ADVANCES IN PROTEIN CHEMISTRY

Volume 34

CONTRIBUTORS TO THIS VOLUME

Eddie Morild

Jane S. Richardson

Ruth Sperling

Ellen J. Wachtel

Donald B. Wetlaufer

ADVANCES IN PROTEIN CHEMISTRY

EDITED BY

C. B. ANFINSEN

National Institute of Arthritis,
Metabolism and Digestive Diseases
Bethesda, Maryland

JOHN T. EDSALL

Biological Laboratories
Harvard University
Cambridge, Massachusetts

FREDERIC M. RICHARDS

Department of Molecular Biophysics
and Biochemistry
Yale University
New Haven, Connecticut

VOLUME 34

1981

ACADEMIC PRESS

A Subsidiary of Harcourt Brace Jovanovich, Publishers

New York   London   Toronto   Sydney   San Francisco

ACADEMIC PRESS, INC.
111 Fifth Avenue, New York, New York 10003

United Kingdom Edition published by
ACADEMIC PRESS, INC. (LONDON) LTD.
24/28 Oval Road, London NW1 7DX

LIBRARY OF CONGRESS CATALOG CARD NUMBER: 44–8853

ISBN 0–12–034234–0

PRINTED IN THE UNITED STATES OF AMERICA

81 82 83 84   9 8 7 6 5 4 3 2 1

CONTENTS

The Histones

RUTH SPERLING AND ELLEN J. WACHTEL

Folding of Protein Fragments

DONALD B. WETLAUFER

The Theory of Pressure Effects on Enzymes

EDDIE MORILD

The Anatomy and Taxonomy of Protein Structure

JANE S. RICHARDSON

CONTRIBUTORS TO VOLUME 34

Numbers in parentheses indicate the pages on which the authors' contributions begin.

EDDIE MORILD,* *Norwegian Underwater Institute, Gravdalsveien 255, N-5034 Ytre Laksevåg/Bergen, Norway* (93)

JANE S. RICHARDSON,† *Department of Anatomy, Duke University, Durham, North Carolina 27710* (167)

RUTH SPERLING,‡ *Department of Chemical Physics, Weizmann Institute of Science, Rehovot, Israel* (1)

ELLEN J. WACHTEL, *Department of Chemical Physics, Weizmann Institute of Science, Rehovot, Israel* (1)

DONALD B. WETLAUFER, *Department of Chemistry, University of Delaware, Newark, Delaware 19711* (61)

* Also at the Department of Chemistry, University of Bergen, N-5014 Bergen, Norway.
† Mailing address: 213 Medical Sciences IA, Duke University, Durham, North Carolina 27710.
‡ Present address: Department of Structural Biology, Stanford University School of Medicine, Stanford, California 94305.

ERRATUM

Advances in Protein Chemistry

Volume 33

Page 252   The text *should read*

of the peptide in the cerebellum. Apparently, the largely motor functions of the cerebellum do not use peptides as neurotransmitters. Cerebellar functions are basic ones which may have evolved before the neuropeptides made their appearance. The cerebral cortex also usually contains relatively low concentrations of neuropeptides, although its large size, in higher animals, makes this brain region a relatively rich source when the total

THE HISTONES

By RUTH SPERLING[1] and ELLEN J. WACHTEL

Department of Chemical Physics, Weizmann Institute of Science, Rehovot, Israel

I. INTRODUCTION

A. Aim of the Article and Summary

The aim of this article is to present evidence and to develop a hypothesis concerning the dynamic structural role of histones in chromatin.

In higher organisms, unlike bacteria, the genetic information is stored in the nucleus. Duplex DNA is found complexed with an

---

[1] Present address: Department of Structural Biology, Stanford University School of Medicine, Stanford, California.

1

equal weight of basic proteins, the histones, and with a smaller amount of nonhistone proteins and RNA. This complex forms a highly compact, organized structure—the chromosome. During interphase, chromosomes expand, revealing the constituent structure called chromatin.

The current view of the histones is that they are primarily structural proteins. Independent of species and tissue, they interact in the same repetitive fashion with the chromosomal DNA to form nucleosomes. Their lack of diversity and non-sequence-specific binding to DNA argue against their playing any explicit role in gene regulation. On the other hand, histone microheterogeneity, arising from post-synthetic modifications and intraspecies sequence variations, still leaves open the possibility that histones are involved, in perhaps a subtle way, in the dynamics of chromosome structure.

The purpose of this article is to correlate the rather unique structural aspects of the five histone molecules—the differences among them as well as their similarities—with their biological function. Such an analysis is best approached, we believe, via a study of the many protein–protein (Section II) and protein–DNA (Section III) interactions in which the histones participate. Emphasis will be placed on the four core histones H2A, H2B, H3, and H4; H1 will be discussed briefly, mainly in relation to its interaction with DNA. In no sense is the bibliography meant to be exhaustive.

During the course of the article, we will present evidence that the histones constitute part of a self-assembly system (Section IV). In the absence of DNA they retain the information to interact with each other to form a hierarchy of structures with dimensions, periodicities, and intermolecular contacts, compatible with what is known about the protein core of chromatin. This strongly suggests that histone–histone interactions have a fundamental role in chromatin structure.

Evidence will be presented in Sections II and IV showing that, under conditions of high salt concentration, histone complexes related to the histone octamer, which is the histone core of the nucleosome, are obtained. Furthermore, higher oligomeric structures are obtained from complexes of H2A·H2B, H3·H4, and all four histones (acid- and salt-extracted). In addition, high ionic strength, while promoting the correct folding and complexing of the core histones, also promotes well-defined reversible histone helical fibers which have axial periodicities similar to those of chromatin. These fibers are formed by assembly of the histone tetrameric and octameric units. The arrangement of the histones in the histone fibers, on the one hand, is related to histone–histone interactions within the nucleosome, and, on the other hand, it also suggests possible histone–histone interactions between nucleosomes. The intranucleosomal interactions of the histones are essential for the formation of the nucleosome. This refers not only to the histone arrangement within the protein core but also to the determination of the periodicity of the DNA fold around it. The internucleosomal interactions may bear relevance to chromatin function.

The similarity of the various histone fibers is probably correlated with the similarity in the distribution of the amino acids in the sequences of the four core histones and reflects their function as the skeleton or backbone of chromatin. However, from the presence of a specific pattern of interactions of the core histones and the existence of histone variants and histone postsynthetic modifications, one can anticipate modulations in the basic general pattern of histone structure. In Section V, a possible mechanism for histone microheterogeneity influencing chromatin structure is suggested. Analogous to other assembly systems, small subunit modifications may be amplified to produce major changes in the assembled superstructure.

For the benefit of those unfamiliar with the field of chromatin research this section continues with background material, characterizing components of the chromatin system which will be pertinent to our discussion.

B. Nucleosomes

It has been shown by biochemical and physical methods that chromatin is built of a repeating unit, the nucleosome (Kornberg, 1974; Olins and Olins, 1974; Woodcock, 1973; Sahasrabuddhe and Van Holde, 1974; Hewish and Burgoyne, 1973; Noll, 1974a; Oudet et al., 1975; Baldwin et al., 1975). This topic has been extensively reviewed (Kornberg, 1977; Felsenfeld, 1978; Chambon, 1978), and therefore only a brief summary will be presented here.

The nucleosome is composed of ~200 base pairs of DNA and an octamer of the histones H2A, H2B, H3, and H4 as well as histone H1 (Kornberg, 1974, 1977). Nucleosomes can be obtained by mild digestion of chromatin with micrococcal nuclease (Noll, 1974a; Axel, 1975), followed by fractionation on a sucrose gradient. Further digestion of the nucleosomes results in the formation of nucleosome core particles composed of 145 base pairs of DNA and an octamer of the histones H2A, H2B, H3, and H4 (Rill and Van Holde, 1973; Sollner-Webb and Felsenfeld, 1975; Axel, 1975; Bakayev et al., 1975; Whitlock and Simpson, 1976; Noll and Kornberg, 1977). The DNA piece thus excised is called "linker" DNA which serves as a link between nucleosomes. It was suggested that histone H1 is bound to the DNA linker (Whitlock and Simpson, 1976; Shaw *et al.*, 1976; Varshavsky *et al.*, 1976; Noll and Kornberg, 1977).

The length of the linker DNA has been shown to vary, resulting in nucleosomes with DNA lengths varying from 154 base pairs in *Aspergillus* to 241 base pairs in sea urchin sperm (for references, see Kornberg, 1977; Chambon, 1978). There are variations among different animal species and also among different cell types in the same organism. Only the linker DNA is variable, while the core DNA length is constant. Changes in H1 have been related to linker variability (Noll, 1976; Morris, 1976a,b).

It has been suggested by Kornberg (1974) that each nucleosome contains two copies of each of the four core histones H2A, H2B, H3, and H4. Immunological studies have shown that histone H2B is present in every nucleosome of rat liver and calf thymus chromatin (Bustin *et al.*, 1976). It was further shown that nucleosomes from HeLa cells contain all four histones (Bustin *et al.*, 1977). This, together with the demonstration of the existence of the histone octamer by cross-linking of chromatin (Thomas and Kornberg, 1975a) and measurements of histone-to-histone (Albright *et al.*, 1979) and histone-to-DNA ratios (Olins *et al.*, 1976; Joffe *et al.*, 1977), supports the model proposed by Kornberg (1974). X-Ray and neutron scattering studies have shown that chromatin has an organized structure with dominant periodicities of 110, 55, 37, and 27 Å (Luzzati and Nicolaieff, 1959; Wilkins *et al.*, 1959; Pardon *et al.*, 1967, 1975; Pardon and Wilkins, 1972; Baldwin *et al.*, 1975). The neutron experiments furthermore determined that the bulk of the histone proteins occupies the central region of the complex to a radius of approximately 32 Å, forming an inner core, and that the DNA is supercoiled on the outside (Baldwin *et al.*, 1975). The packing of the DNA with the histones to form nucleosomes results in an approximate sevenfold increase in the linear packing density of the DNA (Oudet *et al.*, 1975).

Recently, a low-resolution model of the chromatin "core" particle has been derived from a combination of single-crystal X-ray diffraction and electron microscopic data (Finch *et al.*, 1977). The particle is described as a flat cylinder 110 Å in diameter and 57 Å in height. A similar shape and similar dimensions were found to be consistent with the low-angle neutron scattering from core particles in solution (Pardon *et al.*, 1977; Suau *et al.*, 1977). Some conclusions may be drawn concerning the conformation of the DNA. Presumably, the strong 28 Å periodicity apparent in the crystal data (Finch *et al.*, 1977) corresponds to the pitch of the DNA superhelix wound about the histone core. X-Ray and spectroscopic data suggest that the DNA supercoiled around the histone core remains predominantly in the B-conformation (Bram, 1971; Goodwin and Brahms, 1978). The radius of the superhelix must be about 45 Å which, for 145 base pairs of B-form DNA, corresponds to somewhat less than two turns. (1.75 turns have been suggested by Finch *et al.*, 1977). Very little can be deduced from the crystal data concerning the organization of histones in the crystals. However, there is a large body of indirect evidence concerning histone structure and interactions which will be discussed in later sections.

Concerning the path of the DNA around the histone core, a few models have been suggested. Theoretical calculations have shown that the DNA can bend smoothly around the histone core (Levitt, 1978; Sussman and Trifonov, 1978). However, a small change in the number of base pairs per turn relative to that in solution was proposed (Levitt, 1978). A different model was suggested in which the DNA kinks at every tenth or twentieth base pair (Crick and Klug, 1975; Sobell *et al.*, 1976).

Detailed digestion studies of nucleosomes and core particles have shown that a ladder of DNA fragments is obtained, indicating a regular periodic structure of the core particles. The "10-base-pair" repeat obtained by DNase I (Noll, 1974b) digestion of nucleosomes as well as by other nucleases (Sollner-Webb *et al.*, 1978) was recently determined more accurately to be a 10.4-base-pair repeat (Prunell *et al.*, 1979). This may reflect either the periodicity of the DNA double helix in chromatin or the DNA kinking. Experiments with 5'-end labeling of the DNA with ^{32}P followed by DNase I digestion have suggested a more complex structure, in which not all the sites have the same susceptibility for DNase I digestion. Sites at positions 30, 60, and 80 show significantly decreased susceptibility (Simpson and Whitlock, 1976; Lutter, 1978; Noll, 1978). It has been suggested (Finch *et al.*, 1977) that the less susceptible sites for DNase I are the binding sites of the histones which inhibit the digestion.

C. Higher Order Structure of Chromatin

The nucleosome is the building block of chromatin. Chromatin is a dynamic structure and it goes through several levels of compaction, reaching the very compact structure of mitotic chromosomes.

Although the higher order structure of chromatin is still not well understood, two models have been suggested and supported by experimental evidence (for references, see Felsenfeld, 1978; Chambon, 1978). One is the solenoid model of Finch and Klug (1976), obtained by supercoiling the chromatin thread, and the other model is the superbead model (Renz *et al.*, 1977). Each may represent a different state of the same material. Evidence in the literature has suggested that histone H1 is involved in the higher order structure of chromatin (for references, see Felsenfeld, 1978; Chambon, 1978), and some models have been proposed to account for that possibility (Finch and Klug, 1976; Worcel, 1978).

D. The Histones

Since their discovery in 1884 by Kossel, the histones have been the subject of comprehensive study. This work has been summarized in several books and reviews (Phillips, 1971; DeLange and Smith, 1971, 1972; Hnilica, 1972; Elgin and Weintraub, 1975; Li, 1977; Isenberg, 1978, 1979; Von Holt et al., 1979). Therefore, no attempt will be made here to give a comprehensive treatment of the work already reviewed. However, a brief discussion of some special properties of the histones will be presented before we proceed to the treatment of those structural aspects of the histones which are the focus of this article.

Concerning their primary structure, the histones can be divided into three groups: (1) very lysine-rich histone H1, (2) lysine-rich his-

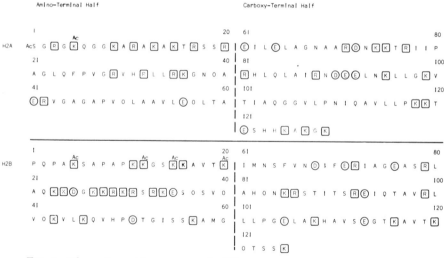

FIG. 1. The amino acid sequences of calf histones H2A (Yeoman et al., 1972; Sautiere et al., 1974) and H2B (Iwai et al., 1972). A one-letter code is used: A, alanine; R, arginine; N, asparagine; D, aspartic acid; C, cysteine; E, glutamic acid; Q, glutamine; G, glycine; H, histidine; I, isoleucine; L, leucine; K, lysine; M, methionine; F, phenylalanine; P, proline; S, serine; T, threonine; O, tyrosine; V, valine. Basic residues are indictated by squares and acidic residues are indicated by circles. Ac denotes acetylation site.

tones H2A and H2B, and (3) arginine-rich histones H3 and H4. This division may also be correlated with the degree of conservation of the histones' primary structure during evolution. On average, the histones are the most highly conserved of all proteins (Dayhoff, 1972; Wilson *et al.*, 1977). However, among the three groups mentioned, there are three degrees of conservation, with H1 being the most variable and H3 and H4 the most highly conserved.

1. Histone H1 is the most variable of the histones (for references, see Cole, 1977; Isenberg, 1978; Von Holt *et al.*, 1979), and the number of its subfractions varies from tissue to tissue and species to species. Three major domains can be distinguished within its primary structure: in the first domain, the amino-terminal region has charged residues and is variable; in the second domain, the central region has hydrophobic residues as in globular proteins and is not variable; and the third domain is the carboxy-terminal end which is charged and variable with respect to sequence conservation.

2. The slightly lysine-rich histones H2A and H2B are composed of two domains—the amino-terminal region, in which are clustered basic residues resulting in a net positive charge; and the carboxy-terminal region, which has a distribution of residues similar to other globular proteins and is rich in hydrophobic residues. This is demonstrated in Fig. 1 which gives the sequences of calf H2A (Yeoman *et al.*, 1972; Sautiere *et al.*, 1974) and H2B (Iwai *et al.*, 1972). H2A and H2B

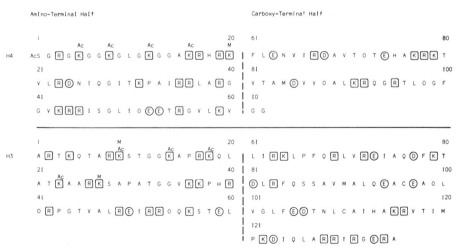

FIG. 2. The amino acid sequences of calf histone H4 (DeLange *et al.*, 1972; Ogawa *et al.*, 1969) and H3 (DeLange *et al.*, 1972; Olson *et al.*, 1972; DeLange *et al.*, 1973). The notation is as in Fig. 1 except that superscript M denotes methylation site.

are conserved to an intermediate extent; there are variations between species. However, there are also subtypes of H2A and H2B which occur in the same species (Franklin and Zweidler, 1977) and whose synthesis is correlated in some species with different developmental stages (Cohen *et al.*, 1975; Newrock *et al.*, 1978). From the sequences known at present the carboxy-terminal region is the more conserved region (for references, see Isenberg, 1978; Von Holt *et al.*, 1979).

3. The arginine-rich histones H3 and H4 are the most highly conserved proteins during evolution (Patthy *et al.*, 1973). Like H2A and H2B, they are composed of two domains—the charged amino-terminal region and the hydrophobic carboxy-terminal region. This is shown in Fig. 2 which gives the sequences of calf H4 (DeLange *et al.*, 1969; Ogawa *et al.*, 1969) and H3 (DeLange *et al.*, 1972, 1973; Olson *et al.*, 1972).

II. HISTONE–HISTONE INTERACTION

A. Introduction

The chromatin repeating unit—the nucleosome core particle—is composed of an octameric unit of the four core histones (two copies each of H2A, H2B, H3, and H4) around which is folded 145 base pairs of DNA (for references, see Kornberg, 1977). This unit is rather compact, and one can *a priori* anticipate histone–histone interactions within the core of the nucleosome. The nucleosomes are further packed in the presence of histone H1 [which is probably bound on the outside of the nucleosome (Baldwin *et al.*, 1975)] to form hierarchies of higher order structures of chromatin. These higher order structures may involve interactions which are additional to those present in the nucleosome, i.e., histone–histone contacts between nucleosomes. A full description of the organization and conformational properties of the histone octamer awaits the three-dimensional solution of the structure of nucleosome core particle crystals (Finch *et al.*, 1977). Our understanding of histone organization and conformation within the higher order structures of chromatin is even more limited at present. Nevertheless, by probing with radiation and chemical cross-linking reagents, it has proved possible to analyze, albeit rather coarsely, the hierarchy of histone–histone contacts within and between nucleosomes. Some of these contacts can be preserved upon removal of the DNA. This permits the isolation of stable protein oligomers which are subcomplexes of the octamer. Similarly, the *in vitro* interaction of purified single histones gives rise to specific complexes reflecting the dominant *in vivo* contacts. These observations point to the possibility that the histones can assemble on their own to form complexes analogous to those found in chromatin. Each histone subunit has sites for interactions with other histones which, under controlled conditions, result in the same well-defined, stoichiometric complexes. As will be discussed later, this is a necessary prerequisite for the histones to constitute an assembly system.

B. In Vivo *Studies of Histone–Histone Interactions and Contacts*

1. *Identification of Cross-Linked Histone Dimers and Trimers*

Histone–histone cross-linking in chromatin and nuclei has been employed in order to study histone–histone contacts in chromatin. The cross-linking agents used can be divided into two groups: (*a*) short-range cross-linkers of length less than 2 Å, which supply information on contact surfaces (this group also includes ultraviolet radiation which induces zero-length cross-links at the binding sites); and (*b*) long-range cross-linkers of length 10–14 Å, where the formation of cross-links does not imply interaction but rather supplies information on neighboring subunits. Although this approach reveals intermolecular contacts, the quantitation of products obtained by cross-linking is not accurate. The cross-linking of histones in chromatin results in the formation of histone dimers and higher oligomers, but in most cases reported in the literature analysis of the cross-linked histones was performed on the cross-linked dimer fraction only. The percentage of dimers in the histone population was not discussed in most cases. Products with molecular weight higher than dimers (in many cases, the major products) do not penetrate the gels in which the results are being analyzed. In such cases the information necessary to determine the percentage which a specific cross-linked dimer constitutes among the total cross-linked histone population is missing. This is even more important in comparing results from various laboratories in which experiments have been performed under different conditions of reagent concentration and cross-linking time. Nevertheless, despite these reservations, data accumulated from various laboratories do give us information on the most abundant dimers.

Inspection of results obtained by cross-linking with short-range cross-linking reagents shows that the most abundant dimers are H4·H2B and H2A·H2B (Van Lente *et al.*, 1975, Martinson and McCarthy, 1975; Martinson *et al.*, 1976). Four slightly different H4·H2B dimers were obtained by cross-linking in nuclei with ultraviolet (UV) radiation, tetranitromethane, and formaldehyde. The H4·H2B dimers were found by CNBr cleavage to be cross-linked in the carboxy-terminal regions of the two molecules (Martinson *et al.*, 1979a). Analysis of the cross-linked H2A·H2B dimer has shown that in the case of UV radiation the site of cross-linking is between tyrosine 37, 40, or 42 of H2B and proline 26 of H2A (DeLange *et al.*, 1979), whereas it occurred in the C-terminus half with methylacetimidate (Martinson *et al.*, 1979a). The H2A·H2B·H4 trimer, obtained with either UV radiation only or tetranitromethane plus UV radiation, has cross-links similar to those found in the H2A·H2B and H4·H2B dimers (Martinson *et al.*, 1979a). The formation of the above dimers and trimers was shown to occur in mononucleosomes as well (Martinson *et al.*, 1979b). However, the formation of an H2A·H2B·H4 trimer and an H2B·H4 dimer was salt-dependent and would not occur at very low salt concentrations.

When nuclei were cross-linked with formaldehyde and the results analyzed by reversal of the cross-links of the dimers obtained (Jackson, 1978) all ten possible dimer combinations of the four core histones were obtained, although most dimers were present in very low yields (<5% of dimer population). The predominant dimer was H4·H2B (two dimers of H4·H2B); H2A·H2B, H3·H2A, and H2A·H2A were also found. The major sites for cross-linking into dimers reside in the carboxy-terminal region of the histone molecules (trypsin resistant region) and therefore indicate proximities of these regions of the histone molecules within the nucleosome (Jackson, 1978). Cross-linking of nuclei and chromatin with 1-ethyl-3-(3-dimethyl-aminopropyl)carbodiimide resulted in formation of an H3·H4 dimer (Bonner and Pollard, 1975), whereas cross-linking of nucleosomal histones with methyl acetimidate yielded the following dimers: H3·H3, H2A·H2B, H3·H4, and small quantities of H4·H4 (Wyns *et al.*, 1978).

To summarize briefly at this point, the fact that all possible histone dimers can be obtained by formaldehyde cross-linking of nuclei (although some at very low yields) indicates that the packing of histones in chromatin is likely to be very compact. Of all the dimer products of the short-range cross-linkers, H2A·H2B and H4·H2B are the most abundant and therefore probably represent strongly interacting pairs. Since these pairs are found not only by cross-linking of chromatin but also of mononucleosomes (Martinson *et al.*, 1979a), there must be H4·H2B and H2A·H2B intranucleosomal interactions. Other intranucleosomal contacts are H3·H3 and H3·H4 (Wyns *et al.*, 1978), while H3·H2A (Jackson, 1978), which occurs in fairly high yields with respect to other dimers in cross-linked nuclei, may reflect intra- and/or internucleosomal contacts. As mentioned above, the lack of information concerning the percentage and identity of higher oligomers present in the cross-linking does not allow us to exclude the possibility of other important contacts.

Cross-linking of histones in chromatin by long-range cross-linkers such as the reversible cross-linker methyl 4-mercaptobutyrimidate which bridges ~ 14 Å was also used in the study of histone packing in chromatin. Analysis of the dimers obtained in this case showed the presence of the following pairs: H3·H3, H2A·H2B, H3·H4, H4· H2A, and H4·H2B (Thomas and Kornberg, 1975b). Similar dimers were obtained by cross-linking with dimethylsuberimidate (Thomas and Kornberg, 1975a). Cross-linking in nuclei with the same reagent but at higher chromatin concentration and at lower pH (pH 7.4) resulted in the following cross-linked dimers: H2A·H2B, H3·H3, H3· H2A, and H2B·H3. H3·H4 and H2B·H4 were also obtained but with lower yields (Hardison et al., 1977). In all the above cases the dimers cross-linked with the long-range cross-linkers may represent intra- as well as internucleosomal histone neighbors.

2. Cross-Linking of Histone Oligomers

In addition to the dimer bands higher oligomeric complexes were obtained by cross-linking chromatin.

a. H1-Oligomers. H1 oligomers were observed by using dimethylsuberimidate (Thomas and Kornberg, 1975a), 1-ethyl-3-(3-dimethylaminopropyl)carbodiimide (Bonner and Pollard, 1975), or glutaraldehyde (Chalkley and Hunter, 1975).

b. Octamer and Higher Oligomers. Cross-linking of chromatin with dimethylsuberimidate at pH 8.0 ionic strength (I) = 0.2, resulted in the formation of a set of aggregates ranging from histone dimers to structures with apparent molecular weight (MW) of ~ 170,000 (Thomas and Kornberg, 1975a). However, when cross-linking with this reagent was performed at pH 9.0, I = 0.1, a histone octamer was the major species. The bands due to 9- and higher mers were absent, but 16- and 24-mer bands were observed. Cross-linking with dithiobissuccinimidyl propionate at pH 9, I = 0.2 (Thomas and Kornberg, 1975a), was more extensive and a histone octamer and multiples of octamer were obtained. Cross-linking with formaldehyde of chromatin, chromatin depleted of H1, and chromatin depleted of H1, H2A, and H2B (Hyde and Walker, 1975a), at salt concentrations of 0.7 M and higher, yielded oligomers of histones with apparent molecular weight of octamer and higher. Higher oligomers were also obtained with glutaraldehyde (Chalkley and Hunter, 1975). These experiments at high pH or high ionic strength with both long- (Thomas and Kornberg, 1975a) and short-range cross-linking reagents

(Hyde and Walker, 1975a) indicate that there are two basic types of histone–histone interactions or contacts: those within the nucleosome which result in the formation of histone octamers and those between nucleosomes which give rise to higher oligomers.

C. Histone–Histone Interactions in Salt-Dissociated Chromatin

Octamers of histones were obtained by cross-linking chromatin or histones removed from chromatin in 2 M NaCl (Thomas and Kornberg, 1975a). In the case of H1-depleted chromatin or (H1, H2A, H2B)-depleted chromatin, when the remaining histones are dissociated by 2 M NaCl and cross-linked, the patterns of cross-linking are similar to those obtained when the histones are still bound to DNA. In all cases dimers are the basic cross-linked unit and octamers and higher oligomers are observed, although the formation of higher histone oligomers is more extensive in the case of the DNA-bound histones (Hyde and Walker, 1975a).

1. Histone Dimer and Tetramer Obtained by Salt Dissociation from Chromatin

Specifically, when histones are dissociated from DNA in 2 M NaCl, and H3·H4 tetramer and H2A·H2B dimer may be identified after fractionation of the histones at pH 5.0 and cross-linking with dimethylsuberimidate (Kornberg and Thomas, 1974). The inference was drawn that these complexes exist as such in the intact nucleosome. Furthermore, since both the un-cross-linked H3·H4 tetramer, and the un-cross-linked H2A·H2B dimer are stable complexes, it has proved possible to characterize their physical properties in solution. Some of these results are summarized here.

Sedimentation velocity measurements on $H3_2 \cdot H4_2$ under a variety of conditions find 2.5–3 S (Kornberg and Thomas, 1974; Roark et al., 1974, 1976; Moss et al., 1976b) which, when combined with MW ~54,000, result in a calculated ratio $f/f_0 \sim 1.7–2.0$, suggestive of a marked shape anisotropy or noncompactness. This suggestion is further supported by ^{13}C (Lilley et al., 1976) and proton resonance spectroscopy (Moss et al., 1976b) (13 mg/ml, 50 mM sodium acetate, 50 mM sodium bisulfate, pH 4.8). Both spectra contain a larger number of sharp resonances than are usually found in native globular proteins with extensive tertiary folding. In particular, the sharpness of glycine resonances (α-CH_2), 59% of which arise from the NH_2-terminal one-third of the H3 and H4 sequences, is seen as evidence for backbone flexibility in the NH_2-terminal region. On the other hand, the ring current shifted peaks of apolar resonances and perturbed aromatic resonances indicate that some regular tertiary folding must take place. The regions of the sequences of H3 and H4 containing apolar and aromatic residues are the COOH-terminal halves, which are seen as the major interacting regions.

One clear-cut piece of evidence of a different type for intermolecular interaction occurring in the COOH-terminal regions was found by investigating the possibility of bonding between pairs of the highly conserved Cys-110 in H3 of avian erythrocyte nucleosomes. It was determined (Camerini-Otero and Felsenfeld, 1977a) that the intermolecular disulfide bridge may be made or broken without disrupting the nucleosome structure as characterized by nuclease digestion, superhelicity of DNA, and sedimentation velocity measurements. This disulfide bridge presumably could also exist in the tetrameric complex.

Infrared and circular dichroism (CD) measurements (Moss et al., 1976b) are both consistent with a sizable fraction of the tetramer being in the α-helical configuration, $\sim 29\%$ α-helix with negligible β structure. This is rather similar to the 25% α-helix and $\sim 0\%$ β structure obtained for the tetramer prepared from acid-extracted histones (D'Anna and Isenberg, 1974b).

As far as the H2A·H2B dimer is concerned, Moss et al. (1976a) have indicated that (at pH 7 in 25 mM phosphate) the complex sediments as a single interacted species ($< 5\%$ aggregate). In water, or buffer at pH 4, the molecules unfold and do not interact. Spectroscopy experiments (Moss et al., 1976a) suggest that the interacting sites are localized in the COOH-terminal two-thirds of the two molecules. This is the location of most of the apolar and aromatic residues. The proton magnetic resonance spectrum of H2A·H2B (in 25 mM phosphate, pH 7, 20 mg/ml) contains characteristic ring shifts of resonances from apolar groups which are not observed for the isolated histones. Secondary structural features present in the folded state were examined (25 mM phosphate pH 7.0, 7 mg/ml), and found by model fitting to CD data to be strongly α-helical, up to 37%, with negligible ($< 2\%$) β-sheet content.

2. The Histone Octamer

Cross-linking of chromatin in 2 M NaCl at pH 8.0 results in the formation of a cross-linked octamer (Thomas and Kornberg, 1975a) which contains the four core histones in equimolar ratios (Thomas and Kornberg, 1975b). A non-cross-linked complex of histones isolated in 2 M NaCl at pH 7.0 also was found to contain an equimolar ratio of the four core histones but had a molecular weight determined to be near that of a tetramer (Weintraub *et al.*, 1975). Additional studies reported a tetramer model for the core histones in 2 *M* NaCl (Campbell and Cotter, 1976; Pardon *et al.*, 1977). However, recently it has been shown that a histone octamer can be obtained free in solution at NaCl concentrations of at least 2 *M* (Thomas and Butler, 1977, 1978; Eickbush and Moudrianakis, 1978). This octamer has a molecular weight of 107,500 and sediments with $s_{20,w}^0$ = 4.95 (Thomas and Butler, 1978), indicative of either asymmetry in shape, or hydration, or both (Thomas and Butler, 1978). The apparent controversy about whether it is an octamer or tetramer which exists at 2 *M* NaCl perhaps may be resolved by noting the tendency of the octamer to dissociate. Various conditions can destabilize the octamer, such as decrease in histone concentration (Thomas and Butler, 1977, 1978; Chung *et al.*, 1978), increase in temperature (Thomas and Butler, 1978; Eickbush and Moudrianakis, 1978), decrease of NaCl concentration (Weintraub *et al.*, 1975; Thomas and Butler, 1977, 1978; Chung *et al.*, 1978; Eickbush and Moudrianakis, 1978; Butler *et al.*, 1979), and values of pH below 7 (Weintraub *et al.*, 1975; Thomas and Butler, 1978; Eickbush and Moudrianakis, 1978) or above 10 (Eickbush and Moudrianakis, 1978). Furthermore, it should be pointed out that in a multicomponent equilibrium system such as this one, dependent on several parameters, more than one local energy minimum may exist. Therefore, if the octamer at 2 *M* NaCl is metastable, approaching the concentration of 2 *M* NaCl by different pathways may lead to the formation of different complexes. Furthermore, hysteresis may also lead to slightly different results when arriving at the same point by different pathways. However, at 4 *M* NaCl, where the octamer is definitely stable (Butler *et al.*, 1979), one can anticipate that the same complex will be obtained independent of the pathway. The important thing to note is that conditions can indeed be found, i.e., 4 *M* NaCl, in which the histone component of the nucleosome—the octameric unit—can be made to reassemble into a stable complex in the absence of DNA. In the following section we will see that this reassembled complex retains its biological activity with respect to association with DNA.

3. Reconstitution of Core Particles with Octamers and DNA

The histone octamer of nucleosome core particles was cross-linked by dimethylsuberimidate and isolated from the DNA by precipitation in 3 *M* NaCl (0.05 *M* sodium phosphate buffer, pH 7.0). The cross-linked octamer, dissolved at low ionic strength, was reconstituted by mixing with DNA at 1.0 *M* NaCl (pH 8.0 Tris buffer) and dialyzed against 0.6 *M* NaCl in the same buffer. The reconstituted particle had properties similar to those of the cross-linked core particle. It sedimented with $s_{20,w}^0 = 10.9$, and had a similar melting curve, CD spectrum, and single-strand DNA fragment size distribution after DNase I digestion. By reconstitution of the cross-linked octamer with SV40 DNA, it was shown that the cross-linked octamer can impose torsional constraints on SV40 DNA and that the structure reconstituted from cross-linked octamer and SV40 DNA is morphologically similar to the SV40 minichromosome as demonstrated by electron microscopy (Stein *et al.*, 1977).

Similar results were obtained from reconstitution experiments with DNA and a non-cross-linked octamer (Thomas and Butler, 1978). Nucleosome-like particles were observed in the EM and a pattern of histone cross-linking comparable to that of native chromatin was obtained. However, only 140-base-pair repeats were obtained upon micrococcal nuclease digestion instead of ~200-base-pair repeats obtained for native rat liver chromatin (Noll and Kornberg, 1977). This indicates that, in the absence of H1, only core particles can be reconstituted. Nevertheless, these studies with both cross-linked and reassembled un-cross-linked histones demonstrate that the octamer is a complete biological functional unit retaining the information for folding the DNA around the histone core.

D. Histone–Histone Interaction in Vitro

1. Histone Pairs

Acid-extracted calf thymus histones can be renatured and mixed pairwise to form strongly associating pairs (D'Anna and Isenberg, 1973, 1974a,b,c; Sperling and Bustin, 1975). Histones in water are unfolded (for references, see Isenberg, 1978). Both the renaturation and complex formation of histones are promoted by the addition of salt, with the former occurring in a short period of time compared with that of the latter (for references, see Isenberg, 1978). The tertiary structure of a protein is determined by its primary structure (Anfinsen, 1973). The quaternary structure of a multicomponent system is affected by the tertiary structure of the components. In such a multiparameter system several structures may exist, each corresponding to a local minimum in the free energy. Thus, the formation of a specific structure may depend on the precise environmental conditions of the preparation. Kinetic as well as thermodynamic factors may affect the final structure obtained.

In view of these considerations, when preparing histone pairs from single histones it is advisable to first equilibrate the individual histones in low salt concentration. This permits folding of the monomeric form. The pairs are then mixed at low salt concentration and given sufficient time to interact. Finally, the salt concentration is increased by dialysis. By preparing histone pairs in this way D'Anna and Isenberg have shown that histones H4 and H2B (D'Anna and Isenberg, 1973) and histones H2A and H2B (D'Anna and Isenberg, 1974a) interact in a 1:1 molar ratio to form strong complexes with an association constant of $10^6\ M^{-1}$. Similarly, H4 and H3 interact with an association constant of $0.7 \times 10^{21}\ M^{-3}$ calculated for a tetramer (D'Anna and Isenberg, 1974b). Other histone pairs form weaker complexes.

a. H2A·H2B. An H2A·H2B dimer obtained from acid-extracted histones was observed by Kelley (1973) in agreement with reports on the strong interaction of this pair (Skandroni *et al.*, 1972; D'Anna and Isenberg, 1974a; Sperling and Bustin, 1975). Measurements of fluorescence anisotropy, relative fluorescence intensity, and CD were employed in characterizing the pairwise interaction of these histones (D'Anna and Isenberg, 1974a). These studies indicate that upon complex formation, the number of α-helical residues increases by about 15.

Renatured H2A and H2B interact to form a dimer at low ionic strength with an $s_{20,w} = 2.2-2.3$ and molecular weight of 28,500–28,900 (Kelley, 1973; Sperling and Bustin, 1975). The formation of a heterodimer was also demonstrated by cross-linking with dimethylsuberimidate (Sperling and Bustin, 1975). The fact that cross-links between H2A and H2B were also obtained in chromatin (for references, see Section II,B) indicates that the histone–histone interaction in the absence of DNA is probably related to that in chromatin. As discussed earlier, this pair was also obtained by salt extraction (Kornberg and Thomas, 1974) and reported to form oligomers. High-resolution proton magnetic resonance (PMR) spectroscopy and CD indicated that H2A·H2B can be renatured from urea-denatured, acid-extracted histones as well as from urea-denatured, salt-extracted histones (Moss *et al.*, 1976a).

b. H3·H4. Cross-linking of the H3·H4 pair and analysis of the results by electrophoresis on SDS-containing polyacrylamide gels as well as by sedimentation velocity and sedimentation equilibrium experiments have indicated that this is a strongly interacting pair. At low ionic strength and protein concentration the major fraction is an assembled heterotetramer accompanied by a fraction of a heterodimer ($K \sim 5 \times 10^4\ M^{-1}$ for the dimer–tetramer equilibrium) and a fraction of high-molecular-weight complexes. The sedimentation value of the low-molecular-weight fraction is 2.6 S (Sperling and Bustin, 1975). This may be compared with the tetrameric complex of H3·H4 obtained by salt extraction (Kornberg and Thomas, 1974) and a dimer in equilibrium with a tetramer reported by Roark *et al.* (1974), also with salt-extracted histones.

c. H4·H2B. In agreement with the binding studies of D'Anna and Isenberg (1973), cross-linking of the H4·H2B pair prepared from acid-extracted histones indicated that at $I = 0.15$ the system consists predominantly of a heterodimer in equilibrium with a heterotetramer. This finding was substantiated by sedimentation velocity and sedimentation equilibrium experiments (Sperling and Bustin, 1975).

d. Other Pairs. The other three combinations of heteropairs of the four core histones, namely, H2A·H3, H2A·H4, and H2B·H3, were studied by fluorescence anisotropy (D'Anna and Isenberg, 1974a,b) and by cross-linking with dimethylsuberimidate and ultracentrifugation (Sperling and Bustin, 1975). H2A·H3 forms a complex of intermediate strength and the remaining two heteropairs interact weakly (D'Anna and Isenberg, 1974b). From cross-linking data H3·H2B is also seen to form a complex of intermediate strength (Sperling and Bustin, 1975).

2. Interchange of Histones upon Addition of Single Histones to Preformed Histone Pairs

Histone H4 forms strong complexes with both histone H3 and H2B (D'Anna and Isenberg, 1973, 1974b; Sperling and Bustin, 1975). Histone H2B also complexes strongly with histone H2A (D'Anna and Isenberg, 1974a). Therefore the question arose whether a histone which can form strong complexes with two types of histones can add to a preformed dimer thus forming a trimer. In the case of the three histones H4, H2B, and H2A, either H4 was added to a preformed H2A·H2B dimer or H2B was added to the preformed H4·H2B pair. No trimer formation was observed (Sperling and Bustin, 1976); rather, a mixture of dimers was obtained. Similar results were obtained for the triplet H3, H4, and H2B. These results indicate that there is exchange of histones with the formation of dimers.

As described in Section II,B, an H2A·H2B·H4 trimer may be obtained by UV cross-linking of chromatin. Therefore, H2B apparently has two distinct binding sites for H2A and H4 (Martinson *et al.*, 1979a). The lack of trimer formation by addition of a third histone to a preformed pair consequently must indicate that a trimeric unit is not stable. The dimer is the basic structural unit in the assembly of histones (Sperling and Bustin, 1976). This idea is supported by the fact that in a mixture of histones only even complexes have been isolated, as will be further discussed below. Therefore in chromatin the H2A·H2B·H4 trimer is probably part of an even higher stable oligomer, such as tetramer, hexamer, or octamer.

3. Dynamic Equilibrium in Assembly of Histone Pairs

H2A·H2B pairs can associate to form higher oligomers (Kornberg and Thomas, 1974) and the H3·H4 tetramer will also form aggregates upon increase in ionic strength and protein concentration (D'Anna and Isenberg, 1974c; Sperling and Bustin, 1975; Moss *et al.*, 1976b; Wachtel and Sperling, 1979). Systematic study of the effect of protein concentration and ionic strength on the six possible combinations of heterodimers has shown (Sperling and Bustin, 1975) that, although the pairwise association of histones leads to formation of dimers and tetramers at $I \sim 0.15$, there is a dynamic equilibrium between these low-molecular-weight complexes and organized periodic histone fibers which are observable by electron microscopy (see Fig. 3). These twisted ribbon structures have a diameter of 40–80 Å, an axial repeat of 330 Å, and the same appearance for all the pairwise combinations of histones (Sperling and Bustin, 1975). Their assembly, which is promoted by increase in protein concentration and ionic strength, is reversible. It should be noted that fiber assembly is facili-

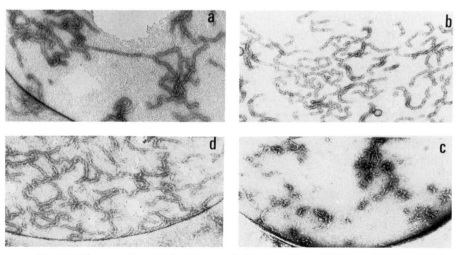

FIG. 3. Electron micrographs of assembled histone fibers: (a) H4, acid-extracted; (b) H3·H4, acid-extracted; (c) H3·H4, salt-extracted; (d) H2A·H2B·H3·H4, acid-extracted (magnification × 78,000).

tated by phosphate ions and fibers can be assembled in the presence of phsophate under physiological conditions (Sperling and Bustin, 1975). Fibers could be obtained from H3·H4 complexes either reconstituted *in vitro* from acid-extracted histones or extracted from chromatin by 2 M NaCl (see Fig. 3b and c). The tendency for fiber formation is most marked for the arginine-rich histones H3 and H4. The lysine-rich histones H2A and H2B polymerize only at higher ionic strength. A pair containing one arginine-rich and one lysine-rich histone will polymerize at ionic strengths intermediate between those of the other two cases.

In order to characterize the intermediates in the assembly of histone pairs to form histone fibers and to prove the formation of heterofibers, the following strong histone pairs — H4·H3, H2A·H2B, and H4·H2B — were each cross-linked with dimethylsuberimidate and fractionated on a Sephadex G-100 column, and the resulting fractions were analyzed (Sperling and Wachtel, 1979). The results indicate that there is a general scheme for histone pairwise assembly, with the major components obtained being a progression of heteroeven complexes: dimers, tetramers, hexamers, and octamers. (It is therefore possible to obtain well-defined octameric complexes of only two histones.) Cross-linking of the H3·H4 pair at various ionic strengths and concentrations (Sperling and Wachtel, 1979) showed that the heterodimers and heterotetramers predominate. The end-products in all these studies were heterofibers obtained either by assembly of the isolated cross-linked heterodimers or by assembly of the mixed pair upon increase in ionic strength. The effect of ionic strength on the isolated cross-linked complexes (dimers, tetramers, hexamers, and octamers) again supported the finding that there is an equilibrium among these complexes which proceeds through dimers, tetramers, hexamers, and octamers to form fibers.

It should be emphasized that the assembly of histone fibers is a completely reversible process (Sperling and Bustin, 1975). We attribute the irreversible aggregates obtained in some pair preparations (Nicola *et al.*, 1978; Lewis, 1976) to failure to follow a preparation technique similar to that described in Section II,D,1.

As noted above, salt-extracted H2A·H2B will form higher oligomers, fibers could be obtained from salt-extracted histones (Fig. 3c), and even octamers are now known to associate into superstructures at very high salt concentration (Wachtel and Sperling, 1979). It would therefore seem that self-assembly is a general property of histones. The pattern of complex formation of the core histones is shown schematically in Fig. 4.

High Order Structures

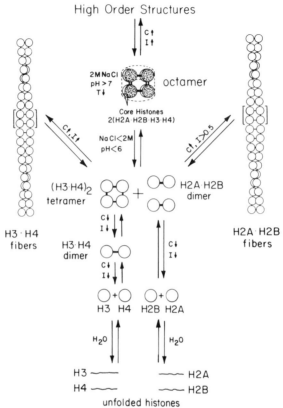

FIG. 4.   Pattern of reversible complex formation of the core histones, H2A, H2B, H3 and H4. In water the histones are unfolded. Upon addition of salt, renaturation occurs, followed by complex formation (Isenberg, 1978). At physiological values of pH and ionic strength (I), H2A and H2B form a strongly interacting dimer (D'Anna and Isenberg, 1973; Sperling and Bustin, 1975) while H3 and H4 form a tetramer (Kornberg and Thomas, 1974) in equilibrium with a dimer (Roark *et al.*, 1974; Sperling and Bustin, 1975). Further increase in ionic strength and protein concentration (C) promotes fiber formation for these pair systems (Sperling and Bustin, 1975). When the four core histones are present in a 1 : 1 : 1 : 1 molar ratio, with I at least 2 M and minimum pH of 7, octamers are formed (Thomas and Butler, 1977, 1978; Chung *et al.*, 1978; Ecikbush and Moudrianakis, 1978; Butler *et al.*, 1979). These octamers will then associate into still higher order structures upon further increase in protein concentration and ionic strength (Wachtel and Sperling, 1979).

4. Assembly of the Four Core Histones

When the four core histones are mixed in equimolar ratio at low ionic strength a single sedimentation velocity peak is observed with $s_{20,w} = 2.0$. Gel filtration and cross-linking with dimethylsuberimidate showed that the peak contained a mixture of dimers (Sperling and Bustin, 1976). Under these conditions, addition of histone H1 does not seem to affect the pattern of dimer formation nor does H1 interact with the other four histones or form dimers.

Upon increase in salt concentration to 2 M, histone octamers were obtained (Thomas and Butler, 1978). The octamer could be assembled from acid-extracted as well as from salt-extracted histones (Thomas and Butler, 1978). In a concentrated solution of the four core histones (prepared by acid extraction) at an ionic strength higher than 2 M NaCl (minimum 10 mg/ml histone concentration), there is a small fraction of assembled fibrous structures which can be observed in the electron microscope (Sperling and Bustin, 1976; Wachtel and Sperling, 1979). These fibers (see Fig. 3d) are ~60 Å in diameter and have a 330 Å axial repeat, and were shown to be composed of the four core histones in an equimolar ratio (Wachtel and Sperling, 1979). The percentage of fibers in the solution of the four core histones is promoted by increase in histone and salt concentrations.

E. Summary

1. Comparison of Histone–Histone Contacts in Chromatin with Histone Complexes in the Absence of DNA

Although the quantitation of the cross-linking data is limited, as discussed above, comparison of the cross-linked histones obtained from chromatin with histone complexes obtained in the absence of DNA can help in developing a more detailed picture of histone–histone interactions. An H2A·H2B dimer was obtained by cross-linking of chromatin with short-range cross-linkers (for references, see Section II,B). This dimer can also be isolated from chromatin by salt extraction (Kornberg and Thomas, 1974) or reconstituted *in vitro* from purified single histones in the presence of low salt (Kelly, 1973; D'Anna and Isenberg, 1974b; Sperling and Bustin, 1975). Therefore, we can conclude that under physiological conditions the H2A·H2B dimer exists as an entity in solution and could be introduced as such (not excluding some local conformational changes) into the nucleosome. An H4·H2B dimer was also obtained by cross-linking of chromatin (for references, see Section II,B) and by *in vitro* reconstitution from single histones (D'Anna and Isenberg, 1973; Sperling and Bustin, 1975). However, it has not been isolated by extraction from chromatin with salt. Nevertheless, it is probably present as a strongly interacting component of the histone octamer and hexamer: $(H3)_2 \cdot (H4)_2 \cdot$ H2A·H2B. It was possible to obtain an H2A·H2B·H4 trimer by cross-linking in nuclei and nucleosomes (Martinson *et al.*, 1979a). Such a trimer was not obtained by *in vitro* reconstitution from a pre-formed H2A·H2B pair with addition of H4 or by adding H2A to H2B·H4 (Sperling and Bustin, 1976). In each case exchange of histones with formation of a mixture of dimers occurred. All this supports the idea that histone dimers are the basic units of histone assembly (Sperling and Bustin, 1976). Furthermore, in a solution of pairs of histones no odd complexes have been observed as major products.

The H3·H4 tetramer has been isolated from chromatin by salt extraction (Kornberg and Thomas, 1974; Roark *et al.*, 1974) and has also been obtained by *in vitro* reconstitution from single histones (D'Anna and Isenberg, 1974; Sperling and Bustin, 1975). The tetramer appears to be an essential structural component of the nucleosome as, on its own, it can fold DNA to form nucleosome-like structures (for references, see Section III). Therefore it seems puzzling that, although the H3·H4 dimer has been obtained by cross-linking of chromatin, the yields are lower than anticipated (for references, see Section II,B). It is possible that $H3_2·H4_2$ serves as the interior backbone of the histone octamer and the H2A·H2B dimer in some way fits into the tetramer, on the one hand making the inner H3·H4 contacts inaccessible to the long-range chemical cross-linking reagents and on the other hand creating on the surface H2B·H4 contacts which are very accessible. It is also possible that due to the compactness of the structure it is hard to isolate an H3·H4 dimer. Rather, H3·H4 cross-links are present in higher oligomers.

The regions involved in the intranucleosomal histone–histone interactions were indicated from cross-linking data on histones in chromatin to be the carboxy-terminal regions (Martinson *et al.*, 1979a; Jackson, 1978). This is in agreement with NMR data suggesting that the interaction of the H3·H4 pair (Böhm *et al.*, 1977) and of the H2A·H2B pair (Moss *et al.*, 1976a) occurs in the carboxy-terminal part of the histone molecules. It should be pointed out that very similar histone cross-linking patterns were obtained with chromatin from various sources. Similarly, the strong *in vitro* association of histones pairs—H2A·H2B, H3·H4, and H4·H2B—is repeated for pea histones (Spiker and Isenberg, 1977), yeast histones (Mardian and Isenberg, 1978), and *Tetrahymena termophila* (Glover and Gorovsky, 1978), and for interkingdom complexes (Spiker and Isenberg, 1978; Isenberg, 1979). This may be correlated with the conservation of the primary sequences of the carboxy-terminal regions of the histones and the indications that these are the sites of histone–histone interaction within the nucleosomes (Spiker and Isenberg, 1978).

2. Histones at High Salt—Similar to Histones in Chromatin

It is apparent from the discussion in this section that conditions exist under which histones, isolated from DNA, or, to varying extents, from each other, will form complexes consistent with the protein–protein cross-linking pattern determined from chromatin. In particular, the H2A·H2B dimer and the H3·H4 tetramer will form in solution at physiological values of pH and ionic strength. However the octamer composed of the H3·H4 tetramer and two H2A·H2B dimers is stable only under conditions of high salt concentration (Thomas and Butler, 1977, 1978; Chung et al., 1978; Eickbush and Moudrianakis, 1978; Butler et al., 1979). At the molecular level this implies that a high-ionic-strength environment, i.e., at least 2 M NaCl, is necessary to preserve native histone secondary and tertiary structure at specific binding sites. Or it could also imply that a minimum amount of electrostatic shielding of charged residues is necessary for complex formation.

There are several types of experiments described in the literature that support the idea that high salt concentration is a necessary, if not sufficient, condition for core histones to acquire and maintain the conformation and complexing pattern native to chromatin. Weintraub et al. (1975) showed that in 2 M NaCl, 1 mg/ml, the complex formed by the core histones H2A, H2B, H3, and H4 was indistinguishable from histones in chromatin with respect to its resistance to digestion by trypsin, its pattern of reactivity with iodine, and its ability to form specific cross-linked products after treatment with formaldehyde. The trypsin limit digest pattern was also observed with a minimum ionic strength of 0.8 M NaCl., pH 7.1, in 2 M NaCl, pH 5.5, and with both salt-extracted and acid-extracted histones. The trypsin digestion probes the folding and complexing of the carboxy-terminal 70–90 residues of the four core histones. The iodination also probes the carboxy-terminal region, while the formaldehyde cross-linking may refer to either amino- or carboxy-terminal regions of the protein. Comparative Raman spectroscopy (Thomas et al., 1977) of salt-extracted core histones in 2 M NaCl (pH 7, 90 mg/ml) and nucleosome core particles consisting of 140-base-pair DNA and eight histones showed that the average local structure of the histones as evidenced by its vibrational spectrum is relatively unchanged. The positions and intensities of the Amide I and III bands of a solution of core histone in D_2O, 2 M NaCl, were fit to a structure containing 51 ± 5% helix, 36 ± 4% disordered, and 13 ± 9% β-sheet.

During nucleosome reconstitution, performed by mixing core histones and DNA in 2 M NaCl with slow back-dialysis to low salt concentration and monitoring the CD spectrum—in particular the ellipticity at 227 nm due to the peptide bond—revealed little variation in protein structure (Wilhelm *et al.*, 1978), in agreement with the above.

In addition, high ionic strength, while promoting the correct folding and complexing of the core histone proteins, also promotes well-defined and reversible higher order association phenomena (Sperling and Bustin, 1974, 1975; Wachtel and Sperling, 1979). Pairs of histones and all the four core histones mixed together all form very similar fibers of 40–80 Å in diameter (see Fig. 3). The structure of the assembled histone fibers and their relation to chromatin will be the subject of Section IV.

III. HISTONE–DNA INTERACTIONS

A. Introduction

The role of histone proteins that is best understood is the compaction of the nuclear DNA. The current picture of histone–DNA interactions which give rise to this compaction is that to a first approximation they may be divided according to (*a*) DNA–core histone interactions involved in the formation of nucleosomes and (*b*) DNA–H1 interactions related to the higher order packaging of the nucleosomes. Little structural detail is available about (*a*) and even less about (*b*). The nucleosome represents the first level of compaction, i.e., 140–200 base pairs of B-form DNA supercoiled around the histone core, with a linear extent of ~55 Å (for references, see Section I). With respect to types and strength of interactions, the core histones can be divided into interactions of the arginine-rich histones H3 and H4 with the DNA and those of the lysine-rich histones H2A and H2B with the DNA. In this regard, the arginine-rich histones are electrostatically more stongly bound than the lysine-rich pairs (Ohlenbusch *et al.*, 1967). Nevertheless, as will be shown, H2A·H2B pairs are preferentially cross-linked to DNA by ultraviolet radiation (Sperling and Sperling, 1978).

As will be discussed in this section, selective degradation and reconstitution of core particles have pointed to H3 and H4 as being both necessary and sufficient for the first level of compaction. Also, only the trypsin-resistant carboxy-terminal regions of all the histones appear to be necessary for nucleosome formation. Exactly what is the primary role of the amino-terminal regions therefore remains somewhat obscure. Due to the special charge distribution in the histone molecules (see Figs. 1 and 2) the amino-terminal regions have a cluster of basic residues resulting in a large net positive charge. It has been suggested that consequently this part of the histone molecule must interact with the negatively charged phosphate groups of DNA. There is as yet no direct evidence for this interaction, although the argument is plausible. Furthermore, enzymic modifications which occur specifically in the amino-terminal region affect the net charge. Such modifications are correlated with nuclear events in which there is an alteration in histone–DNA interaction (such as during transcription). These facts suggest that the interaction of the amino-terminal regions of the histones with DNA, although currently not well understood, may play an important role in the dynamic aspect of chromosome structure, varying during the cell cycle. This possibility will be discussed in Section V.

B. DNA–Core Histone Interactions Involved in Formation of the Nucleosome

1. Interactions in the Intact Nucleosome—Cross-Linking Studies

In contrast to the extensive use of cross-linking to study histone–histone interaction, few histone–DNA cross-links have been identified. Nevertheless, protein–nucleic acid cross-links have been shown to be significant products of UV irradiation of prokaryotic and eukaryotic organisms (Smith, 1976; Todd and Han, 1976). Such cross-links occur also *in vitro* upon irradiation of native nucleoprotein complexes and have been found to be major products in irradiated chromatin (Strniste and Rall, 1976; Sperling and Havron, 1977). Photochemical cross-linking experiments can provide detailed information regarding protein–nucleic acid contacts due to (*a*) the ability of both purines and pyrimidines to form UV-induced covalent adducts with a major number of amino acids (for references, see Sperling and Havron, 1977) and (*b*) the formation of covalent bonds *only* between neighboring residues in the native nucleoprotein structure (Sperling and Havron, 1976; Havron and Sperling, 1977). In some cases, it has proved possible to identify the residues located in the binding site (Havron and Sperling, 1977).

An attempt was made to produce and identify such "zero-length" protein–DNA cross-links in the intact nucleosome (Sperling and Sperling, 1978). Rat liver mononucleosomes containing 185 base pairs of DNA and a full complement of the histones including H1 were irradiated for 5 minutes with UV light of wavelength $\lambda > 290$ nm in the presence of acetone as a photosensitizer. Under those irradiation conditions H1, H2A, and H2B were the first proteins involved in UV-induced cross-linking. The cross-linking of H1 to DNA will be discussed later. The cross-linked nucleosomes were separated into DNA and protein fractions. The histones bound to

DNA in the protein fraction were identified by direct analysis. Using two independent labeling experiments, it was shown that of the four core histones, histones H2A and H2B were preferentially cross-linked to the DNA. Equimolar amounts of H2A and H2B were cross-linked to the DNA. Furthermore, an H2A·H2B dimer, cross-linked by a short piece of DNA, was also identified. This is in agreement with the finding of a piece of DNA 40–50 base pairs long, bound to H2A and H2B in the digestion products of chromatin (Nelson et al., 1977). At short times, the UV-induced disappearance of histones H3 and H4, as well as part of histones H2A and H2B, from their normal migration position on polyacrylamide gels, can be accounted for by histone– histone cross-linking with aggregate formation (Martinson et al., 1976; Sperling, 1976). It appears that histone H4, although capable of forming covalent adducts with ATP (Sperling, 1976), does not cross-link to DNA under the short irradiation period used in the above studies (Sperling and Sperling, 1978).

As discussed in Section I digestion of nucleosome particles with DNase I produces a set of DNA fragments. However, not all the cutting points show the same frequency of cutting. It has been suggested that the low-frequency cutting sites in the DNA are protected by neighboring histones. The photochemical cross-linking studies described here (Sperling and Sperling, 1978) indicate that within the nucleosome core histones H2A and H2B are in close proximity to the DNA. It is possible, therefore, that some of the more infrequent cutting sites of DNase I are at the contact points of DNA with histones H2A and H2B.

As will be described later in this section, reconstitution and salt effect studies have indicated that histones H3 and H4 have a central role in the folding of the DNA, and that electrostatically they are more tightly bound to the DNA than H2A or H2b. Thus, the absence of cross-links between DNA and histone H3 and H4 may appear surprising. However, the results showing that histones H2A and H2B cross-link preferentially to the DNA are not necessarily contradictory to these studies. There are at least two possible explanations. On the one hand, it is possible that after the DNA has been folded into its nucleosomal conformation by the H3·H4 tetramer, binding sites for H2A and H2B are created in which H2A and H2B are in closer proximity to the DNA. On the other hand, although proximity is a necessary requirement for the occurrence of cross-linking, steric factors may also have to be considered in order to explain the preferential photochemical cross-linking of H2A and H2B. For instance, the interacting residues may have to be appropriately oriented at the contact points prior to thir cross-linking. In any event photochemical cross-linking shows that within the nucleosome core the four core histones are not equivalently positioned with respect to the DNA. This histone–DNA arrangement leads to the preferential cross-linking of the lysine-rich histones H2A and H2B to the DNA.

Histones H1 and H2A·H2B were cross-linked to the DNA by UV irradiation of calf thymus nuclei (Kunkel and Martinson, 1978), in agreement with the above results (Sperling and Sperling, 1978). Kunkel and Martinson's further observation of H3 cross-linked to DNA may be due to a higher dosage of UV light employed for cross-linking as compared to that of Sperling and Sperling (1978). When chromatin and isolated nucleosomes were irradiated with higher energy UV light, i.e., $\lambda = 254$ nm, all histones were interpreted as being cross-linked to the DNA and at an equal rate, as judged by their disappearance from SDS–polyacrylamide gels (Mandel *et al.*, 1979).

Another method used to study histone–DNA interaction was the treatment of DNA with dimethyl sulfate. This results in the methylation of the N-7 atom of guanine in the major groove and the N-3 atom of adenine in the minor groove of DNA (for references, see Singer, 1975). The comparative kinetic measurements of methylation of DNA and chromatin have shown that in isolated chromatin, in chromatin reconstituted from core histones and DNA, nuclei, mononucleosomes, as well as intact cells (at low temperatures), the minor groove of the DNA is well exposed and not shielded by histones. The major groove of the DNA is rather exposed and shielded by the histones against methylation of the DNA to the extent of 14–20% (Mirzabekov *et al.*, 1977, 1978a). However, when ^{32}P-terminally labeled nucleosomal and naked DNA were methylated, and the patterns of strand cleavage at methylation sites of both cases were compared, no significant difference was detected between the reactivity of N-7 of guanines in nucleosomal DNA and of that in naked DNA (McGhee and Felsenfeld, 1979). These results indicate that in nucleosome DNA the bases are nearly as accessible to solvent as they are in naked DNA.

A cross-linking method was used to map the arrangement of histones along the length of the DNA. In this approach (Levina and Mirzabekov, 1975) core particles were methylated at the N-7 position of guanine and the N-3 position of adenine and the alkylated bases were depurinated by heating. The aldehyde group formed would react with the ε-amino group of a lysine residue of neighboring histone to form a Schiff base. This cross-linking was further stabilized by reducing the double bond by sodium borohydride. Cross-linking causes the specific scission of one DNA strand at the point of cross-linking and the attachment of only the 5'-terminal DNA fragment to histones (Mirzabekov *et al.*, 1978a). Using this method and 5'-end labeling of the DNA with polynucleotidyl kinase, histone H3 was shown to be cross-linked to the 5' end of core particle DNA (Simpson, 1978b).

Recently a map of cross-linking sites of histones to DNA in core particles was obtained by using the same approach and analyzing the results with 5'-end labeling of the DNA after cross-linking and iodination of the histones (Mirzabekov *et al.*, 1978b). The results of the cross-linking of ϵ-amino groups of lysines with the neighboring aldehyde groups in randomly distributed depurinated sites of the DNA (Mirzabekov *et al.*, 1978b) led to the suggestion that the two molecules of each histone type in the core are likely to be bound similarly, one to each of the antiparallel strands of the DNA, in agreement with the proposed dyad axis in the core (Finch *et al.*, 1977).

2. Stepwise Salt Dissociation and Association

There is a strong electrostatic contribution to the interaction of histones with DNA. Increasing concentration of salt induces the stepwise dissociation of histones from chromatin and nucleosomes. There are three stages in this process. In the first stage, up to 0.7 *M* NaCl, H1 is selectively dissociated. Between 0.7 and 1.2 *M* NaCl, H2A and H2B are removed, whereas the arginine-rich histones H3 and H4 are dissociated only above 1.2 *M* NaCl. At 2.0 *M* NaCl the DNA of chromatin is free of histones (Ohlenbusch *et al.*, 1967; Burton *et al.*, 1975).

The binding of the core histones to the DNA is a reversible process. The interaction of lysine-rich histones with DNA is noncooperative whereas H3 and H4 bind to the DNA in a cooperative manner (Burton *et al.*, 1978). The reversibility of the interaction is demonstrated by the stepwise reconstitution of nucleosomes from histones plus DNA upon gradual decrease in salt concentration. This stepwise reconstitution has been studied using several procedures: by mixing the four core histones (Wilhelm *et al.*, 1978) with DNA at 2 *M* NaCl; by bringing chromatin to 2 *M* NaCl and then decreasing the salt to low ionic strength either in the presence (Burton *et al.*, 1978) or absence of H1. In each of these cases the arginine-rich histones H3 and H4 bind first to the DNA in the range of 2 *M* to 1.2 *M* NaCl. The binding of H3 and H4 imposes a nucleosome-like structure observable in the electron microscope (Wilhelm *et al.*, 1978). Nuclease digestion studies have

shown that the bound H3·H4 tetramer protects 145-base-pair fragments of DNA, comparable to the DNA component of the complete
core particle, and therefore the tetramer spans the nucleosome core
particle (Burton *et al.*, 1978). The proper binding of H2A and H2B requires the presence of bound H3 and H4 and occurs below ionic
strength of 1.2 *M*. It seems therefore that the binding of H3 and H4 to
DNA folds the DNA and results in the formation of a binding site for
H2A and H2B.

3. Subnucleosomal Particles

Two types of subnucleosomal particles which retain many, if not all,
of the properties of the intact nucleosome have been identified. The
first type contains only H3 and H4, either as a tetramer (Bina-Stein
and Simpson, 1977) or an octamer (Simon *et al.*, 1978; Stockley and
Thomas, 1979), while the second contains all core histones, each
lacking up to 30 amino-terminal residues which have been digested
away by trypsin (Whitlock and Simpson, 1977). The fact that other
subnucleosomal particles have not been isolated does not necessarily
mean that they cannot exist; it indicates only that the proper reconstitution or dissociation conditions have not been found. Nevertheless,
results to date point to H3·H4 on the one hand, and the trypsin-
resistant carboxy-terminal regions of all the core histones on the other
hand, as playing controlling structural roles in the formation of the nucleosome and the consequent folding of the DNA.

a. H3·H4–DNA. Reconstitution experiments performed with the
arginine-rich histones H3 and H4 (either acid- or salt-extracted) and
DNA resulted in the formation of nucleoprotein complexes which resembled chromatin with respect to the digestion patterns obtained
using micrococcal nuclease, DNase I, DNase II, trypsin, and chymotrypsin (Camerini-Otero *et al.*, 1976; Sollner-Webb *et al.*, 1976).
Other combinations of histones did not result in these characteristic
patterns. The nuclease digestion experiments indicate that in the
complex reconstituted with H3 and H4 alone the DNA is folded in
such a way that the periodicity of cleavage of nucleosomal DNA by
nucleases is very closely retained. Mild proteolytic cleavage experiments of core histones in nucleosomes results in cleavage of their
amino-terminal region, keeping the carboxy-terminal region intact
(Weintraub and Van Lente, 1974; Weintraub *et al.*, 1975). Comparable experiments with complexes reconstituted from DNA and H3 and
H4 revealed that these complexes resemble chromatin with respect to
protection from digestion of the carboxy-terminal region.

The reconstituted H3·H4–DNA complexes have a nucleosome-like appearance in the electron microscope (Bina-Stein and Simpson, 1977; Oudet *et al.*, 1978) and the low-angle X-ray diffraction of the reconstituted complex reveals the same intensity maxima at 6.0, 3.5, and 2.8 nm that are seen in experiments with intact nucleosomes (Boseley *et al.*, 1976; Moss *et al.*, 1977; Bradbury *et al.*, 1978). In several independent experiments it was shown that H3 and H4 can induce supercoiling in closed circular DNA (Bina-Stein and Simpson, 1977; Camerini-Otero and Felsenfeld, 1977b; Oudet *et al.*, 1978).

As to the stoichiometry of the H3·H4–DNA particle, two complexes were identified: an H3·H4 tetramer and an H3·H4 octamer, each associated with about 140 base pairs of DNA. The complexing of 140 base pairs of DNA with H3 and H4 resulted in the formation of nucleosome-like particles, as observed by the EM, and reported to have an $s_{20,w} \sim 9.8$ and contain a tetramer of H3 and H4 per 140 base pairs (Bina-Stein and Simpson, 1977; Bina-Stein, 1978). These results differ from those of Simon *et al.* (1978) who report that at least two complexes of H3·H4–DNA can be obtained upon reconstitution of H3, H4, and 150 bp DNA. In this experiment both an octamer and a tetramer of H3·H4 were found bound to 150 base pairs of DNA, having $s_{20,w}$ equal to 10.4 and 7.5 for the octamer and tetramer, respectively. The stoichiometry of the complexes obtained is dependent on the histone-to-DNA ratio. At low ratios of histone to DNA the predominant species contains an H3·H4 tetramer per 150 base pairs of DNA. At a histone-to-DNA ratio of 1:1 the octamer prevails. The nuclease and protease digestion experiments (Camerini-Otero *et al.*, 1976; Sollner-Webb *et al.*, 1976) were performed at a histone-to-DNA ratio of 0.5, conditions which for 140-base-pair DNA would lead primarily to a tetrameric complex. Therefore, it seems that a tetramer of H3·H4 is sufficient for the generation of nuclease-resistant fragments similar to those of complete nucleosomes. Upon addition of H2A and H2B to the tetrameric complex, nucleosomes are formed. Addition of H3·H4 to the tetrameric complex resulted in an octameric complex which is similar in compaction to nucleosomes. H3·H4 tetramers and octamers were similarly found complexed with about 140 base pairs of DNA upon reconstitution of H3·H4 with SV40 DNA. Both complexes were reported to be able to fold 140 base pairs of DNA (Thomas and Oudet, 1979).

In experiments performed with chromatin or core particles depleted of histones except for H3 and H4 (Stockley and Thomas, 1979), two complexes were obtained, one containing an octamer and one a tetramer of H3 and H4 per 140 base pairs of DNA. The physical properties of the two core complexes were similar to those observed by

Simon *et al.* (1978). The complex with an H3·H4 octamer more closely resembled the nucleosome core particle with respect to compaction, suggesting that H3 and H4 could replace H2A and H2B in the compaction of the DNA into nucleosome core particles.

b. Trypsin-Digested Nucleosomes. Trypsin digestion of core particles removes up to 30 amino acid residues from the amino-terminal end of each histone, leaving approximately the carboxy-terminal two-thirds intact. This treatment results in an increase of the CD peak at 280 nm from 2000 deg · cm^2/dmol PO_4 to 4000 deg · cm^2/dmol PO_4 (Whitlock and Simpson, 1977; Simpson *et al.*, 1978). The DNA must still be partly folded as the normal value for DNA in solution is 8000 deg · cm^2/dmol PO_4. Furthermore, the nucleosomes sedimented similarly to intact particles and DNase I digestion revealed $\sim n \times 10$ periodicity of susceptible sites, although base pairs 60–80 from the 5' end showed an increased susceptibility to digestion. The 5' end of the nucleosomal DNA also became more susceptible to nuclease digestion after treatment with trypsin. These positions were therefore presented as possible binding sites for the amino-terminal peptides (Whitlock and Simpson, 1977; Simpson *et al.*, 1978).

The inverse of the above experiments gave similar results (Whitlock and Stein, 1978). Trypsin-digested histones removed from HeLa core particles can subsequently fold DNA, although DNase I digests the resulting particles more rapidly than the untreated ones. Parallel experiments were performed for chicken erythrocyte core particles (Lilley and Tatchell, 1977). In all cases it could be concluded that it is the trypsin-insensitive carboxy-terminal regions of the histones which are responsible for the folding of the DNA in the nucleosome.

It is interesting to note that nucleohistone lacking amino termini does not aggregate as readily in the presence of $MgCl_2$ as do untreated particles (Whitlock and Stein, 1978). It might be hypothesized from this result that, although the amino termini are not necessary for the maintenance of nucleosome structure, they are involved *in vivo* in internucleosomal interactions.

C. H1–DNA Interactions

1. The Binding of H1 to Mononucleosomes

H1 forms a microheterogeneous group of very lysine-rich histones. In nucleated avian erythrocytes H5 is also present (for references, see Cole, 1977), bearing a similarity in sequence and stoichiometry to H1. Histone H1 does not interact with the other core histones in solution

(Sperling and Bustin, 1976) and does not self-associate *in vitro*, yet in chromatin poly(H1) cross-linking is observed (Thomas and Kornberg, 1975a; Bonner and Pollard, 1975; Chalkley and Hunter, 1975). H1 is the first histone to dissociate from DNA upon increase in salt concentration (Ohlenbusch *et al.*, 1967; Burton *et al.*, 1975), i.e., at about $I = 0.45-0.7$. As already discussed in Section I, H1 is removed upon micrococcal nuclease digestion of nucleosomes to form nucleosome core particles. This led to the suggestion that histone H1 is bound to the DNA linker. The core histones alone are sufficient to maintain the fold of the DNA and form nucleosome core particles, which have a low-angle X-ray diffraction pattern similar to that of chromatin (Finch *et al.*, 1977). This fact, together with the compaction of chromatin obtained by increasing the concentration of salt and magnesium ions in the presence of H1, gave further support to the idea that H1 is involved in the higher order structure of chromatin (Littau *et al.*, 1965; Mirsky *et al.*, 1968; Bradbury *et al.*, 1973; Oudet *et al.*, 1975; Finch and Klug, 1976; Noll and Kornberg, 1977; Renz *et al.*, 1977). However, in order to understand this effect, a better understanding of the mode of interaction of H1 with the nucleosome is required.

Ultraviolet irradiation of mononucleosomes composed of 185 base pairs of DNA and a full complement of the histones including H1 resulted in the photochemical cross-linking of histone H1. Histone H1 was the first to disappear from its normal migration position on SDS–polyacrylamide gels. Fractionation of the cross-linked products yielded a water-soluble covalent complex of DNA and histones containing 15% protein. This complex, which had a mobility very similar to that of 185-base-pair DNA on agarose gels, was partially resistant to mild micrococcal nuclease digestion. It exhibited a CD spectrum similar to that of chromatin, had the appearance of holey circles in the EM, and exhibited a positive reaction with antibodies against histone H1 (Sperling and Sperling, 1978). This led to the conclusion that histone H1, although anchored to the DNA linker, spans the whole length of the nucleosome and binds to the DNA at additional sites. As shown in the schematic model in Fig. 5, by doing so histone H1 clamps the folds of the DNA molecule, once it has been coiled around the nucleosome core (Sperling and Sperling, 1978). Therefore, it seems that although H1 is not required for folding the DNA around the histone core, H1 may stabilize the fold. A similar model of H1 binding to the nucleosome subparticle containing 160 base pairs of DNA was suggested by Simpson (1978a). The stabilizing effect of H1 binding suggested by the photochemical cross-linking (Sperling and Sperling, 1978) may explain the results obtained in

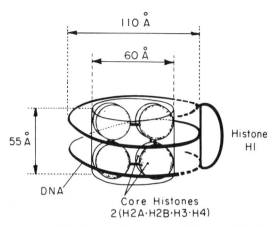

FIG. 5. Schematic model of the nucleosome, with histone H1 shown as stabilizing the fold of the DNA molecule around the core histones [based on results of Sperling and Sperling (1978)]. The nucleosome dimensions are derived from X-ray (Finch *et al.*, 1977) and neutron (Baldwin *et al.*, 1975; Pardon *et al.*, 1977; Suau *et al.*, 1977) scattering experiments. The histone core dimensions are derived from electron microscopic and X-ray studies (Sperling and Amos, 1977; Wachtel and Sperling, 1979; Sperling and Wachtel, 1979). The regions of the DNA molecule indicated by dashed lines indicate those base pairs which are not present in nucleosome core particles.

digestion studies of chromatin with DNase II which yielded repeats of 100 base pairs in the presence of H1 in contrast to the 200-base-pair repeats obtained by micrococcal nuclease digestion (Altenburger *et al.*, 1976) by DNase II in the absence of H1.

This mode of binding of H1 to the nucleosome is consistent with the findings that at low histone-to-DNA ratios and/or at low ionic strength histone H1 binds preferentially to supercoiled DNA (Vogel and Singer, 1975, 1976). This preferential binding has been shown to be the property of the H1 peptide fragment, amino acid residues (73–106).

A similar model for the binding of H1 to the nucleosome was suggested by Thoma *et al.* (1979) on the basis of electron microscopic studies of H1-containing and H1-depleted chromatin. At very low ionic strength nucleosomes were observed in chromatin but not in H1-depleted chromatin. At somewhat higher ionic strength nucleosomes were visible in both samples. However, in chromatin containing H1 the DNA enters and leaves the nucleosome on the same side, whereas in H1-depleted chromatin the entrance and exit points are more randomly distributed and tend to occur on opposite sites of the nucleosome. A regular "zig-zag" pattern of "beads on a string" was obtained from chromatin at 1 mM NaCl in the presence of H1, suggesting that H1 binds at that region where the DNA molecule enters and leaves the nucleosome, i.e., at the linker region.

2. Histone H1 and Higher Order Structures of Chromatin

The interaction of H1 with relaxed DNA and with superhelical DNA at high histone-to-DNA input ratios results in a cooperative salt-promoted binding with the concomitant formation of aggregates (Böttger et al., 1976b; Singer and Singer, 1978). Some of these aggregates have been observed to have a doughnut shape in the electron microscope (Hsiang and Cole, 1977). For complexes formed from relaxed DNA, but not from superhelical DNA, the CD spectrum appears to be characteristic of ψ DNA (Böttger et al., 1976a). In Section III,C,1 analogy was drawn between the mode of binding of H1 to individual nucleosomes and its preferential binding to superhelical DNA. A possible relationship may also be drawn between the above-mentioned aggregates and the ability of H1 to condense chromatin to form higher order structures. Upon increase in ionic strength, chromatin folds up progressively to form higher order structures (Thoma and Koller, 1977; Renz et al., 1977; Thoma et al., 1979). Several models of higher order structure in chromatin have been proposed. We mention two: the solenoid model (Finch and Klug, 1976) and the superbead model (Renz et al., 1977; Hozier et al., 1977; Vengerov and Popenko, 1977), both of which may coexist representing different states of the same material (Olins, 1978). Both models require the presence of H1 and magnesium ions and invoke H1–H1 interactions. A possible way in which H1 binding to the nucleosome may influence the higher order structure of chromatin is by the formation of H1 aggregates along the chromatin superstructure inside the solenoid hole (Finch and Klug, 1976) or outside the superstructure (Worcel, 1978).

3. Effect of H1 Phosphorylation on Interaction with DNA

The phosphorylation of histone H1 has been reviewed several times in recent years (Dixon et al., 1975; Elgin and Weintraub, 1975; Gurley et al., 1978; Langan, 1978; Isenberg, 1978). Therefore, we will not give a full discussion of the subject here. Only points relevant to H1–DNA interaction will be mentioned.

Histone H1 can be modified through phosphorylation (Langan and Hohmann, 1975). A change in the extent of H1 phosphorylation was correlated with the onset of mitosis (Lake and Salzman, 1972) and it has been suggested that this serves as a trigger for chromosome condensation and mitosis (Bradbury *et al.*, 1974). It was also associated with the onset of cellular DNA synthesis (Balhorn *et al.*, 1972). Site-specific phosphorylation of H1 occurs in rat liver upon glucagon administration (Langan, 1969). Phosphorylation occurring during interphase involves 1–3 serines on the carboxyl side of Tyr-73. Another type of phosphorylation involves 3–6 serine and threonine sites distributed throughout the molecule and occurs during mitosis (Gurley *et al.*, 1978).

It has been suggested that H1 phosphorylation may alter H1–DNA interactions (Adler *et al.*, 1972). The comparative study of binding of phosphorylated and nonphosphorylated H1 to superhelical circular DNA has shown that the four phosphorylated species studied (Ser-37; Ser-106; Ser-37 and Ser-106; five sites of serine and threonine) bind DNA and discriminate between superhelical and relaxed DNA with the same efficiency as nonphosphorylated H1. However, qualitative differences were observed in the effect of salt on H1 binding to DNA. Whereas H1 would bind maximally to superhelical DNA at low ionic strength, phosphorylated H1 (Ser-37, Ser-106) binds and recognizes superhelical DNA more efficiently than H1 in the absence of salt (Singer and Singer, 1978). Results supporting this finding were also obtained with *Col*E1 superhelical DNA (Knippers *et al.*, 1978).

IV. Histones as the Chromatin Skeleton

A. Biological Assembly Systems

In biological systems built from many copies of a few components, the principles of self-assembly are utilized for the formation of organized repeating superstructures (Caspar and Klug, 1962). In such structures the constituent subunits occupy equivalent or semiequivalent positions in space and therefore have the same kind of interactions with neighboring subunits. Depending on the symmetry of the assembled structure, each subunit may have two or more specific binding sites for neighbors at the surfaces of the subunit. The formation of the assembled structure does not involve major changes in the secondary and tertiary structure of the subunit, and to a first approximation one can assume that only small, local changes occur at the binding sites. The intersubunit interaction energy is of the order of 5–10 kcal/mol, and the assembly is reversible.

It has been shown in several biological assembly systems that, due to the fact that tertiary structure determines quaternary structure, a major component of the system can be assembled on its own to form structures similar to the intact organelle (Caspar and Klug, 1962). Examples are coat proteins of viruses such as tobacco mosaic virus (TMV) (for references, see Butler and Durham, 1977), and the different protein components of the muscle system (for references, see Clark and Spudich, 1977; Cohen, 1979). It should be pointed out, however, that the assembly of a major component on its own usually requires conditions substantially different from that of the assembly of the complete biological structure. This would seem logical in order for the system to compensate for the missing component and also to safeguard against the formation of partial superstructures under conditions in which the intact superstructure is required.

B. Chromatin as an Assembly System

Histones were initially assumed to form part of an assembly system (Sperling and Bustin, 1974, 1975) for two reasons. First, they are present in chromatin as many copies of a few components. [The mass of the five histones equals that of the DNA (for references, see Phillips, 1971; Hnilica, 1972).] Second, fiber X-ray diffraction studies of chromatin (Luzzati and Nicolaieff, 1959; Wilkins *et al.*, 1959; Pardon *et al.*, 1967; Pardon and Wilkins, 1972) indicated that they are involved in a regular periodic structure. The fact that the histones are highly conserved with respect to their primary structure (Dayhoff, 1972; Wilson *et al.*, 1977), especially the arginine-rich histones H3 and H4, also indicates that they may play a primarily structural role in which each residue is crucial.

Since 1974, evidence has accumulated in the literature which indicates that chromatin itself may be considered as an assembly system. It is true that chromatin is more complex than assembly systems analyzed to date, both with respect to the size of the nucleic acid involved and therefore the amount (and variety) of protein complexed with it and with respect to the dynamic aspect of the multilevel higher order structure. Nevertheless, at least at the lower levels of organization, the interpretation of chromatin as an assembly system may be valid. Evidence for this derives from three basic lines of research described in previous sections: (1) the reconstitution of the nucleosome, (2) the self-assembly of the octamer, and (3) the putative self-organization of nucleosomes into higher order structures.

1. Reconstitution of Nucleosomes

The fact that nucleosome-like particles can be reversibly reconstituted from histones and DNA, without any additional factor, exemplifies the principle of self-assembly of biological structures. The reconstituted particles have the characteristic beaded appearance of nucleosomes in the electron microscope (Germond *et al.*, 1975; Oudet *et al.*, 1975), and give typical periodic digestion fragment patterns when digested with nucleases (Axel *et al.*, 1974; Camerini-Otero *et al.*, 1976; Sollner-Webb *et al.*, 1976). These reconstitutions can be obtained using both acid- and salt-extracted histones. Chromatin core particles may also be reconstituted from core histones and 145-base-pair DNA (Tatchell and Van Holde, 1977). The resulting particles appear identical to native core particles with respect to physical properties such as sedimentation velocity, histone content, CD spectrum, melting profile, and the DNase I digestion pattern.

2. Assembly of the Histone Octamer

The histone octamer is the histone unit of the nucleosome. As discussed in Section II, it has been shown that at high salt concentration ($I > 2\ M$) the core histones can assemble on their own, in the absence of DNA, to form histone octamers (this assembly occurs with both acid- and salt-extracted histones). Furthermore, the secondary and tertiary structures of core histones at high salt concentration are similar to the structures they have in the intact nucleosome. The basic units of the assembly of the four core histones are histone dimers which are obtained at low salt concentration. Upon increase in salt concentration, tetramers, hexamers, and octamers are obtained. The cross-linking pattern of histones in high salt concentration is similar to that in chromatin, again supporting the idea that the assembly of core histones at high salt concentration is similar to that in chromatin.

3. Higher Order Structures of Chromatin and Histone Fibers

The evidence that the nucleosome behaves like part of an assembly system is rather clear both from the point of view of reconstitution and from the ability of the histones to assemble on their own into structures similar to those of the intact organelle. To extend this analogy to the assembly of nucleosomes to form higher order structures is less straightforward. Our knowledge and understanding of the higher order structure of chromatin is rather limited at present. However, as described in Section III, we do have some information concerning the type of structure(s) which may be responsible for the higher levels of organization.

At low ionic strength, chromatin has the appearance of "beads-on-a-string" as observed in the EM (Olins and Olins, 1974; Woodcock, 1973). However, it is more compact in its native environment in the nucleus (for references, see Chambon, 1978). Several levels of compaction beyond the nucleosome are required to account both for the dimensions of the chromosome and for the structural changes that it undergoes during the cell cycle. Electron microscopy of chromatin has so far revealed two basic structures, "thin fibers" 100 Å in diameter and "thick fibers" 200–300 Å in diameter (Ris and Kubai, 1970; Dupraw, 1970; Ris, 1974; Solari, 1974; Davies and Haynes, 1975; Finch and Klug, 1976; Worcel and Benyajati, 1977; Rattner and Hamkalo, 1979). A solenoid model was proposed for the 300-Å structure (Finch and Klug, 1976). The solenoid is formed by coiling the 100-Å fiber into a shallow helix. Physical studies have also suggested that nucleosomes could be arranged in helical arrays to form higher order structures (Carpenter et al., 1976; Pardon et al., 1978).

A second model to explain the 200- to 300-Å fiber is the "superbead." In this model the thick fiber is discontinuous and is made up of 200-Å beads containing 8 nucleosomes per globule (Renz et al., 1977).

A low-energy in vitro form of nucleosome packing was observed in nucleosome core particle crystals (Finch et al., 1977). Two variants of these crystals occurred. (a) Wavy columns of nucleosomes stacked one on top of each other with an axial repeat of 340 Å were obtained upon crystallization of nucleosomes containing proteolytically cleaved histones (Finch et al., 1977). (b) Straight columns of closely packed nucleosomes, 110 Å in diameter, were obtained upon crystallization of nucleosomes with intact histones (Finch and Klug, 1978). In both these structures histone–histone contacts between nucleosomes are implied. Similar face-to-face packing of nucleosomes in arcs and helical patterns was observed in the EM by Dubochet and Noll (1978).

Obviously, several of the model structures described here could accommodate a continuous core of histones inside a closely packed array of nucleosomes. That such a continuous histone core might exist is independently suggested by the observation that histones not only can self-assemble into the octameric unit but that, under similar conditions of high salt concentration, they can further assemble into extended fibrous structures possessing helical symmetry (Sperling and Bustin, 1974; 1975; Sperling and Amos, 1977; Wachtel and Sperling, 1979; Sperling and Wachtel, 1979). These fibers have intra- and interoctamer complexing patterns similar to those found in chromatin [determined by cross-linking experiments (Sperling and Bustin, 1975; Sperling and Wachtel, 1979)]. The relationship of the histone fibers to the histone core of chromatin is further strengthened by the fact that the fibers have axial periodicities (Sperling and Amos, 1977; Sperling and Wachtel, 1979) similar to those found in X-ray diffraction patterns of nucleosome crystals and chromatin. In this sense, histones behave like part of a classical assembly system.

C. Self-Assembly of Histones to Form Histone Fibers

1. Self-Assembly of Single Histones

As described earlier, the assembly pathway of pairs of histones proceeds from dimers through tetramers, hexamers, and octamers to form fibers (Sperling and Wachtel, 1979). In the EM these fibers as well as those composed of all four histones are observed to be helical, 40–80 Å in diameter with a 330-Å pitch, and to contain an even number of strands (Sperling and Bustin, 1975; Wachtel and Sperling, 1979; Sperling and Wachtel, 1979). The fact that morphologically identical structures (see Fig. 3a) can be reversibly obtained also with single histones (Sperling and Bustin, 1974, 1975; Sperling and Amos, 1977; Wachtel and Sperling, 1979) emphasizes the general structural similarities and perhaps even interchangeability of the histones. As with the other fibers, several intermediates can be observed. There is a transition from single histone monomers and dimers to single histone fibers, the process being favored by increase in protein concentration, ionic strength, and temperature. The cross-linking of solutions of single histones gave a series of multimers from mono-, di-, tri-, and tetra- up to octamer and above. Cross-linking with glutaraldehyde and dimethylsuberimidate under several different sets of conditions has shown that single histone dimers and tetramers are predominant intermediates on the pathway to higher multimers (Sperling and Bustin, 1975).

2. Structure of Histone Fibers from Optical Diffraction

Optical diffraction from electron micrographs of single histone fibers and bundles (which are better specimens than the single fibers for optical diffraction studies) gave similar diffraction patterns. The striking observation is that the main reflections obtained from histone fibers are at axial spacings of 55, 37, and 27 Å (Sperling and Amos, 1977; Sperling and Wachtel, 1979). These are the dominant low-angle rings obtained by X-ray diffraction of chromatin (Luzzati and Nicolaieff, 1959; Wilkins et al., 1959; Pardon et al., 1967; Pardon and Wilkins, 1972; Boseley et al., 1976). However, in the case of the histone fibers, there are other reflections at intermediate spacings which lead to a diffraction pattern with 12 layer lines which are subdivisions of a structure with a 330-Å repeat. Calculation of the Fourier transforms of the single fibers from their digitized images and comparison

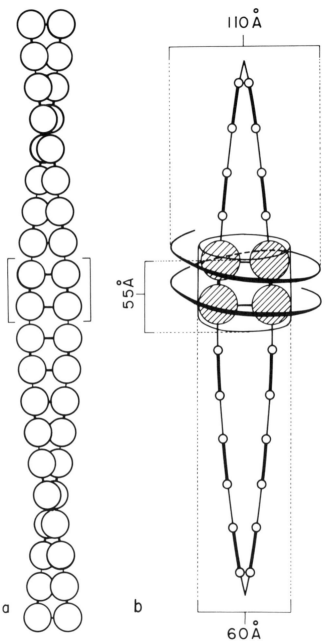

FIG. 6. (a) A schematic model of the helical, double-stranded, unstaggered, H4 fiber (Sperling and Amos, 1977). The asymmetric unit is an axial dimer and there are six such dimers per strand per repeat. The repeat distance is 330 Å. The two different types of axial bonds—within and between dimers—are denoted by a thick and thin line, respectively. The tetrameric grouping is indicated. (b) A model of (a) upon which is superimposed a schematic representation of a nucleosome core particle

of the phases on opposite sides of the meridian led to the conclusion that the Bessel functions on the main layer lines are even functions. The diffraction pattern obtained from electron micrographs was rather noisy and it was difficult to index. However, it was possible to limit the lattice schemes to two very similar ones. Two related schematic models were proposed for the arrangement of subunits in histone H4 fibers (Sperling and Amos, 1977). In both models the fiber is a double-stranded unstaggered fiber in which the asymmetric unit is an axial histone dimer. The best way to explain this is by "head-to-head" and "tail-to-tail" assembly of the histone subunits. Therefore, there are different axial bonds within the dimer and between dimers. Due to the double strandedness of the fibers, units of dimers and tetramers are grouped, in agreement with the cross-linking data (Sperling and Bustin, 1975). The two models differ in the number of subunits per strand per repeat, having six (see Fig. 6) or four histone dimers per repeat per strand, respectively.

Structural studies were also performed on other histone fibers, in particular H3·H4 fibers and fibers prepared from all four core histones mixed in equimolar ratios. Bundles of fibers from both systems have also been obtained. The optical diffraction patterns from electron micrographs again showed dominant axial spacings of 55, 37, and 27 Å, indicating a fundamental similarity of organization for all the histone fibers (Sperling and Wachtel, 1979).

3. X-Ray Diffraction of Histone Fibers

To complement and extend the optical diffraction work, X-ray diffraction data from self-assembled histone fibers were analyzed for three systems: H4, H3·H4, and the four core histones H2A, H2B, H3, and H4 (Wachtel and Sperling, 1979). These data were obtained under conditions of high ionic strength and high protein concentration. Three main types of specimen geometry were investigated: gel, fiber, and film. The third type of specimen, the thin film, gave the best orientation. All systems gave qualitatively similar results.

Fourier analysis of the low-angle equatorial data from semioriented specimens of H4 and H3·H4 indicates that the data are consistent with the scattering from a two- or four-stranded structure, where each strand is about 12–14 Å in radius, the outer diameter is 60 Å, and the

(Sperling and Wachtel, 1979) containing a tetramer of histones and 1.75 turns of smoothly bent DNA (Finch et al., 1977). The superposition illustrates how an H3·H4 double-stranded fiber could form the core of the H3·H4 subnucleosomal particle as well as serving as the arginine-rich kernel of the histone core of closely packed nucleosomes.

strands are separated by a 10-Å central region of low electron density. It is the interference of the protein subfibrils across this 10-Å "hole" which gives rise to the characteristic equatorial diffraction peak at ~36 Å (Wachtel and Sperling, 1979).

Furthermore, there is a striking parallelism between these data and the neutron diffraction data from nucleosomes in 100% D_2O (Pardon *et al.*, 1977; Suau *et al.*, 1977), where scattering from the histone protein dominates, and from core protein in 2 *M* NaCl solution (Pardon *et al.*, 1978). The above interference phenomenon may well be the explanation for the protein-dominated scattering maximum between 35 and 37 Å observed for chromatin and nucleosomes in solution (Pardon *et al.*, 1977; Suau *et al.*, 1977).

4. Histone Fibers Are Related to the Histone Core of Chromatin

The similarity between the layer line spacings found in the optical diffraction patterns of histone fibers (Sperling and Amos, 1977; Sperling and Wachtel, 1979) and the spacings found in X-ray diffraction patterns of chromatin, as well as features of the low-angle equatorial scattering (Wachtel and Sperling, 1979), reinforces the suggestion that the structure of the assembled histone fibers is closely related to a substructure of chromatin, most probably to 100-Å fibers consisting of continuously associated nucleosomes, or to those parts of the higher order structure of chromatin in which close packing of nucleosomes exists. It should be noted that a fibrous model for the histone core of chromatin has also been suggested by Hyde and Walker (1975b). Although the most detailed structural analysis has been carried out for H4 (Sperling and Amos, 1977), the pattern of linear aggregation seems to be a common property of histones (Sperling and Wachtel, 1979). It is likely that H2A, H2B, H3, and H4 are all homologous with respect to fiber formation, but that differences among them are necessary for other functions, for example, for differentiating the continuous 100-Å fibers into beadlike nucleosomes, or for the formation of higher order structures. The interchangeability of the histones is also suggested by recent work of Stockley and Thomas (1979) who found that H3 and H4 could replace H2A and H2B in the octameric histone cores of nucleosomes and chromatin stripped of lysine-rich histones and by the reconstitution experiments of Simon *et al.* (1978). The commensurate nature of the periodicities obtained from optical diffraction of the histone fibers (Sperling and Amos, 1977; Sperling and Wachtel, 1979) and X-ray diffraction of nucleosome core crystals (Finch *et al.*, 1977) is even more striking. A 27-Å meridional reflection is obtained in both cases as well as a strong 55-Å layer line. It is worth noting that one dimension of the crystal unit cell is 340 Å while the repeat of the histone fibers is 330 Å, although this agreement may be accidental. From the point of view of symmetry, the histone model with six histone pairs per repeat per strand (Fig. 6a) is in better accord with the crystallographic data (Finch *et al.*, 1977) and is more likely to be correct for the histone core in chromatin. Furthermore, the model of the histone fibers can easily fit within the volume assigned to the histone core of closely packed nucleosomes.

Figure 6b shows the model of a double-stranded histone fiber with six axial dimers per repeat on which is superimposed a schematic representation of the nucleosome core particle. This schematic model consists of a histone core (derived from the electron microscopic and X-ray studies) with a diameter of 60 Å and a height of 55 Å (Sperling and Amos, 1977; Wachtel and Sperling, 1979), wrapped by 1.75 turns of smoothly bent DNA (Finch *et al.*, 1977). The fiber shows the 27 Å axial spacing of the histone subunits. In the region of superposition, a simplified tetrameric unit is shown, obtained by axial pairing of the histone subunits and best explained by "head-to-head" and "tail-to-tail" axial assembly of histones. The axial pairing results in two kinds of bonds, within the dimer and between dimers. This bipartite nature of the histone tetrameric unit is in very good agreement with the bipartite nature of the nucleosome core particle found by Finch *et al.* (1977). Although the analysis of subunit arrangement in the histone fiber was performed on H4 (Sperling and Amos, 1977), H3·H4 fibers show very similar dimensions and axial periodicities (Wachtel and Sperling, 1979; Sperling and Wachtel, 1979). Therefore it is plausible to assume that at low resolution the model of the H4 fiber could characterize the H3·H4 fiber as well. As discussed in Section III, H3 and H4 constitute a biological functional unit, as on their own they can be reconstituted with DNA to form nucleosome-like particles. The superposition illustrates how an H3·H4 double-stranded fiber could form the core of the H3·H4-reconstituted subnucleosomal particle as well as serving as the arginine-rich kernel of the histone core of closely packed nucleosomes.

In arrays of closely packed nucleosomes composed of all four core histones, strands of H2A·H2B dimers could be incorporated in the grooves between the two H3·H4 strands, producing a four-stranded polymer. Alternatively, they could bind to the H3·H4 double-stranded fiber to give an octamer of the histones per nucleosome. This latter model is supported by the photochemical cross-linking of histones to DNA which have shown that within the nucleosome core the four core histones are not equivalently positioned with respect to the DNA (Sperling and Sperling, 1978). From the three-dimensional reconstruction from electron micrographs (Sperling and Amos, 1977) and low-angle X-ray diffraction of histone fibers (Wachtel and Sperling, 1979), we suggest that the histone core of the nucleosome can be regarded as a truncated fibrous structure of 60 Å maximum outer diameter, 55 Å in height, and having an internal region of low electron density of 10 Å diameter (as illustrated schematically in Fig. 6b).

V. Histones—Not Only a Static Skeleton?

A. Introduction

Although nucleosomes appear to constitute a homogeneous population with respect to composition of histones (for references, see Kornberg, 1977) as observed by SDS–polyacrylamide gel electrophoresis, inspection of histone composition on acid urea gels (Panyim and Chalkley, 1969) or urea triton gels (Zweidler and Cohen, 1972; Alfageme et al., 1974) reveals the presence of microheterogeneity in each group of histone molecules. There are two major sources of this microheterogeneity: (a) the presence of postsynthetic modifications such as acetylation, phosphorylation, methylation, and ADP-ribosylation; and (b) the presence of histone variants whose synthesis is correlated in some systems with different stages of development.

B. Histone Modifications

The topic of histone modifications has been reviewed recently (Elgin and Weintraub, 1975; Dixon, 1976; Allfrey, 1977; Isenberg, 1979), and no attempt will be made here to give a comprehensive review. However, we will present a brief summary of results which can be correlated with the structural role of histones.

1. Acetylation

The topic of histone acetylation has been the subject of extensive reviews (Dixon et al., 1975; Dixon, 1976; Allfrey, 1977; Isenberg, 1979). Acetylation occurs at specific lysine residues. In the case of the four core histones the acetylation sites are all located in the amino-terminal half of the molecule.

Two types of modifications occur (see Figs. 1 and 2): (a) amino-terminal acetylation of histones H1, H2A, and H4 with the formation of α-acetylserine. This modification occurs in the cytoplasm before the histones are transported to the nucleus (Liew et al., 1970), and is apparently an irreversible modification; (b) the formation of N^6-

acetyllysine at specific positions in the amino-terminal region of H4, H3, and H2A and H2B. This type of modification has a very rapid turnover (for references, see Dixon, 1976; Allfrey, 1977). The positions of N^6-acetylation of lysines are precise in the histone sequence and are also conserved. Their location in calf thymus are at lysines 5, 8, 12, and 16 for H4; lysines 9, 14, 18, 23 for H3; lysines 5, 12, 15, and 20 for H2B; and at position 5 for H2A (for references, see Dixon, 1976; Allfrey, 1977). Isolation of histone H4 from calf thymus or rat liver yields two major equal populations with (a) no acetylated ε-lysine residues, and (b) monoacetylated ε-lysine at position 16. [It should be noted that, in contrast, H4 isolated from sea urchin sperm cells contains no ε-acetyllysine residues (Wangh et al., 1972).] The di-, tri-, and tetraacetylated species have a higher turnover.

Histone acetylation has been correlated in the literature with increase of transcriptional activity in several systems (for references, see Allfrey, 1971, 1977; Dixon, 1976), although direct evidence for involvement of acetylation with gene activity is missing. Acetylated H4 was shown to be preferentially associated with a chromatin fraction enriched in actively transcribing genes (Davie and Candido, 1978). Recently, use has been made of the enhancement of multiacetylated histones in HeLa cells and Friend erythroleukemia cells grown in the presence of sodium butyrate (Riggs et al., 1977, 1978). n-Butyrate apparently inhibits deacetylase activity, thereby permitting the study of the effect of modification enrichment on chromatin structure (Perry et al., 1979; Cousens et al., 1979; Boffa et al., 1978; Sealy and Chalkley, 1978; Candido et al., 1978). Although highly acetylated chromatin is degraded faster by DNase I as compared to chromatin from cells grown in the absence of n-butyrate (Simpson, 1978c; Nelson et al., 1978; Vidali et al., 1978; Mathis et al., 1978), the rate of digestion by DNase I of mononucleosomes from cells grown with or without sodium butyrate was comparable (Simpson, 1978c). Similarly, no major differences could be detected in mononucleosomes reconstituted with histones isolated from butyrate or non-butyrate-treated cells (Mathis et al., 1978). Either the degree of enrichment of the 3- and 4-acetylated histones is not sufficient to exhibit the structural differences, which are probably small, or the effect of this modification is not reflected at the level of mononucleosomes but rather at a higher order of chromatin structure, e.g., at the level of interaction between nucleosomes.

The site of acetylation in the amino-terminal region has prompted suggestions to explain how such modifications might modulate histone binding to the DNA (for references, see Allfrey, 1971, 1977;

Dixon et al., 1975; Dixon, 1976). However, there is still not enough experimental evidence to support any one particular model.

2. Phosphorylation

Phosphorylation occurs in all histones. This topic has been reviewed extensively (Dixon et al., 1975; Elgin and Weintraub, 1975; Gurley et al., 1978; Langan, 1978; Isenberg, 1979) and will be discussed only briefly here. The phosphorylation of H1 has already been discussed in Section III. Phosphorylation of the four core histones occurs at serine residues in the amino-terminal region of the histones. One type of phosphorylation occurs at the hydroxy group of the α-acetylated serine 1 of H4 and H2A (Sung and Dixon, 1970; Sung et al., 1971). A second type of phosphorylation occurs at serine residues which follow a lysine residue: at Ser-6 for H2B and Ser-10 and Ser-28 for H3 (Sung and Dixon, 1970; Marzluff and McCarthy, 1972). The occurrence of the phosphorylation sites of the core histones in the amino-terminal end suggests, as in the case of histone acetylation, correlation of this modification with the alteration of interaction of the histones with the DNA.

3. Methylation

Methylation of histone lysines has been seen only in H3 and H4: at position 9, 27 for H3 and at position 20 of H4. Sometimes a lysine may have more than one methyl group (Ogawa et al., 1969; Patthy et al., 1973; DeLange et al., 1969, 1973). Methylation of histidines was reported for H1 and H5 in duck erythroid cells. This modification, which occurs late in S phase or in G_2, is found in the amino-terminal region of H3 and H4 and is irreversible (for references, see Allfrey, 1977; Dixon, 1976).

4. ADP-Ribosylation

Covalent links of histones H1, H2A, H2B, and H3 with poly(ADP-ribose) have been reported (for references, see Hayaishi and Ueda, 1977). Furthermore, a histone H1 dimer linked by poly(ADP-ribose) has been reported. The increase in ADP-ribosylation is concomitant with cellular replication and ADP-ribosylation has been proposed as a trigger for in vivo replication in eukaryotic cells.

5. Protein A24

Protein A24 is histone H2A into which the peptide ubiquitin is covalently attached via its carboxy-terminal end to the ϵ-amino of lysine 119 of H2A (Goldknopf et al., 1977). A24 is a constituent of nucleosomes and is present to the extent of 10% of calf thymus H2A (Busch *et al.*, 1978). A24 content in chromatin is decreased upon increase of cellular activity. In contrast to acetylation, phosphorylation, and methylation, all of which occur in the amino-terminal end, this modification is confined to the carboxy-terminal end of the histone molecule and may be involved with changes in histone–histone interaction.

C. Histone Variants

The use of urea triton gel electrophoresis (Zweidler and Cohen, 1972; Alfageme *et al.*, 1974) has permitted the further resolution of histone fractions and the identification of histone variants. Microcomponents or variants of histone H1 have been known for a long time (for references, see Cole, 1977). Recently, it has been shown that microcomponents or variants of histones H2B, H2A, and H3 occur in many systems (Marzluff *et al.*, 1972; Laine *et al.*, 1976; Garrard, 1976; Blankstein and Levy, 1976; Zweidler, 1976; Franklin and Zweidler, 1977). Furthermore, in sea urchin embryo the synthesis of new histone variants has been correlated with development (Cohen *et al.*, 1975; Newrock *et al.*, 1978; Von Holt *et al.*, 1979).

1. H2A

Table I represents the sequence changes that have been identified in H2A variants. The ratio of the different variants changes from species to species. Furthermore, the ratio $H2A_1/H2A_2$ changes from 3:1 to ~1:1 in normal cells or Friend cells after proliferation induced by DMSO treatment (Blankstein and Levy, 1976).

2. H2B

H2B variants are known to occur in mammalian nuclei. These variants are characterized by the changes summarized in Table II. Variants 1 and 3 have been found in all tissues of all mammals exam-

TABLE I
H2A Variants

Variants	Residue number			Reference
	16	51	99	
Calf				
1	Thr	Leu	Lys	Franklin and Zweidler (1977)
2	Ser	Met	Lys	
Rat leukemia				
α	Ser	Gly	Lys (60%)	Laine *et al.* (1976)
β_1	Thr	Gly	Arg (20%)	
β_2	Ser	Gly	Arg	

ined, although at different ratios. Variant 2 was found only in mouse (Franklin and Zweidler, 1977).

The carboxy-terminal region of the sequences of H2B studied until now is highly conserved except for a region of $\simeq 20$ amino acids from position 75 to position 94 (in calf) which is polar and was found to be variable. Therefore, the region 75–94 is variable among different species and also among different histone variants of the same species. This type of variation occurring in the carboxy-terminal region may affect the histone–histone interactions occurring in this region (for references, see Von Holt et al., 1979). In sea urchin embryo and sperm six H2B variants have been identified (for references, see Von Holt et al., 1979). Sperm H2B has two variants with 143 and 144 amino acid residues in contrast to calf H2B with 125 amino acid residues. Therefore, the amino-terminal region of H2B is highly variable. This extra piece is composed of a repeating pentapeptide added at the amino terminus, which, as in other histones, is charged. In contrast to other H2B histones which are lysine rich, the sperm $H2B_1$ and $H2B_2$ are arginine-rich.

3. H3

H3 variants were identified in mammalia nuclei at the positions recorded in Table III.

D. Clustering of Like Nucleosomes

The fact that all the histone variants which are synthesized during different stages of development of sea urchin embryo remain in the system led Cohen et al. (1975) to suggest that there must exist multiple forms of nucleosomes. Using specific anti-histone antibodies heterogeneity was detected in rat liver nucleosomes with respect to the availability of the antigenic determinants of histones H2A and H4, yielding two populations of nucleosomes, one which exhibited an increase in diameter upon reaction with anti-H4 and anti-H2A and one which did not (Goldblatt et al., 1978a; Bustin et al., 1978). Furthermore, experiments with fractionated long pieces of rat liver chromatin (Goldblatt et al., 1978b; Goldblatt et al., 1981) have shown that

TABLE II
H2B Variants

| Variant | Residue number | | Reference |
	75	76	
1	Gly	Glu	Franklin and Zweidler (1977)
2	Ser	Glu	
3	Gly	Gln	

TABLE III

H3 Variants

Variant	Residue number			Reference
	89	90	96	
1	Val	Met	Cys	Franklin and Zweidler (1977)
2	Val	Met	Ser	
3[a]	Ile	Gly	Ser	

[a] Some uncertainty.

there is clustering of like nucleosomes along the chromatin array. These nucleosomes are similar with respect to the availability of the antigenic determinants of histones H2A and H4. Furthermore, super-coiled chromatin arrays which were fully reactive with anti-H4 were observed with chromatin arrays of about 100 nucleosomes. The most plausible source of this observed heterogeneity is the presence of his-tone subfractions: histone variants in the case of H2A (Zweidler, 1976; Franklin and Zweidler, 1977) and histone monoacetylation in the case of H4, as rat liver H4 is composed of two equal populations of zero-ϵ-lysine and mono-ϵ-lysine acetylated H4 (Wangh *et al.*, 1972). If this is indeed the case, these results would support the idea of chromatin organization into domains (Worcel and Benyajati, 1977; Laemmli *et al.*, 1978; Igo-Kemenes and Zachau, 1978). Each domain would con-tain like nucleosomes which would be characterized by the same his-tone subfractions. This concept is also in agreement with the sugges-tion that in replication the new histone subfractions will be assembled consecutively on chromatin (Weintraub *et al.*, 1978).

E. Histones Are Associated with DNA during Transcription

Digestion studies using micrococcal nuclease have indicated that actively transcribing genes have a periodic structure similar to that of bulk chromatin (Mathis and Gorovsky, 1978; Allfrey *et al.*, 1978; Reeves, 1978; Bellard *et al.*, 1978). On the other hand, there is increasing evidence that the structure of active chromatin is different from that of nontranscribing chromatin (for reviews, see Felsenfeld, 1978; Franke *et al.*, 1978; Paul *et al.*, 1978; Chambon, 1978), as shown by preferential digestion with DNase I (Weintraub and Groudine, 1976; Garel and Axel, 1976; Mathis and Gorovsky, 1978) and micro-coccal nuclease (Bellard *et al.*, 1978; Bloom and Anderson, 1978; Reeves and Jones, 1976; Reeves, 1978). Limited digestion of *Phy-sarum* chromatin with micrococcal nuclease results in enrichment of ribosomal DNA sequences, known to be engaged in transcription, in chromatin subunits that differ in sedimentation velocity and protein composition from that of nonactive mononucleosomes (Allfrey *et al.*,

1978). Using electron microscopy spreading techniques it was demonstrated that transcribing ribosomal genes are not compacted into nucleosomes, but rather are extended to the same length as the corresponding B-form DNA (Franke *et al.*, 1976, 1978; Foe, 1978; McKnight *et al.*, 1978; Reeder *et al.*, 1978; Woodcock *et al.*, 1976). In addition, structural transitions between extended (transcriptionally active) and compacted (transcriptionally inactive) states have been demonstrated (Foe, 1978; Scheer, 1978; Scheer and Zentgraf, 1978; Franke *et al.*, 1978).

Biochemical evidence tends to suggest that histones are present in transcriptionally active chromatin (Higashinakagawa *et al.*, 1977; Reeder *et al.*, 1978; Gottesfeld and Butler, 1978). Using immunoelectron microscopy it has been shown that histones H2B and H3 are associated with actively transcribing nonribosomal genes (McKnight *et al.*, 1978). Recently, immunoelectron microscopy has also revealed that histone H2B is associated with the DNA of actively transcribing ribosomal genes of newt oocytes (Goldblatt *et al.*, 1979; Sperling *et al.*, 1981). The distribution of the antibody molecules (detected by ferritin labeling) along the axis of the transcribing ribosomal DNA genes is not continuous; rather, clusters of ferritins are observed at approximately uniform intervals of ~1000 Å. As the DNA of the ribosomal DNA genes observed by this technique is not compacted, one can conclude that the core histones are associated with the DNA of actively transcribing ribosomal genes, but that the structure of this histone–DNA complex differs from that of nonactive chromatin.

The mechanism of the transition between the transcriptionally active and nonactive states is not known, nor are the detailed structural changes. It is not known how this transition is induced or how it is maintained. Although there is evidence that HMG proteins are correlated with active or potentially active regions (Levy *et al.*, 1978), there are also indications that an increase in histone acetylation is associated with an increase of transcriptional activity (for references see Section V,B,1). Therefore, it is plausible to assume that histones have some role in the structure of transcriptionally active chromatin. It is not clear whether this role is in promoting the structural change or in maintaining it; neither is it clear whether this is an active or passive role. However, if a repeating structure is required in active chromatin (and the indications are that some periodic structure is present) the histones probably are involved in creating the repeating unit.

F. A Possible Dynamic Role for Histones

Although histones have been regarded as the most highly conserved proteins during evolution (Dayhoff, 1972; Wilson *et al.*, 1977), and although they have been shown to be an important constituent of chromatin in all species, a large degree of microheterogeneity has been observed. There are histone postsynthetic modifications which have a rapid turnover and vary during the cell cycle. On the other hand, there are histone variants which introduce changes in the composition of histone subfractions. These changes have been correlated with developmental stages.

As has been emphasized in this article, histones play a structural role in chromatin, particularly the core octamer which contains the information for folding the DNA into its nucleosomal conformation. Now, can we anticipate what might happen upon histone modification or upon the introduction of histone variants? It has been shown in other assembly systems of regular subunits that a slight modification in each subunit can cause modification in the entire superstructure. Since the histones appear to form an assembly system, it is reasonable to postulate that histone modifications and histone variants can affect the chromosomal superstructure. As already discussed, it has been shown that the histones are associated with DNA during transcription and replication and that the structure of the nucleoprotein complex undergoes changes during these stages. It is also known that histone modifications occur at specific times during the cell cycle. Therefore, it is possible that these modifications will affect the structure of the histone skeleton which, in turn, will affect the fold of the DNA around the histone core. Preliminary evidence supporting the idea that histone modifications can affect the histone skeleton were obtained from highly acetylated H4 histones. Decrease in histone H4 assembly was correlated with increased acetylation (R. Sperling, unpublished).

The fundamental idea is that histone–histone interactions within and between nucleosomes as well as histone–DNA interactions can be modulated by histone variants and histone modification and lead to conformational changes in chromatin. Histone modifications occurring at the NH_2-terminal region may affect both histone–DNA interaction and histone–histone interaction between nucleosomes. Histone variants in systems like that of rat liver differ in the carboxy-terminal region (Zweidler, 1976). This may affect histone–histone interaction within the nucleosomes. However, in other systems histone variants also involve major alterations in the NH_2-terminal region. Therefore, this may involve structural changes in the nucleosome structure as well as in the higher order structure of the chromatin. The postulated existence of domains containing a clustering of similar histone subtypes is suggestive of a type of long-range "communication" of information. This communication could take place via histone–histone interaction between nucleosomes, possibly in a mode related to the interactions present in the histone fibers. Any or all of these considerations could be relevant to the structural changes which the nucleohistone complex is believed to undergo during stages of transcription or replication.

Whatever the details of the model, it is clear that histone–histone interactions play a central role in chromatin structure. We know that chromatin is a dynamic structure and that the histones serve as its skeleton, containing the information for folding of the DNA. The microdiversity of the histone variants and histone modifications introduces into the system the potential for dynamic change. The signals for structural change probably come from other cellular components such as hormones and nonhistone proteins. However, once the signal is given, it could be communicated to the DNA in different regions of the chromosome via the histone skeleton. Furthermore, the system could mediate the signal in a controlled, yet diversified way.

NOTE ADDED IN PROOF: Klug *et al.* (1980) obtained regular fibers by the assembly of histone octamers at high salt. From image reconstruction of these histone fibers a 22-Å resolution model was proposed for the histone octamer, which has a 2-fold axis of symmetry and is wedge-shaped. From this structure and the results of various cross-linking data an arrangement of the individual histones within the octamer has been proposed.

There is good agreement between the overall dimensions of the histone octamer found by Klug *et al.* and data obtained from other types of histone fibers discussed here. Similarity of cross-linking data of histone octamer fibers, octamer free in solution, and octamer in nucleosomes makes the extrapolation from the octamer model in the fibers to the octameric core of nucleosome valid (Klug *et al.*, 1980). This further substantiates the idea that histones are part of an assembly system, and therefore the histone core of the nucleosome can be regarded as a truncated histone fiber (see Section IV).

ACKNOWLEDGMENTS

We are grateful to Drs. R. D. Kornberg and J. Sperling for helpful discussions and suggestions concerning the manuscript. The work of the authors' laboratory was partially supported by grants from the United States–Israel Binational Science Foundation (BSF), the Jewish Agency, and the Israeli Commission for Basic Research.

REFERENCES

Adler, A. J., Langan, T. A., and Fasman, G. D. (1972). *Arch. Biochem. Biophys.* **153**, 769–777.

Albright, S., Nelson, P. P., and Garrard, W. T. (1979). *J. Biol. Chem.* **254**, 1065–1073.

Alfageme, C. R., Zweidler, A., Mahowald, A., and Cohen, L. H. (1974). *J. Biol. Chem.* **249**, 3729–3736.

Allfrey, V. G. (1971). *In* "Histones and Nucleohistones" (D. M. Phillips, ed.), pp. 241–294. Plenum, New York.

Allfrey, V. G. (1977). *In* "Chromatin and Chromosome Structure" (H. J. Li and R. A. Eckhardt, eds.), pp. 167–191. Academic Press, New York.

Allfrey, V. G., Johnson, E. M., Sun, I. Y.-C., Littau, V. C., Matthews, H. R., and Bradbury. E. M. (1978). *Cold Spring Harbor Symp. Quant. Biol.* **42**, 505–514.

Altenburger, W., Horz, W., and Zachau, H. G. (1976). *Nature (London)* **264**, 517–522.

Anfinsen, C. B. (1973). *Science* **181**, 223–230.

Axel, R. (1975). *Biochemistry* **14**, 2921–2925.

Axel, R., Melchoir, W., Jr., Sollner-Webb, B., and Felsenfeld, G. (1974). *Proc. Natl. Acad. Sci. U.S.A.* **71**, 4101–4105.

Bakayev, V. V., Melnickov, A. A., Osicka, V. D., and Varshavsky, A. J. (1975). *Nucleic Acids Res.* **2**, 1401–1419.

Baldwin, J. P., Boseley, P.G., Bradbury, E. M., and Ibel, K., (1975). *Nature (London)* **253**, 245–249.

Balhorn, R., Boodwell, J., Sellers, L., Granner, D., and Chalkley, R. (1972). *Biochem. Biophys. Res. Commun.* **43**, 1326–1333.

Bellard, M., Gannon, F., and Chambon, P. (1978). *Cold Spring Harbor Symp. Quant. Biol.* **42**, 779–791.

Bina-Stein, M. (1978). *J. Biol. Chem.* **253**, 5213–5219.

Bina-Stein, M., and Simpson, R. T. (1977). *Cell* **11**, 609–618.

Blankstein, L. A., and Levy, S. B. (1976). *Nature (London)* **260**, 638–640.

Bloom, K. S., and Anderson, J. W. (1978). *Cell* **15**, 141–150.

Boffa, L. C., Vidali, G., Mann, G. S., and Allfrey, V. G. (1978). *J. Biol. Chem.* **253**, 3364–3366.

Böhm, L., Hayashi, H., Cary, P. D., Moss, T., Crane-Robinson, C., and Bradbury, E. M. (1977). *Eur. J. Biochem.* **77**, 498–493.

Bonner, W. M., and Pollard, H. B. (1975). *Biochem. Biophys. Res. Commun.* **64**, 282–288.

Boseley, P. G., Bradbury, E. M., Butler-Browne, G. S., Carpenter, B. G., and Stephens, R. M. (1976). *Eur. J. Biochem.* **62**, 21–31.

Böttger, M., Becker, M., Fenske, H., and Scherneck, S. (1976a). *Acta Biol. Med. Ger.* **35**, 679–681.

Böttger, M., Scherneck, S., and Fenske, H. (1976b). *Nucleic Acids Res.* **3**, 419–429.

Bradbury, E. M., Carpenter, B. G., and Rattle, H. W. E. (1973). *Nature (London)* **241**, 123–125.

Bradbury, E. M., Inglis, K. J., and Matthews, H. R. (1974). *Nature (London)* **247**, 257–261.

Bradbury, E. M., Moss, T., Hayashi, H., Hjelm, R. P., Suau, P., Stephens, R. M., Baldwin, J. P., and Crane-Robinson, C. (1978). *Cold Spring Harbor Symp. Quant. Biol.* **42**, 277–286.

Bram, S. (1977). *J. Mol. Biol.* **58**, 277–288.

Burton, D. R., Hyde, J. E., and Walker, I. O. (1975). *FEBS Lett.* **55**, 77–80.

Burton, D. R., Butler, M. J., Hyde, J. E., Phillips, D., Skidmore, C. J., and Walker, I. O. (1978). *Nucleic Acids Res.* **5**, 3643–3663.

Busch, H., Ballal, N. R., Busch, R. K., Choi, Y. C., Davies, F., Goldknopf, I. L., Matsui, S. I., Rao, M. S., and Rothblum, L. I. (1978). *Cold Spring Harbor Symp. Quant. Biol.* **42**, 665–683.

Bustin, M., Goldblatt, D., and Sperling, R. (1976). *Cell* **7**, 297–304.

Bustin, M., Simpson, R. T., Sperling, R., and Goldblatt, D. (1977). *Biochemistry* **16**, 5381–5385.

Bustin, M., Kurth, P. D., Moudrianakis, E. N., Goldblatt, D., Sperling, R., and Rizzo, W. B. (1978). *Cold Spring Harbor Symp. Quant. Biol.* **42**, 379–388.

Butler, A. P., Harrington, R. E., and Olins, D. E. (1979). *Nucleic Acids Res.* **6**, 1509–1519.

Butler, P. J. G., and Durham, A. C. H. (1977). *Adv. Protein Chem.* **31**, 187–251.

Camerini-Otero, R. D., and Felsenfeld, G. (1977a). *Proc. Natl. Acad. Sci. U.S.A.* **74**, 5519–5523.

Camerini-Otero, R. D., and Felsenfeld, G. (1977b). *Nucleic Acids Res.* **4**, 1159–1181.

Camerini-Otero, R. D., Sollner-Webb, B., and Felsenfeld, G. (1976). *Cell* **8**, 333–347.

Campbell, A. M., and Cotter, R. I. (1976). *FEBS Lett.* **70**, 209–211.

Candido, E. P. M., Reeves, R., and Davie, J. R. (1978). *Cell* **14**, 105–113.

Carpenter, B. G., Baldwin, J. P., Bradbury, E. M., and Ibel, K. (1976). *Nucleic Acids Res.* **3**, 1739–1746.

Caspar, D. L. D., and Klug, A. (1962). *Cold Spring Harbor Symp. Quant. Biol.* **27**, 1–23.

Chalkley, R., and Hunter, C. (1975). *Proc. Natl. Acad. Sci. U.S.A.* **72**, 1304–1308.

Chambon, P. (1978). *Cold Spring Harbor Symp. Quant. Biol.* **42**, 1209–1234.

Chung, S.-Y., Hill, W. E., and Doty, P. (1978). *Proc. Natl. Acad. Sci. U.S.A.* **75**, 1680–1684.

Clark, M., and Spudich, J. A. (1977). *Annu. Rev. Biochem.* **46**, 797–822.

Cohen, C. (1979). *Trends Biochem. Sci.* **4**, 73–77.

Cohen, L. H., Newrock, K. M., and Zweidler, A. (1975). *Science* **190**, 994–997.

Cole, R. D. (1977). *In* "The Molecular Biology of the Mammalian Genetic Apparatus" (P. O. P. Ts'o, ed.)., Vol. 1, pp. 93–104. Elsevier/North-Holland Biomedical Press, Amsterdam.

Cousens, L. S., Gallwitz, D., and Alberts, B. M. (1979). *J. Biol. Chem.* **254**, 1716–1723.

Crick, F. H. C., and Klug, A. (1975). *Nature (London)* **255**, 530–533.

D'Anna, J. A., Jr., and Isenberg, I. (1973). *Biochemistry* **12**, 1035–1043.

D'Anna, J. A., Jr., and Isenberg, I. (1974a). *Biochemistry* **13**, 2098–2104.

D'Anna, J. A., Jr., and Isenberg, I. (1974b). *Biochemistry* **13**, 4992–4997.

D'Anna, J. A., Jr., and Isenberg, I. (1974c). *Biochem. Biophys. Res. Commun.* **61**, 343–347.

Davie, J. R., and Candido, E. P. M. (1978). *Proc. Natl. Acad. Sci. U.S.A.* **75**, 3574–3577.

Davies, H. G., and Haynes, M. R. (1975). *J. Cell Sci.* **17**, 263–285.

Dayhoff, M., ed. (1972). "Atlas of Protein Sequences and Structure," Vol. 5. Natl. Biomed. Res. Found., Silver Spring, Maryland.

DeLange, R. J., and Smith, E. L. (1971). *Annu. Rev. Biochem.* **40**, 279–314.

DeLange, R. J., and Smith, E. L. (1972). *Acc. Chem. Res.* **5**, 368–373.

DeLange, R. J., Fambrough, D. M., Smith, E. L., and Bonner, J. (1969). *J. Biol. Chem.* **244**, 319–334.

DeLange, R. J., Hooper, J. A., and Smith, E. L. (1972). *Proc. Natl. Acad. Sci. U.S.A.* **69**, 882–884.

DeLange, R. J., Hooper, J. A., and Smith, E. L. (1973). *J. Biol. Chem.* **248**, 3261–3274.

DeLange, R. J., Williams, L. C., and Martinson, H. G. (1979). *Biochemistry* **18**, 1942–1946.

Dixon, G. H. (1976). *In* "Organization and Expression of Chromosomes" (V. G. Allfrey, ed.), pp. 197–207. Dahlem Konferenzen, Abakon Verlagsgesellschaft, Berlin.

Dixon, G. H., Candido, E. P. M., Honda, B. M., Louie, A. J., Macleod, A. R., and Sung, M. T. (1975). *Ciba Found. Symp.* **28** (new ser.), 229–258.

Dubochet, J., and Noll, M. (1978). *Science* **202**, 280–286.

Dupraw, E. J., (1970). "DNA and Chromosomes." Holt, New York.

Eickbusch, T. H., and Moudrianakis, E. N. (1978). *Biochemistry* **17**, 4955–4964.

Elgin, S. C. R., and Weintraub, H. (1975). *Annu. Rev. Biochem.* **44**, 725–774.

Felsenfeld, G. (1978). *Nature (London)* **271**, 115–122.

Finch, J. T., and Klug, A. (1976). *Proc. Natl. Acad. Sci. U.S.A.* **73**, 1897–1901.

Finch, J. T., and Klug, A. (1978). *Cold Spring Harbor Symp. Quant. Biol.* **42**, 1–9.

Finch, J. T., Lutter, L. C., Rhodes, D., Brown, R. S., Rushton, B., Levitt, M., and Klug, A. (1977). *Nature (London)* **269**, 29–36.

Foe, V. E. (1978). *Cold Spring Harbor Symp. Quant. Biol.* **42**, 723–740.

Franke, W. W., Scheer, U., Trendelenberg, M. F., Spring, H., and Zentgraf, H. (1976). *Cytobiologie* **13**, 401–434.

Franke, W. W., Scheer, U., Trendelenberg, M. F., Zentgraf, H., and Spring, H. (1978). *Cold Spring Harbor Symp. Quant. Biol.* **42**, 755–772.

Franklin, S. G., and Zweidler, A. (1977). *Nature (London)* **266**, 273–275.

Garel, A., and Axel, R. (1976). *Proc. Natl. Acad. Sci. U.S.A.* **73**, 3966–3970.

Garrard, W. T. (1976). *FEBS Lett.* **64**, 323–325.

Germond, J. E., Hirt, B., Oudet, P., Gross-Bellard, M., and Chambon, P. (1975). *Proc. Natl. Acad. Sci. U.S.A.* **72**, 1843–1847.

Glover, C. V. C., and Gorovsky, M. A. (1978). *Biochemistry* **17**, 5705–5713.

Goldblatt, D., Bustin, M., and Sperling, R. (1978a). *Exp. Cell Res.* **112**, 1–14.

Goldblatt, D., Polacow, I., Litowsky, I., Bustin, M., and Sperling, R. (1978b). *Annu. Meet. Isr. Biochem. Soc.*

Goldblatt, D., Sheer, U., Bustin, M., and Sperling, R. (1979). *Annu. Meet. Isr. Biochem. Soc.*

Goldblatt, D., Polacow, I., Litowsky, I., Bustin, M., and Sperling, R. (1981). In preparation.

Goldknopf, I. L., French, M. F., Musso, R., and Busch, H. (1977). *Proc. Natl. Acad. Sci. U.S.A.* **74**, 5492–5495.

Goodwin, D. C., and Brahms, J. (1978). *Nucleic Acids Res.* **5**, 835–850.

Gottesfeld, J. M., and Butler, P. J. G. (1978). *Nucleic Acids Res.* **4**, 3155–3173.

Gurley, L. R., D'Anna, J. A., Jr., Braham, S. S., Deaven, L. L., and Tobey, R. A. (1978). *Eur. J. Biochem.* **84**, 1–15.

Hardison, R. C., Zeitler, D. P., Murphy, J. M., and Chalkley, R. (1977). *Cell* **12**, 417–427.

Havron, A., and Sperling J. (1977). *Biochemistry* **16**, 5631–5635.

Hayaishi, O., and Ueda, K. (1977). *Annu. Rev. Biochem.* **46**, 95–116.

Hewish, D. R., and Burgoyne, L. A. (1973). *Biochem. Biophys. Res. Commun.* **52**, 504–510.

Higashinakagawa, T., Wahn, H., and Reeder, R. H. (1977). *Dev. Biol.* **55**, 375–386.

Hnilica, L. S. (1972). "The Structure and Function of Histones," CRC Press, Cleveland, Ohio.

Hozier, H., Renz, M., and Nehls, R. (1977). *Chromosoma* **62**, 301–317.

Hsiang, M. W., and Cole, R. D. (1977). *Proc. Natl. Acad. Sci. U.S.A.* **74**, 4852–4856.

Hyde, J. E., and Walker, I. O. (1975a). *FEBS Lett.* **50**, 150–154.

Hyde, J. E., and Walker, I. O. (1975b). *Nucleic Acids Res.* **2**, 405–421.

Igo-Kemenes, T., and Zachau, H. G. (1978). *Cold Spring Harbor Symp. Quant. Biol.* **42**, 109–118.

Isenberg, I. (1978). *In* "The Cell Nucleus" (H. Busch, ed.), Vol. 4, pp. 135–154. Academic Press, New York.

Isenberg, I. (1979). *Annu. Rev. Biochem.* **48**, 159–191.

Iwai, K., Hayashi, H., and Ishikawa, K. (1972). *J. Biochem. (Tokyo)* **72**, 357–367.

Jackson, V. (1978). *Cell* **15**, 945–954.

Joffe, J., Keene, M., and Weintraub, H. (1977). *Biochemistry* **16**, 1236–2238.

Kelley, R. I. (1973). *Biochem. Biophys. Res. Commun.* **54**, 1588–1594.

Klug, A., Rhodes, D., Smith, J., Finch, J. T., and Thomas, J. O. (1980). *Nature (London)* **287**, 509–516.

Knippers, R., Otto, B., and Böhme, R. (1978). *Nucleic Acids Res.* **5**, 2113–2131.

Kornberg, R. D. (1974). *Science* **184**, 868–871.

Kornberg, R. D. (1977). *Annu. Rev. Biochem.* **46**, 931–954.

Kornberg, R. D., and Thomas, J. O. (1974). *Science* **184**, 865–868.

Kossel, A. (1884). *Hoppe-Seyler's Z. Physiol. Chem.* **8**, 511–515.

Kunkel, G. R., and Martinson, H. G. (1978). *Nucleic Acids Res.* **5**, 4263–4272.

Laemmli, U. K., Cheng, S. M., Adolph, K. W., Paulson, J. R., Brown, J. A., and Baumbach, W. L. (1978). *Cold Spring Harbor Symp. Quant. Biol.* **42**, 351–360.

Laine, B., Sautiere, P., and Biserte, G. (1976). *Biochemistry* **15**, 1640–1645.

Lake, R. S., and Salzman, N. P. (1972). *Biochemistry* **11**, 4817–4826.

Langan, T. A. (1969). *Proc. Natl. Acad. Sci. U.S.A.* **64**, 1276–1283.

Langan, T. A. (1978). *Methods Cell Biol.* **19**, 127–142.

Langan, T. A., and Hohmann, P. (1975). *In* "Chromosomal Proteins and Their Role in the Regulation of Gene Expression" (G. S. Stein and L. J. Kleinsmith, eds.), pp. 113–126. Academic Press, New York.

Levina, E. S., and Mirzabekov, A. D. (1975). *Dokl. Akad. Nauk SSSR* **221**, 1222–1225.

Levitt, M. (1978). *Proc. Natl. Acad. Sci. U.S.A.* **75**, 640–644.

Levy, B. W., Wong, N. C. W., Watson, D. C., Peters, E. H., and Dixon, G. H. (1978). *Cold Spring Harbor Symp. Quant. Biol.* **42**, 793–802.

Lewis, P. N. (1976). *Biochem. Biophys. Res. Commun.* **68**, 329–335.

Li, H. J. (1977). *In* "Chromatin and Chromosome Structure" (H. J. Li and R. A. Eckhardt, eds.), pp. 1–35. Academic Press, New York.

Liew, C. C., Haslett, G. W., and Allfrey, V. G. (1970). *Nature (London)* **226**, 414–417.

Lilley, D. M., and Tatchell, K. (1977). *Nucleic Acids Res.* **4**, 2039–2055.

Lilley, D. M., Howarth, O. W., Clark, V. M., Pardon, J. F., and Richards, B. M. (1976). *FEBS Lett.* **62**, 7–10.

Littau, V. C., Burdick, C. J., Allfrey, V. G., and Mirsky, A. E. (1965). *Proc. Natl. Acad. Sci. U.S.A.* **54**, 1204–1212.

Lutter, L. (1978). *Cold Spring Harbor Symp. Quant. Biol.* **42**, 137–147.

Luzzati, V., and Nicolaieff, A. (1959). *J. Mol. Biol.* **1**, 127–133.

McGhee, J. D., and Felsenfeld, G. (1979). *Proc. Natl. Acad. Sci. U.S.A.* **76**, 2133–2137.

McKnight, S. L., Bustin, M., and Miller, D. L. (1978). *Cold Spring Harbor Symp. Quant. Biol.* **42**, 741–754.

Mandel, R., Kolomijtseva, G., and Brahms, G. (1979). *Eur. J. Biochem.* **96**, 257–265.

Mardian, J. K. W., and Isenberg, I. (1978). *Biochemistry* **17**, 3825–3833.

Martinson, H. G., and McCarthy, B. J. (1975). *Biochemistry* **14**, 1073–1078.

Martinson, H. G., Shetlar, M. D., and McCarthy, B. J. (1976). *Biochemistry* **15**, 2002–2007.

Martinson, H. G., True, R., Lau, C. K., and Mehrabian, M. (1979a). *Biochemistry* **18**, 1075–1082.

Martinson, H. G., True, R. J., and Burch, J. B. E. (1979b). *Biochemistry* **18**, 1082–1089.

Marzluff, W. F., Jr., and McCarthy, K. S. (1972). *Biochemistry* **11**, 2677–2681.

Marzluff, W. F., Jr., Sanders, L. A., Miller, D. M., and McCarthy, K. S. (1972). *J. Biol. Chem.* **247**, 2026–2033.

Mathis, D. J., and Gorovsky, M. A. (1978). *Cold Spring Harbor Symp. Quant. Biol.* **42**, 773–778.

Mathis, D. J., Oudet, P., Wasylyk, B., and Chambon, P. (1978). *Nucleic Acids Res.* **5**, 3523–3547.

Mirsky, A. E., Burdick, C. J., Davidson, E. H., and Littau, V. C. (1968). *Proc. Natl. Acad. Sci. U. S. A.* **61**, 592–597.

Mirzabekov, A. D., Sanko, D. F., Kolchinsky, A. M., and Melnikova, A. F. (1977). *Eur. J. Biochem.* **75**, 379–389.

Mirzabekov, A. D., Schick, V. V., Belyavsky, A. V., Krapov, V. L., and Barykin, S. G. (1978a). *Cold Spring Harbor Symp. Quant. Biol.* **42**, 149–155.

Mirzabekov, A. D., Schick, V. V., Belyavsky, A. V., and Barykin, S. G. (1978b). *Proc. Natl. Acad. Sci. U. S. A.* **75**, 4184–4188.

Morris, N. R. (1976a). *Cell* **8**, 357–364.

Morris, N. R. (1976b). *Cell* **9**, 627–637.

Moss, T., Cary, P. D., Abercrombie, B. D., Crane-Robinson, C., and Bradbury, E. M. (1976a). *Eur. J. Biochem.* **71**, 337–350.

Moss, T., Cary, P. D., Crane-Robinson, C., and Bradbury, E. M. (1976b). *Biochemistry* **15**, 2261–2267.

Moss, T., Stephens, R. M., Crane-Robinson, C., and Bradbury, E. M. (1977). *Nucleic Acids Res.* **4**, 2477–2485.

Nelson, D. A., Oosterhof, D. K., and Rill, R. L. (1977). *Nucleic Acids Res.* **4**, 4223–4233.

Nelson, D. A., Perry, W. M., and Chalkley, R. (1978). *Biochem. Biophys. Res. Commun.* **82**, 356–363.

Newrock, K. M., Alfageme, C. R., Nardi, R. V., and Cohen, L. H. (1978). *Cold Spring Harbor Symp. Quant. Biol.* **42**, 421–432.

Nicola, N. A., Fulmer, A. W., Schwartz, A. M., and Fasman, G. D. (1978). *Biochemistry* **17**, 1779–1785.

Noll, M. (1974a). *Nature (London)* **251**, 249–261.

Noll, M. (1974b). *Nucleic Acids Res.* **1**, 1573–1578.

Noll, M. (1976). *Cell* **8**, 349–355.

Noll, M. (1978). *Cold Spring Harbor Symp. Quant. Biol.* **42**, 77–85.

Noll, N., and Kornberg, R. D. (1977). *J. Mol. Biol.* **109**, 393–404.

Ogawa, Y., Quagliarotti, G., Jordan, J., Taylor, C. W., Starbuck, W. C., and Busch, H. (1969). *J. Biol. Chem.* **244**, 4387–4392.

Ohlenbusch, H. H., Olivera, B. M., Tuan, D., and Davidson, N. (1967). *J. Mol. Biol.* **25**, 299–315.

Olins, A. L. (1978). *Cold Spring Harbor Symp. Quant. Biol.* **42**, 325–329.

Olins, A. L., and Olins, D. E. (1974). *Science* **183**, 330–332.

Olins, A. L., Carlson, R. D., Wright, E. B., and Olins, D. E. (1976). *Nucleic Acids Res.* **3**, 3271–3291.

Olson, M. O. J., Jordan, J., and Busch, H. (1972). *Biochem. Biophys. Res. Commun.* **46**, 50–55.

Oudet, P., Gross-Bellard, M., and Chambon, P. (1975). *Cell* **4**, 281–300.

Oudet, P., Germond, J. P., Sures, M., Gallwitz, D., Bellard, M., and Chambon, P. (1978). *Cold Spring Harbor Symp. Quant. Biol.* **42**, 287–300.

Panyim, S., and Chalkley, R. (1969). *Arch. Biochem. Biophys.* **130**, 337–346.

Pardon, J. F., and Wilkins, M. H. F. (1972). *J. Mol. Biol.* **68**, 115–124.

Pardon, J. F., Wilkins, M. H. F., and Richards, B. M. (1967). *Nature (London)* **215**, 508–509.

Pardon, J. F., Worcester, D. L., Wooley, J. C., Tatchell, K., Van Holde, K. E., and Richards, B. M. (1975). *Nucleic Acids Res.* **2**, 2163–2176.

Pardon, J. F., Worcester, D. L., Wooley, J. C., Cotter, R. I., Lilley, D. M. J., and Richards, B. M. (1977). *Nucleic Acids Res.* **4**, 3199–3214.

Pardon, J. F., Cotter, R. I., Lilley, D. M. J., Worcester, D. L., Campbell, A. M., Wooley, J. C., and Richards, B. M. (1978). *Cold Spring Harbor Symp. Quant. Biol.* **42**, 11–22.

Patthy, L., Smith, E. L., and Johnson, J. (1973). *J. Biol. Chem.* **248**, 6834–6840.

Paul, J., Zollinger, E. J., Gilmour, R. S., and Birnie, G. D. (1978). *Cold Spring Harbor Symp. Quant. Biol.* **42**, 597–603.

Perry, M. E., Nelson, D., Moore, M., and Chalkley, R. (1979). *Biochim. Biophys. Acta* **561**, 517–525.

Phillips, D. M. P., ed. (1971). "Histones and Nucleohistones," Plenum, New York.

Prunell, A., Kornberg, R. D., Lutter, L., Klug, A., Levitt, M., and Crick, F. H. C. (1979). *Science* **204**, 855–858.

Rattner, J. B., and Hamkalo, B. A. (1979). *J. Cell Biol.* **81**, 453–457.

Reeder, R. H., Wahn, H. L., Botchan, P., Hipskind, R., and Sollner-Webb, B. (1978). *Cold Spring Harbor Symp. Quant. Biol.* **42**, 1167–1177.

Reeves, R. (1978). *Cold Spring Harbor Symp. Quant. Biol.* **42**, 709–722.

Reeves, R., and Jones, A. (1976). *Nature (London)* **260**, 495–500.

Renz, M., Nehls, P., and Hozier, J. (1977). *Proc. Natl. Acad. Sci. U.S.A.* **74**, 1879–1883.

Riggs, M. G., Whittaker, R. G., Neumann, J. R., and Ingram, V. M. (1977). *Nature (London)* **268**, 462–464.

Riggs, M. G., Whitakker, R. G., Neumann, J. R., and Ingram, V. M. (1978). *Cold Spring Harbor Symp. Quant. Biol.* **42**, 815–818.

Rill, R., and Van Holde, K. E. (1973). *J. Biol. Chem.* **248**, 1080–1083.

Ris, H. (1974). *Ciba Found. Symp.* **28**, (new ser.), 7–23.

Ris, H., and Kubai, D. F. (1970). *Annu. Rev. Genet.* **4**, 263–294.

Roark, D. E., Geoghegan, T. F., and Keller, G. H. (1974). *Biochem. Biophys. Res. Commun.* **59**, 542–547.

Roark, D. E., Geoghegan, T. F., Keller, G. H., Matter, K. V., and Engle, R. L. (1976). *Biochemistry* **15**, 3019–3025.

Sahasrabuddhe, C. G., and Van Holde, K. E. (1974). *J. Biol. Chem.* **249**, 152–156.

Sautiere, P., Tyrou, D., Laine, B., Mizon, J., Ruffin, P., and Biserte, G. (1974). *Eur. J. Biochem.* **41**, 563–576.

Scheer, U. (1978). *Cell* **13**, 535–549.

Scheer, U., and Zentgraf, H. (1978). *Chromosoma* **69**, 243–254.

Sealy, L., and Chalkley, R. (1978). *Cell* **14**, 115–121.

Shaw, B. R., Herman, T. M., Kovacic, R. T., Beaudreau, G. S., and Van Holde, K. E. (1976). *Proc. Natl. Acad. Sci. U. S. A.* **73**, 505–509.

Simon, R. H., Camerini-Otero, R. D., and Felsenfeld, G. (1978). *Nucleic Acids Res.* **5**, 4805–4818.

Simpson, R. T. (1978a). *Biochemistry* **17**, 5524–5531.

Simpson, R. T. (1978b). *Nucleic Acids Res.* **5**, 1109–1119.

Simpson, R. T. (1978c). *Cell* **13**, 691–699.

Simpson, R. T., and Whitlock, J. P., Jr. (1976). *Cell* **9**, 347–353.

Simpson, R. T., Whitlock, J. P., Jr., Bina-Stein, M., and Stein, A. (1978). *Cold Spring Harbor Symp. Quant. Biol.* **42**, 127–136.

Singer, D. S., and Singer, M. F. (1978). *Biochemistry* **17**, 2086–2095.

Singer, G. (1975). *Prog. Nucleic Acid Res. Mol. Biol.* **15**, 219–284.

Skandrani, E., Mizon, J., Sautiere, P., and Biserte, G. (1972). *Biochimie* **54**, 1267–1272.

Smith, K. C. (1976). *In* "Aging, Carcinogenesis and Radiation Biology" (K. C. Smith, ed.), pp. 67–82. Plenum, New York.

Sobell, H. M., Tsai, C. C., Gilbert, S. G., Jain, S. C., and Sakore, T. D. (1976). *Proc. Natl. Acad. Sci. U. S. A.* **73**, 3068–3072.

Solari, A. J. (1974). *In* "The Cell Nucleus" (H. Busch, ed.), Vol. 1, pp. 494–537. Academic Press, New York.

Sollner-Webb, B., and Felsenfeld, G. (1975). *Biochemistry* **14**, 2915–2920.

Sollner-Webb, B., Camerini-Otero, R. D., and Felsenfeld, G. (1976). *Cell* **9**, 179–193.

Sollner-Webb, B., Melchior, W., Jr., and Felsenfeld, G. (1978). *Cell* **14**, 611–627.

Sperling, J. (1976). *Photochem. Photobiol.* **23**, 323–326.

Sperling, J., and Havron, A. (1976). *Biochemistry* **15**, 1489–1495.

Sperling, J., and Havron, A. (1977). *Photochem. Photobiol.* **26**, 661–664.

Sperling, J., and Sperling R. (1978). *Nucleic Acids Res.* **5**, 2755–2773.

Sperling, R., and Amos, L. A. (1977). *Proc. Natl. Acad. Sci. U. S. A.* **74**, 3772–3776.

Sperling, R., and Bustin, M. (1974). *Proc. Natl. Acad Sci. U. S. A.* **71**, 4625–4629.

Sperling, R., and Bustin, M. (1975). *Biochemistry* **14**, 3322–3331.

Sperling, R., and Bustin, M. (1976). *Nucleic Acids Res.* **3**, 1263–1275.

Sperling, R., and Wachtel, E. J. (1979). *In* "Molecular Mechanisms of Biological Recognition" (M. Balaban, ed.), pp. 183–200. Elsevier/North-Holland Biomedical Press, Amsterdam.

Sperling, R., Scheer, U., Goldblatt, D., and Bustin, M. (1981). In preparation.

Spiker, S., and Isenberg, I. (1977). *Biochemistry* **16**, 1819–1826.

Spiker, S., and Isenberg, I. (1978). *Cold Spring Harbor Symp. Quant. Biol.* **42**, 157–163.

Stein, A., Bina-Stein, M., and Simpson, R. T. (1977). *Proc. Natl. Acad. Sci. U. S. A.* **74**, 2780–2784.

Stockley, P. G., and Thomas, J. O. (1979). *FEBS Lett.* **99**, 129–135.

Strniste, G. F., and Rall, S. C. (1976). *Biochemistry* **15**, 1712–1719.

Suau, P., Kneale, G. G., Braddock, G. W., Baldwin, J. P., and Bradbury, E. M. (1977). *Nucleic Acids Res.* **3**, 3769–3786.

Sung, M. T., and Dixon, G. H. (1970). *Proc. Natl. Acad. Sci. U. S. A.* **67**, 16–1623.

Sung, M. T., Dixon, G. H., and Smithies, O. (1971). *J. Biol. Chem.* **246**, 1358–1364.

Sussman, J. L., and Trifonov, E. (1978). *Proc. Natl. Acad. Sci. U. S. A.* **75**, 103–107.

Tatchell, K., and Van Holde, K. E. (1977). *Biochemistry* **16**, 5295–5303.

Thoma, F., and Koller, T. (1977). *Cell* **12**, 101–107.

Thoma, F., Koller, T., and Klug, A. (1979). *J. Cell Biol.* **83**, 403–427.

Thomas, G. J., Jr., Prescott, B., and Olins, D. (1977). *Science* **197**, 385–388.

Thomas, J. O., and Butler, P. J. G. (1977). *J. Mol. Biol.* **116**, 769–781.

Thomas, J. O., and Butler, P. J. G. (1978). *Cold Spring Harbor Symp. Quant. Biol.* **42**, 119–125.

Thomas, J. O., and Kornberg, R. D. (1975a). *Proc. Natl. Acad. Sci. U. S. A.* **72**, 2626–2630.

Thomas, J. O., and Kornberg, R. D. (1975b). *FEBS Lett.* **58**, 353–358.

Thomas, J. O., and Oudet, P. (1979). *Nucl. Acids Res.* **7**, 611–623.

Todd, P., and Han, A. (1976). *In* "Aging, Carcinogenesis and Radiation Biology" (K. C. Smith, ed.), pp. 83–104. Plenum, New York.

Van Lente, F., Jackson, J. F., and Weintraub, H. (1975). *Cell* **5**, 45–50.

Varshavsky, A. J., Bakayev, V. V., and Georgiev, G. P. (1976). *Nucleic Acids Res.* **3**, 477–492.

Vengerov, Y. Y., and Popenko, V. I. (1977). *Nucleic Acids Res.* **4**, 3017–3027.

Vidali, G., Boffa, L. C., Bradbury, E. M., and Allfrey, V. G. (1978). *Proc. Natl. Acad. Sci. U. S. A.* **75**, 2239–2243.

Vogel, T., and Singer, M. F. (1975). *J. Biol. Chem,* **250**, 796–798.

Vogel, T., and Singer, M. F. (1976). *J. Biol. Chem.* **251**, 2334–2338.

Von Holt, C., Strickland, W. N., Brandt, W. F., and Strickland, M. W. (1979). *FEBS Lett.* **100**, 201–218.

Wachtel, E. J., and Sperling R. (1979). *Nucleic Acids Res.* **6**, 139–151.

Wangh, L., Ruiz-Carrillo, A., and Allfrey, V. G. (1972). *Arch. Biochem. Biophys.* **150**, 44–56.

Weintraub, H., and Groudine, M. (1976). *Science* **193**, 848–856.

Weintraub, H., and Van Lente, F. (1974). *Proc. Natl. Acad. Sci. U. S. A.* **71**, 4249–4253.

Weintraub, H., Palter, K., and Van Lente, F. (1975). *Cell* **6**, 85–110.

Weintraub, H., Flint, S. J., Leffak, I. M., Groudine, M., and Grainger, R. M. (1978). *Cold Spring Harbor Symp. Quant. Biol.* **42**, 401–407.

Whitlock, J. P., Jr., and Simpson, R. T. (1976). *Biochemistry* **15**, 3307–3314.

Whitlock, J. P., Jr., and Simpson, R. T. (1977). *J. Biol. Chem.* **252**, 6516–6520.

Whitlock, J. P., Jr., and Stein, A. (1978). *J. Biol. Chem.* **253**, 3857–3861.

Wilhelm, F. X., Wilhelm, M. L., Erard, M., and Daune, M. P. (1978). *Nucleic Acids Res.* **5**, 505–521.

Wilkins, M. H. F., Zubay, G., and Wilson, H. R. (1959). *J. Mol. Biol.* **1**, 179–185.

Wilson, A. C., Carlson, S. S., and White, T. J. (1977). *Annu. Rev. Biochem.* **46**, 573–640.

Woodcock, C. L. F. (1973). *J. Cell Biol.* **59**, 368a.

Woodcock, C. L. F., Frado, L.-L. Y., Hatch, C. L., and Ricciardiello, L. (1976). *Chromosoma* **58**, 33–39.

Worcel, A. (1978). *Cold Spring Harbor Symp. Quant. Biol.* **42**, 313–324.

Worcel, A., and Benyajati, C. (1977). *Cell* **12**, 83–100.

Wyns, L., Lasters, I., and Hamers, R. (1978). *Nucleic Acids Res.* **5**, 2345–2358.

Yeoman, L. C., Olson, W. O., Sugano, N., Jordan, J. J., Taylor, C. W., Starbuck, W. C., and Busch, H. (1972). *J. Biol. Chem.* **247**, 6018–6023.

Zweidler, A. (1976). *In* "Organization and Expression of Chromosomes" (V. G. Allfrey, ed.), pp. 187–196. Dahlem Konferenzen, Abakon Verlagsgesellschaft, Berlin.

Zweidler, A., and Cohen, L. H., (1972). *Fed. Proc., Fed. Am. Soc. Exp. Biol.* **31**, 926.

FOLDING OF PROTEIN FRAGMENTS

By DONALD B. WETLAUFER

Department of Chemistry, University of Delaware, Newark, Delaware

I. INTRODUCTION

The mechanism of acquisition of three-dimensional structure by proteins (protein folding) has drawn the attention of numerous investigators over the past two decades and has been the subject of frequent reviews (Anfinsen and Scheraga, 1975; Baldwin, 1975; Creighton, 1978; Nemethy and Scheraga, 1977; Wetlaufer and Ristow, 1973). The idea that a globular protein might fold in subassemblies of peptide chain before forming the final native structure emerged independently from two different laboratories. We will begin with a sketch of the context from which this work on the folding of protein fragments has developed.

In the mid-1960s, several attempts were made to regenerate oxidatively specific binding from reduced immunoglobulins and their proteolytic fragments (Haber, 1964; Whitney and Tanford, 1965; see also the reviews of Friedman and Beychok, 1979; Wetlaufer and Ristow, 1973). While these varied greatly in the particular systems studied and in experimental detail, the results agreed in the finding that specific binding could be regenerated from immunoglobulin fragments, often in good yield. Because there were many indications that immunoglobulins are structurally unique, there was no inclination to generalize fragment folding as something to expect of globular proteins in general.

Attempts to prepare synthetic insulin repeatedly encountered the

ADVANCES IN
PROTEIN CHEMISTRY, Vol. 34

problem of low yields in the attempt to link oxidatively the A and B chains via disulfides (Dixon and Wardlaw, 1960; Du *et al.*, 1965; Zahn *et al.*, 1966). The view of this problem was broadened by Givol *et al.* (1965) who showed that insulin is apparently metastable and loses its activity and native structure in the presence of a thiol–disulfide shuffling system. Similar but less detailed studies were carried out with chymotrypsinogen and its two-chained daughter, chymotrypsin. The inference of Givol *et al.* that insulin is synthesized as a single chain precursor was soon validated by studies from Steiner's laboratory demonstrating proinsulin as the precursor. With *in vitro* studies, Steiner and Clark (1968) also demonstrated the oxidative renaturation of reduced proinsulin in high yield, in marked contrast with the poor yields in forming insulin from its two chains. These results taken altogether seemed to indicate that the immunoglobulins indeed were a special case, and that protein fragments would probably not form structures resembling the native structure.

Studies of proteolytic fragments of staphylococcal nuclease (Taniuchi and Anfinsen, 1969) and RNase A (Taniuchi, 1970) seemed to support this view. Taniuchi (1970), in summary remarks, said "Thus, the minimum information of the specific folding of a protein requiring almost the entire amino acid sequence is observed with both staphyloccocal nuclease and bovine pancreatic ribonuclease."

Another set of studies, in the Pasteur Institute, led to a different perspective on these questions. These studies concerned β-galactosidase, a large, inducible enzyme in *Escherichia coli*. Early termination mutant forms of β-galactosidase provide a number of enzymically inactive, incomplete peptide chains, which can combine *in vitro* and *in vivo* with a peptide containing an extensive C-terminal sequence of the enzyme, to yield enzymic activity (Ullmann and Perrin, 1970). This, of course, is an example of protein complementation (see the excellent review of Zabin and Villarejo, 1975). From determinations of the sedimentation coefficient and the molecular weight of the C-terminal peptide, Goldberg (1969) inferred that both the N-terminal and the C-terminal fragment have compact native-like structures. He further suggested that these "globules" are folded around independent nucleation centers. It seems fair to say that in the early 1970s there was no agreement on the question, "Is all (or nearly all) of a peptide chain necessary for folding to native-like structure?"

Unaware of Goldberg's (1969) publication on β-galactosidase "globules," we had been studying the structures of globular proteins solved by X-ray crystallography, hoping to find in those structures some clue to the assembly process. We saw that not only are most globular proteins composed of multiple domains, but also that those domains are characterized by chain continuity. We advanced the hypothesis that these domains resulted from independent assembly processes (Wetlaufer, 1973):

> Distinct structural regions have been found in several globular proteins composed of single polypeptide chains. The existence of such regions and the continuity of peptide chain within them, coupled with kinetic arguments, suggests that the early stages of the three-dimensional structure formation (nucleation) occur independently in separate parts of these molecules. A nucleus can grow rapidly by adding peptide chain segments that are close to the nucleus in amino acid sequence. Such a process would generate three-dimensional (native) protein structures that contain separate regions of continuous peptide chain. . . . One of the most searching experimental tests for an independent continuous region would be to demonstrate self-assembly of just that region, with a high degree of fidelity.

A substantial amount of experimental evidence has since been developed that, on balance, supports this prediction. In the following section, we will outline experimental findings, protein by protein, roughly in chronological order. Finally, we will summarize and explore the significance of these findings.

II. Experimental Studies on Specific Proteins

A. β-Galactosidase

This enzyme from *E. coli* is a tetramer of four identical subunits, each of molecular weight ~116,500. *Amber* and *ochre* (premature termination) mutants of the enzyme provide a number of enzymically inactive, incomplete peptide chains, identical in sequence with the N-terminal part of the wild-type chains. A subset of these N-terminal peptides, called acceptor peptides, can combine with so-called ω-peptides identical in sequence with the C-terminal part of the wild-type chain, to restore enzymic activity (Ullmann *et al.*, 1965, 1967; Ullmann and Perrin, 1970; see also the review by Zabin and Villarejo, 1975). Goldberg (1969) suggested that the acceptor peptides and the ω-peptides "are presumably folded around independent nucleation centers as evidenced by the following facts:

> (i) The acceptor, in the absence of ω-protein, cross-reacts with pure anti-galactosidase serum (D. Perrin, personal communication) and therefore must share some structural features with the wild-type protomer.
> (ii) The ω-protein, even though it does not react with anti-galactosidase serum, is certainly not a random coil: from the value of its sedimentation coefficient, $S_{20,w} = 3.15$ (Ullmann *et al.*, 1967), and of its molecular weight, one can compute an f/f_0 value of 1.1 compatible with a globular structure."

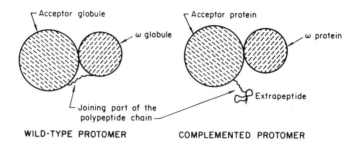

FIG. 1. Scheme for the "globule" structure of β-galactosidase monomer. Reprinted with permission from Goldberg (1969).

Goldberg illustrated his idea schematically (Fig. 1). As will be seen shortly, Goldberg's view was as prescient as his evidence was limited. First, some comment on the evidence he cited. A frictional ratio (f/f_0) as small as 1.1 is not only compatible with a globular structure, it *demands* a shape approaching spherical. Of course, that is not proof of any but the grossest structural resemblance to the ω-region of the native protein, and it is unsettling that the ω-protein was reported not to react with anti-galactosidase serum. Ullmann and Monod's studies of the kinetics of ω-complementation (1970) showed processes that required hours for completion of a complementation reaction. This rather slow rate of reaction raises the suspicion that complementation requires rather more structural rearrangement than is implied by Fig. 1.

In a brief report, Villarejo and Zabin (1973) investigated the ability of a number of peptide fragments of β-galactosidase to bind to an affinity column. The column was prepared by coupling the substrate analog p-aminophenyl-β-D-thiogalactopyranoside to agarose. Several of the large peptide fragments were retarded on the column under conditions of salt gradient elution. The fragments were assayed in the effluent by their ability to complement appropriate mutant species of galactosidase to produce enzyme activity. The same set of peptides that were retarded on the affinity column were also inactivated by the substrate analog N-bromoacetyl-β-D-galactopyranosylamine. Inactivation in this case means loss of ability to form active complemented enzyme. This set of peptides included all of the β-region of the wild-type sequence, which extends approximately from residues 200 to 800. It has since been shown that the site-specific alkylation

occurs on methionine 500 (Fowler *et al.*, 1978), in the middle of the
β-region.

We suggest that the specificity of the chromatographic adsorption
might have been more strongly demonstrated by specific elution by a
substrate for the native enzyme. However, the evidence obtained is
consistent with the conclusion that the retarded fragments contain
structures similar to the substrate-binding site of the native enzyme.

We noted above that, in preliminary experiments, ω-protein appeared not to cross-react with antibody to native β-galactosidase.
This obstacle to the overall argument was removed by the immunochemical studies of Celada *et al.* (1974). They tested the reactivity of
the ω and acceptor fragments with anti-galactosidase antibodies, as
well as the immunogenicity of these fragments. They observed
strong cross-reactivity between anti-galactosidase and the two fragments, and found that both anti-ω and anti-acceptor sera bind the
wild-type enzyme. These investigators note that in both the complementation and in the combination of anti-galactosidase with the fragments, specific stabilizations are involved, and that it is possible that
wild-type structure does not exist in this fragment in the absence of
specific interactions. They proceed (correctly, we believe) to stress
the importance of the finding that both ω and acceptor elicit formation
of antibodies that strongly bind to the wild-type enzyme. This shows
near identity of the fragment structures to the corresponding parts of
wild-type enzyme in the absence of the specific stabilizing interactions suggested above.

Celada *et al.* (1974) were fully aware of the significance of these
findings. They point out that the ω-fragment was prepared by procedures requiring denaturation in 8 *M* urea, so that the ω-fragment, as
finally purified and in nondenaturing solvent, had passed through an
in vitro denaturation–renaturation cycle. To show the perspective in
which Celada *et al.* viewed their results, we quote from their con-
cluding remarks:

> In summary, on the basis of these findings, we are led to conclude that the ω poly-
> peptide, even though it corresponds to only 30% of the complete wild-type poly-
> peptide, is able by itself (i.e., in the absence of any interactions with other seg-
> ments or parts of the wild-type molecule) to fold into the correct wild-type struc-
> ture, or one very close to it. It seems reasonable, on this basis, to assume that the
> normal *in vivo* mechanism of folding involves the stepwise activation of several vir-
> tually independent centers of nucleation.

The underlying idea of the above experiments has been extended in
the recent work of Celada *et al.* (1978), who determine the immuno-

DONALD B. WETLAUFER

TABLE I
Binding of Anti-Peptide Antibodies to β-Galactosidase[a]

Antibody to peptide	Size (amino acid residues)	Binding capacity (nmol of antigen per ml of serum)	Avidity (liters/mol)	Relative binding capacity[b] (%)
CNBr2	90	none		
T8	81	none		
CNBr3	95	none		
CNBr4	15	none		
T16	20	none		
T28-30	36	trace		
CNBr10	41	0.11	1×10^8	11
CNBr14	59	0.043	1.4×10^8	9.5
CNBr15	40	0.01	1.6×10^8	0.5
CNBr16	61	0.2	4×10^7	12
CNBr18	90	0.4	1.4×10^8	7
CNBr19	23	none		
CNBr20	96	1.6	2×10^7	13
CNBr20B	62	0.14	1.3×10^8	1
CNBr21	61	none		
CNBr22	43	none		
CNBr23	23	0.022^c	2×10^7	4
CNBr24	32	1.0	4.5×10^7	20

[a] Reprinted with permission from Celada *et al.* (1978).

[b] Binding capacity for β-galactosidase compared to homologous antigen.

[c] The binding to control serum was three-fourths of the experimental value.

genicity of a number of smaller β-galactosidase peptides. Eighteen tryptic and CNBr peptides were used to raise antibodies in rabbits. The peptides ranged in size from 15 to 96 amino acid residues. All 18 peptides were effective immunogens, and half of the anti-peptide antibodies bound strongly to β-galactosidase. Determinations of the "avidities" or association constants of the latter group of antibodies with native β-galactosidase gave values ranging from 2×10^7 to 1.6×10^8 liters mol^{-1} (Table I). These values are surely large enough to indicate strong binding.

Inspection of Fig. 2 shows the distribution (along the β-galactosidase amino acid sequence) of the relative binding capacity of the sera raised against the several peptides. No simple relationship was seen between peptide size and the amount of antibody evoked or its avidity for β-galactosidase. We take these results as substantial evidence that many relatively small peptide fragments in β-galactosidase have the capability for folding to structures resembling those of their counterpart sequences in the native enzyme. The re-

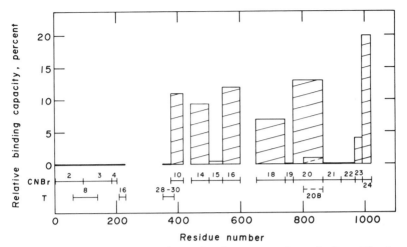

FIG. 2. Relative binding of β-galactosidase by antipeptide antibodies. The figure shows a linear map of the polypeptide chain of β-galactosidase and the cyanogen bromide (CNBr) and tryptic (T) peptides which were used to prepare antibodies. Solid lines and hatched segments represent the binding of antibodies to β-galactosidase as compared with the homologous peptide. Reprinted with permission from Celada *et al.* (1978). Copyright 1978 by the American Chemical Society.

sults also naturally suggest the existence of several subdomains within the acceptor and ω-domains of the native protomer. These conclusions would not be valid if the antibodies evoked by the above set of peptides are specific for particular sequences but indifferent to their conformations. No test for this situation appears to exist. In our view, the high affinities determined for the antibody–peptide combinations suggest a specificity that would be hard to achieve in the absence of specific three-dimensional structures in the peptides.

Taken altogether, the available evidence makes a strong case for independent folding of several domains of β-galactosidase, both *in vitro* and *in vivo*.

B. Ribonuclease A

This enzyme (RNase A) is a single chain protein of 124 amino acid residues, cross-linked by four intrachain disulfide bonds. Limited proteolysis of the enzyme cuts a single peptide bond between residues 20 and 21 (Richards and Vithayathil, 1959). The derived protein, RNase S, retains enzymic activity although the N-terminal peptide of 20 amino acids (S-peptide) is no longer covalently attached to the balance of the molecule (S-protein). Removal of S-peptide from

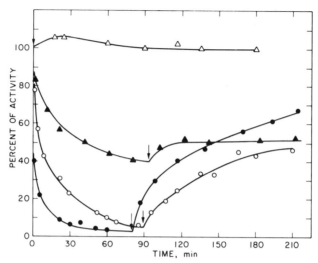

Fig. 3. Inactivation of S-protein, by disulfide interchange in the absence, and subsequent reactivation in the presence, of S-peptide. The reaction was carried out with different concentrations of the interchange enzyme. ●——●, 134 μg/ml; ○——○, 67 μg/ml; ▲——▲, none. Stabilization against denaturation by S-peptide, 134 μg/ml, interchange enzyme (△——△). All reactions were carried out in 0.1 M Tris–chloride buffer, pH 7.4, β-mercaptoethanol, 10^{-3} M, with 180 μg/ml of S-protein. The arrows indicate the times of addition of S-peptide (1.3 eq relative to S-protein) to the reaction mixture. Reprinted with permission from Kato and Anfinsen (1969). Copyright by The American Society of Biological Chemists, Inc.

S-protein leads to loss of enzymic activity. Activity is rapidly restored upon mixing of S-protein and S-peptide.

Employing the disulfide interchange enzyme, Kato and Anfinsen (1969) studied the stability of S-protein relative to RNase S. The interchange enzyme led to rapid "scrambling" of disulfide bonds of S-protein. This was accompanied by loss of activity (when assayed in the presence of added S-peptide). Loss of activity and scrambling did not occur when S-peptide was present in slightly greater than stoichiometric concentration. Moreover, the scrambled S-protein is reactivated (in high yield) on addition of S-peptide and the interchange enzyme. At much higher concentrations, C-peptide (sequence 1–13) of RNase showed the same ability as S-peptide in reactivating scrambled S-protein. Both inactivation and reactivation are shown in Fig. 3. Kato and Anfinsen suggest that the residual activity (~20%) remaining after treatment with the reduced interchange enzyme may be due to a fraction of only partly scrambled molecules. This conclusion appears to be supported by their results from mapping the disulfide peptides by the "off-diagonal" method of Brown and Hartley (1966). They found the enzymically scrambled S-protein appeared to contain a significant amount of natively paired cystine peptides, although one of the native disulfide peptides was completely missing. Kato and Anfinsen concluded that the information in the N-terminal 20 residues of RNase A is required for the stability of the S-protein, although a covalent connection linking S-peptide and S-protein is unnecessary.

The issue of stability of another fragment of RNase A was further examined by Taniuchi (1970). He studied the structures formed in de-(121–124) ribonuclease after reduction and reoxidation. This enzymically inactive derivative results from proteolytic removal of the C-terminal sequence of four amino acids from RNase A (Anfinsen, 1956). The derivative [which we will abbreviate as de-(121–4)RNase] was shown to contain no internal peptide bonds cut. Its circular dichroism (CD) spectrum was shown to be highly similar to that of RNase A. De-(121–124)RNase was subjected to reduction and air reoxidation under conditions that are optimal for yield of enzymic activity when applied to RNase A, the parent protein. Reoxidized de-(121–4)RNase was shown to be largely monomeric by gel permeation chromatography and sedimentation equilibrium studies. The CD spectrum of reoxidized de-(121–4)RNase was substantially different from that of the derivative before reduction, suggesting gross conformational differences. RNase A and reoxidized de-(121–4)RNase were subjected to CNBr cleavage (both molecules have four methionine residues) and the resulting peptides were compared by gel permeation chromatography. Different patterns were obtained for the two molecules. The chromatograms were interpreted as indicating randomly paired disulfide bonds in reoxidized de-(121–4)RNase. Taniuchi drew the overall conclusion that the four residues at the C-terminus of RNase A carry information essential for the formation of the native three-dimensional structure.

In a study very similar to the study of RNase S by Kato and Anfinsen (1969), Andria and Taniuchi (1978) examined the stability of RNase peptide 1–118 under disulfide exchange conditions. Enzymically inactive peptide 1–118 can be complemented to generate enzymic activity by fragment 105–124, and this complemented species does not lose activity in the presence of the disulfide interchange enzyme. Peptide 1–118 contains all of the four disulfide bonds of RNase A, and is rapidly converted to a noncomplementing form in the presence of the disulfide exchange enzyme. This inactivated 1–118 can be partially reactivated by the simultaneous presence of the interchange enzyme and peptide 105–124. The inactivation and reactivation reactions were also monitored by gel permeation chromatography and CD measurements. The overall conclusion was that fragment 1–118 with its native disulfides is thermodynamically unstable compared with the available alternative set of disulfides.

Ten of the thirteen residues in N-terminal tridecapeptide of RNase A are helical in the parent molecule (Kartha *et al.*, 1967; Wyckoff *et al.*, 1970). In a thoughtful and careful study, Brown and Klee (1971) investigated the CD spectrum and molecular weight of this peptide in various solvents. The CD spectrum was sensitive to both temperature and ionic strength under conditions in which the peptide was monomeric by sedimentation equilibrium. The CD could be fitted to a model mixture composed of a small fraction of α-helix and a large fraction of random coil structure. Brown and Klee were careful not to make a numerical estimate of the apparent helical content of the peptide, rather, commenting that the helical structure showed only marginal stability under physiological conditions. We have no quarrel with these experiments or their conclusions. However, we must point out that, while a mixture of helix and random structure is a plausible model, it is an assumption; other conformations or mixtures might fit the data equally well. We have earlier discussed this work in comparison with similar but less careful studies of the conformation of fragments of other proteins (Wetlaufer and Ristow, 1973).

We cannot finish our discussion of RNase without mentioning the work of Gutte (1975a,b, 1977, 1978). He has synthesized (Merrifield solid-phase methodology) several shortened analogs of RNase A and of S-protein. The position of the deletions was "decided from a careful inspection of the x-ray structure" (Gutte, 1975a). The gaps resulting from these deletions were filled by glycyl or alanyl residues. Considering the subjective criteria for the deletions, it seems remarkable that some of these analog preparations did show low levels of catalytic and/or substrate binding activity. This work, although certainly provocative, is very difficult to evaluate. One obvious difficulty, which Gutte acknowledges, is due to the sequence heterogeneity of the peptides he has prepared. Another problem comes from an expansion of the original objective "to study the importance of several such exterior loops for the correct folding of RNase to give an active enzyme" (Gutte, 1975a). The results and discussion in that and subsequent publications are not focused on the original objective.

C. Staphylococcal Nuclease

This enzyme consists of a single polypeptide chain of 149 residues containing neither cysteine nor cystine. It contains one tryptophan residue near the C-terminus and, in its active form, contains bound Ca^{2+}. Limited proteolysis provides a number of large fragments (Taniuchi and Anfinsen, 1969), which will here be designated according to the sequence positions of their N- and C-terminal amino acids.

Fragment 6–149 has full enzymic activity and optical rotatory dispersion (ORD), CD, and fluorescence emission spectrum closely resembling those properties of the native enzyme (1–149). However, when fragments 1–126 and 127–149 are examined by the same optical criteria, both fragments behave as "loose and disorganized structures very different from that of nuclease." No activity results when these two fragments are mixed; but when 1–126 is mixed with 50–149, enzymic activity is generated, along with the ability to form a precipitate with antinuclease serum. The various optical properties of 1–126 mixed with 50–149 change in a nonadditive way to resemble more nearly the properties of the native enzyme.

Fragments 6–48 and 49–149 also will combine on mixing to generate enzymic activity and a conformation similar to that of the native enzyme. It is worth noting that the derivative formed by combining 6–48 with 49–149 showed only 8–10% of the specific activity of the native enzyme. This was also the case with the activity generated by the association of peptides 1–126 and 49–149. Taniuchi and Anfinsen concluded that almost the entire amino acid sequence of this enzyme was necessary to generate native-like structure. We agree that, in the aggregate, their conclusion is consistent with their data. However, there is some question (Wetlaufer and Ristow, 1973) whether the CD results offer substantial support.

In a further study, Taniuchi et al. (1977) have shown that in the association of overlapping fragments of staphylococcal nuclease, two different species of active enzyme are formed. On the basis of the products of limited proteolysis, structures for the two species were deduced. In one case a structure is proposed in which fragment 1–126 assumes native-like structure over the sequence 1–48, and all of fragment 50–149 assumes native-like structure. In the other case the structure is one in which fragment 1–126 assumes native-like structure over the sequence 1–110, while that part of fragment 50–149 in the sequence interval 111–149 assumes native-like structure. The interest of these results is enhanced by the finding that the two active species initially form in relative concentrations substantially different from their equilibrium concentrations. Thus, both a mobile equilibrium and substantial kinetic control of the early products are evident. Taniuchi et al. did not reach a clear-cut mechanistic conclusion from their studies.

D. Hen Egg White Lysozyme

An early study on native-like structure formation in a fragment of lysozyme was carried out immunochemically by Arnon and Sela (1969). The native molecule is a single peptide chain of 129 amino acid residues and contains four disulfide bonds. One links cysteinyl residues at sequence positions 64 and 80; a second links residues 76 and 94. By peptic digestion and reduction (followed by air reoxidation) of the peptide mixture, a so-called loop peptide was obtained. This was probably a mixture of peptides 64–83 and 57–83, containing one undetermined disulfide bond. This loop peptide was coupled to a synthetic water-soluble branched poly(DL-alanine), and the resulting conjugate was used to immunize rabbits. Antibodies reactive both with the immunogen and with native lysozyme were isolated from the immune sera. Reduced, carboxymethylated loop peptide did not react with the antibodies elicited by the conjugate, nor with antibodies raised against native lysozyme. Arnon and Sela (1969) concluded that antibodies against the conjugate were directed against a conformation-dependent determinant whose structure resembled its structure as part of native lysozyme.

A later paper from the same laboratory (Arnon et al., 1971) employed a synthetic peptide whose sequence was the same as 64–82 of lysozyme except that residue 76 was Ala rather than Cys. The authors point out that this replacement removes the possibility, inherent in the earlier work, of forming various intramolecular disulfide-pairing isomers. Air oxidation of the synthetic peptide for 20 hours gave a 65% yield of oxidized monomer, with the balance aggregates. Oxidized monomer was coupled as before to synthetic poly(DL-alanine) and this conjugate was again used as an antigen. The results essentially duplicate those obtained with the fragment isolated from lysozyme. Performic acid oxidation of the synthetic peptide opened the "loop" and converted the two half-cystines to cysteic acid residues. This open form of the peptide did not react with antibodies to the closed loop. The foregoing experiments are attractive and support the idea that the loop peptide by itself can form native-like structure. However, there is still the question of whether the experiments with the synthetic loop peptide show an intrinsic tendency of the fragment to fold to native-like structure, or whether they show that that structure can be formed under forcing conditions. The opened loop peptides, whose half-cystines have been converted to cysteic acid or carboxymethyl cysteinyl residues, also raise a question. With bulky, charged groups on the partner residues of the former cross-link, can we fairly say that the principal difference is conformational? These derivatives of half-cystine must alter the steric and electrostatic picture, as well as the conformation. We wonder whether a loop peptide, in which all three cysteinyl residues were replaced by α-aminobutyryl residues, would cross-react with antibodies evoked by the conjugates of Arnon *et al.* A further question is raised by the results of Mathyssens *et al.* (1974). These investigators found that lysozyme fragment 57–83 + 91–107 (the two peptides are linked by native —S—S— bonds) did not react with antilysozyme antibodies. This apparent contradiction with the findings of Arnon's group may only reflect differences between the animals that produced the antibodies used by the two groups—nonetheless, some additional uncertainty is introduced into the issue.

The oxidative folding of reduced peptide 13–105 of hen egg lysozyme was studied by Johnson *et al.* (1975, 1978). This fragment contains five half-cystines, and is of special interest since it has been independently shown (Anderson and Wetlaufer, 1976) that oxidative regeneration of reduced lysozyme preferentially involves residues 64, 76, 80, and 94 in the initial stages. It also contains the residues comprising the continuous chain region 39–100, earlier identified (Wetlaufer, 1973) as a candidate for an independent self-assembly region. Johnson *et al.* (1975, 1978) employed the glutathione oxidoreduction buffer under conditions previously shown (Saxena and Wetlaufer, 1970) to be optimal for oxidative regeneration of reduced lysozyme. The folding processes for reduced fragment 13–105 and for reduced lysozyme were compared (*a*) by the kinetics of protein disulfide formation, (*b*) by reaction with antilysozyme antibodies, and (*c*) by affinity chromatography.

The extent of formation of protein disulfides with time was determined by withdrawing aliquots which were acidified to pH 5.5 and alkylated with N-ethylmaleimide. The disulfide content of the peptide was determined after its isolation. Formation of two intrapeptide disulfide bonds proceeded at the same rate (within experimental error) as formation of the first two disulfides in reduced lysozyme. The first-order rate constant for these two processes (0.5 min^{-1}) was eight times that describing the rate of oxidation of reduced lysozyme in the presence of 6 M guanidinium chloride, suggesting substantial specificity in the process in absence of denaturant. An additional indication of specificity was the finding that 13–105 reached its maximum of two —S—S— bonds in less than 20 minutes, retaining one reduced thiol from 20 to 240 minutes. For subsequent studies this material was S-alkylated with N-ethylmaleimide.

Early course antisera elicted by native lysozyme were obtained from two goats. These antisera showed 32 and 56% of the maximum reaction obtained with native lysozyme, respectively. Precipitin and immunoabsorption estimates of the percentage of cross-reactivity were in close agreement. The precipitin curves indicate that regenerated 13–105 has substantially lower affinity for the antisera than does native lysozyme. This is qualitatively consistent with the observation of a CD spectrum indicating considerably less α-helix and β structure in regenerated 13–105 than is found in the X-ray diffraction structure of the corresponding part of native lysozyme. Although there are differences in the detailed localization of the antigenic sites of lysozyme (Fujio et al., 1968; Arnon et al., 1971; Lee and Atassi, 1975; Atassi et al., 1976), there appears to be general agreement that at least one major antigenic determinant of native lysozyme is found in the three-dimensional structure comprising the sequence 57–105. The immunochemical evidence for regeneration of native-like structure from reduced 13–105 is subject to some of the same qualifications we have previously noted in the discussion of the synthetic loop peptide studies of Arnon et al. It should be noted that with its content of five half-cystines, the 13–105 regeneration is very permissive in the opportunities it provides for forming nonnative intramolecular disulfides. This clearly differs from synthetic loop analog in which there were only two half-cystines, Cys-76 having been replaced by alanine "in order to avoid ambiguous disulfide bond formation" (Arnon et al., 1971).

Affinity chromatography was carried out on columns prepared with lightly carboxymethylated chitin, which is known to be a poor substrate for lysozyme. Both native lysozyme and regenerated 13–105 were bound to the column at pH 7 and eluted at pH 3. As controls, the basic proteins cytochrome c and pancreatic RNase A, as well as concanavalin A and α-amylase, were not bound from the same solvent at pH 7. These findings constitute a third line of evidence for formation of native-like structure in regenerated 13–105.

The kinetics of disulfide formation, the demonstration of specific binding, and the immunochemical results all support the conclusion that native-like structure results from the oxidative folding of reduced peptide 13–105. These three independent lines of evidence support the conclusion that lysozyme has a continuous chain independent assembly region somewhere in the sequence 13–105.

In a more limited study, Johnson (1974) studied the —S—S— pairing in the oxidative regeneration of reduced lysozyme peptide 57–83. By limited proteolysis and fractionation of the resultant pep-

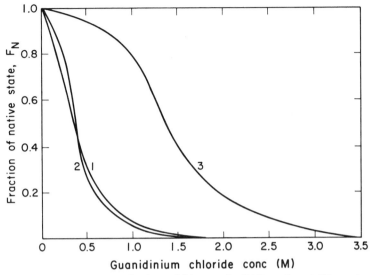

FIG. 4. Equilibrium curves for the unfolding and refolding of penicillinase in guanidinium chloride measured by viscosity (1), difference spectroscopy (2), and mean residue rotation $[m']_{234}$ (3). Redrawn with permission from Robson and Pain (1976b).

tides, Johnson separated a major disulfide-containing peptide. Compositional analysis showed that the disulfide of this peptide linked half-cystines 64 and 80, which is a native pairing. Limited amounts of the reduced 57–83 made confirmatory experiments such as material balances impossible, so the presence of nonnative disulfides cannot be ruled out.

E. Penicillinase

This enzyme from *Staphylococcus aureus* is a single-chain globular protein whose molecular weight is 28,000. It contains neither cysteinyl nor cystinyl residues. Careful equilibrium studies of the guanidinium chloride denaturation of penicillinase by Robson and Pain (1973, 1976a,b), showed nonparallel changes in the viscosity and optical rotation of penicillinase solutions as a function of denaturant concentration (see Fig. 4). They interpreted these results as being inconsistent with a two-state model for the denaturation, $N \rightleftharpoons U$, but requiring at least a third species in the reaction scheme. Kinetic studies of the denaturation and renaturation showed that the third state, designated H, lies on both the unfolding and folding pathways, thus supporting the reaction scheme $N \rightleftharpoons H \rightleftharpoons U$. Species H predominates in 0.8 *M* guanidinium chloride, and has been further characterized by viscosity, sedimentation, and CD measurements (Carrey and Pain, 1979). To summarize, they found the frictional ratio, f/f_0, to be 2.09 for species H, compared with 1.02 for native penicillinase, indicating a substantially larger effective volume for H than for N. The CD spectra were taken to mean that penicillinase (state N) contains 27% α-helix while state H contains 16%, with no change in β-structure from N to H. Although an exact interpretation of CD spectra is not presently possible (Saxena and Wetlaufer, 1971), we can say with confidence that the state H spectrum reflects a substantially ordered structure. Carrey and Pain (1977) have reported that penicillinase fragments resulting from CNBr cleavage aggregate to form a structure very similar to that of the native enzyme, but lacking enzymic activity. Three fragments whose molecular weights are 10,500, 9,500, and 8,500 are formed and tend to associate as an aggregate whose properties resemble the native enzyme. Sedimentation equilibrium results showed the molecular weight for the aggregate to be the same as that of the native enzyme. The CD spectra of the aggregate and the native enzyme are nearly identical in the peptide absorption region 205–240 nm. Reversible guanidinium chloride denaturation of the aggregate was demonstrated, although the aggregate was less stable than the native enzyme.

The same laboratory has reported (Adams *et al.*, 1980) immunochemical studies of the fragments. Antiserum raised against penicillinase cross-reacts with the specific aggregate, showing complete identity in the Ouchterlony gel diffusion method. The peptides were separated by chromatography under denaturing conditions (8 *M* urea), each showing partial immunochemical identity with native penicillinase. All the foregoing results support the conclusion that the separate fragments can undergo independent folding to structures closely resembling their counterparts in the native molecule.

F. Plasma Albumin

Fifteen years ago plasma albumin was one of the most thoroughly studied proteins. Today, although X-ray crystallographers have solved structures for a large number of proteins, providing information in exquisite detail, the three-dimensional structure of plasma albumin is still unknown. However, the human and the bovine proteins, both single-chain proteins, have been sequenced (Brown, 1974). On the basis of internal sequence homologies and the pattern of —S—S— bonding, Brown (1975) has proposed that the protein has arisen by gene triplication. Figure 5 shows schematically the S—S— bonding and the proposed domains. A much earlier suggestion for in-

BOVINE SERUM ALBUMIN

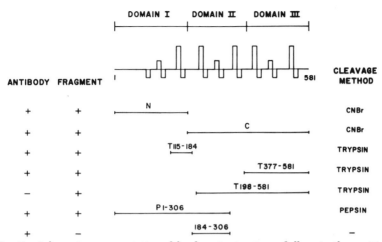

FIG. 5. Schematic representation of the domain structure of albumin, the position of the disulfide loops, and the fragments of the molecule that have been isolated. Included in the figure are the cleavage methods used to obtain the fragments plus the regions to which the restricted antibody populations used in these experiments are directed. Reprinted, with permission, from Teale and Benjamin (1976). Copyright by the American Society of Biological Chemists, Inc.

ternal homology and a kind of domain structure for plasma albumin was made by Foster (1960) (see Fig. 6). Foster's proposal was made as an attempt to rationalize a wide range of pH-dependent physical properties of the protein. It is remarkable that Foster's view is still serviceable after 20 years and much new information. We see that even in the absence of an X-ray crystallographic structural solution there are strong reasons to believe that serum albumin is a multiple-domain protein, and that the domains have chain continuity. Reed *et al.* (1975a) showed that limited proteolysis of serum albumin generates fragments that retain specific bonding functions of the native molecule. The following year, reports from two groups gave evidence of independent folding of those fragments.

Native bovine serum albumin has the surprising property of catalyzing the decomposition of the Meisenheimer complex: 1,1-dihydro-2,4,6-trinitrocylohexadienate. Taylor and Silver (1976) prepared albumin fragments 1–306 and 307–581 and separated them without disulfide reduction. These were called "native fragments." Neither of these fragments alone has catalytic activity to decompose the Meisenheimer complex, but when mixed together stoichiometrically, ~35%

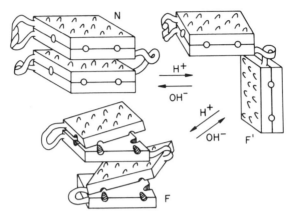

FIG. 6. A highly idealized model for the plasma albumin molecule to account for the N–F transformation and its relationship to the titration anomaly, the cooperative detergent binding, and the altered solubility behavior of the low pH form. The model contains four folded amphipathic subunits, the hydrophobic surfaces being buried in the N form and exposed in the F form. Holes around the periphery of the molecule represent the 10 to 12 strong binding sites for detergent ions which are destroyed, upon isomerization, with the exposure of a large number of weaker sites. Reprinted with permission from Foster (1960). Copyright by Academic Press, Inc.

of the specific catalytic activity of albumin is obtained. Reduction of each fragment, followed by air reoxidation, led to substantial potential activity. The yield was $45 \pm 5\%$ when oxidatively refolded 307–582 was assayed in the presence of equimolar native fragment 1–306. When the same kind of experiment was carried out to test the oxidative refolding of 1–306, assaying catalytic activity in the presence of equimolar native fragment 307–582, the yield was $15 \pm 2\%$. When refolded 1–306 was added to refolded 307–581, $7 \pm 2\%$ activity was recovered. The authors had previously demonstrated that the catalytic activity in question is very sensitive to protein conformation, and is especially sensitive to disulfide pairing in the protein. They concluded that "a very high degree of native structure is present in the refolded fragments."

That conclusion is supported and extended by the virtually simultaneous publication of Teale and Benjamin (1976). These investigators studied the oxidative regeneration of bovine serum albumin, assaying the extent of refolding immunochemically. Their results showed clearly that some parts of the molecule fold faster than others. Two fragments of albumin were tested for oxidative regeneration in the same way. Substantial return of native structure was seen in both fragments.

To carry out these experiments, specific peptides derived from albumin were covalently coupled to agarose. These materials served as immunoadsorbents for the preparation of restricted populations of antibodies. Figure 5 shows the relationship of the peptides used to the disulfide and one-dimensional structure of albumin. The regenerations of reduced protein were facilitated either with the disulfide exchange enzyme system (Goldberger et al., 1964) or with glutathione oxidoreduction buffers (Saxena and Wetlaufer, 1970). Aliquots were withdrawn from the regenerating protein solutions and incubated with antiserum and ^{125}I-labeled albumin at pH 6.0 for 30 minutes, then chilled and precipitated with ammonium sulfate. The precipitates were collected and washed, and the radioactivity was determined. The extent of folding to native-like structures was estimated by measuring the extent of inhibition of precipitation of ^{125}I-labeled albumin by rabbit antibodies elicited by native albumin. Restricted populations of antibody were used to determine whether different parts of reduced albumin acquire antigenic structure at the same rate.

Results of several experiments in which the return of native-like structure to various parts of reduced albumin was obtained are shown in Fig. 7. It is clear that different antigenic determinants are formed

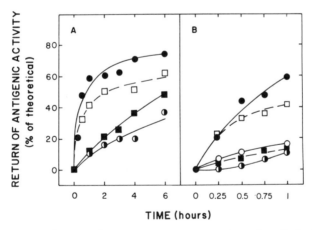

FIG. 7. Kinetics of return of native antigenic structure in various limited regions of reduced albumin as determined with restricted populations of antialbumin. Reduced protein was reoxidized in the presence of optimal amounts of rat liver disulfide-exchange enzyme (0.5 mg/ml of RII/2.5 ml of reaction mixture). The reoxidation buffer and other conditions were as described in Section II. (A) ●, Anti-$T_{115-184}$; □, anti-$T_{377-581}$; ◐, anti-N-fragment; ■, antialbumin, anti-C-fragment and anti anti-$T_{184-306}$. (B) ○, Anti-bovine serum albumin; ■, anti-C-fragment and anti-$T_{184-306}$; ◐, anti-N-fragment. Reprinted, with permission, from Teale and Benjamin (1976). Copyright by the American Society of Biological Chemists, Inc.

at substantially different rates during the refolding of the whole molecule.

Similar experiments with fragments $T_{198-581}$ and $T_{377-581}$ showed the ability of fragments of albumin to form native-like structure independently of the remainder of the molecule. From a comparison of the rates of fragment folding with that of the whole molecule, it appears that these two fragments both fold faster than the whole molecule. Results of another experiment, relevant to this point, are shown in Fig. 8. Here with antibodies specific for the native structure of peptide $T_{377-581}$ it is seen that the rates of formation of native structure are related as $T_{377-581} > T_{198-581} >$ albumin (1–581). In other words the C-terminal third of the molecule finds its native structure more slowly as part of the C-terminal two-thirds, and more slowly still as part of the whole molecule. Thus, while fragments can assemble independently, some interdependence is seen in the assembly of larger fragments and the whole albumin peptide. Teale and Benjamin suggest that the observed decrease in rate for the whole molecule may be due to the greater chance for mispairing in disulfide bonding, or perhaps an effect of the extra thiol of Cys-34.

In a subsequent paper, Teale and Benjamin (1977) carried out further experiments of the same kind with additional fragments of albumin. They concluded that the C-terminal one-third of each of the

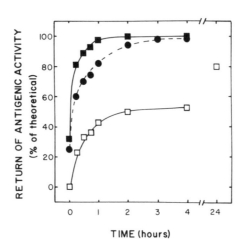

FIG. 8. Interdomain influence on the rate of refolding of Domain III of albumin (residues 377–381), later time points. All conditions were the same as for Fig. 7. ■——■, $T_{377-581}$ (Domain III); ●——●, $T_{198-581}$ (Domains II and III); □——□, albumin (Domains I, II, and III). Reprinted, with permission, from Teale and Benjamin (1976). Copyright by the American Society of Biological Chemists, Inc.

three domains refolds faster than the remainder of the respective domain, and presented further evidence for interdomain interference in folding rates.

The foregoing studies of Teale and Benjamin relied entirely on immunochemical criteria for measuring the rates of return of native-like structure. For us this raised the inevitable question, "Does the antibody assist the folding process by binding one of several species in a mobile equilibrium?" Another question of concern was, "Is folding and disulfide shuffling stopped in the mildly acidic (pH 6.0) antibody reaction medium?" Another question was raised by Habeeb and Atassi (1976) in a paper whose title carries its conclusion, "A Fragment Comprising the Last Third of Bovine Serum Albumin Which Accounts for Almost All the Antigenic Reactivity of the Native Protein." Without attempting to address the relative merits of this conclusion and those implicit in the work of Teale and Benjamin (1976), a substantial disagreement is apparent.

With these concerns in mind, a reexamination of the folding-by-parts issue using other criteria for native-like structure was undertaken (Johanson et al., 1977, 1981). Albumin has a strong fatty acid binding site in the C-terminal third of its sequence, and a strong bilirubin binding site in its N-terminal half. Binding of ligands to proteins and fragments (sampled during oxidative regeneration) was measured with a system that exploits the ability of soluble protein to remove ligand from albumin immobilized on agarose (Reed et al., 1975b). These assays require only 10 minutes equilibration time, limiting the extent of protein folding during the assay itself. Oxidative regenerations were carried out with oxidoreduction buffers of glutathione or β-mercaptoethylamine.

Palmitate binding was regained more rapidly and more completely in the smallest of the three peptides studied: fragment 377–582 > fragment 307–582 > albumin (1–582); similarly, bilirubin binding returned faster in fragment 1–306 than in the whole molecule (1–582). These results are consistent with the findings of Teale and Benjamin. Thus, folding studies evaluated by specific ligand binding and by immunochemical criteria lead to the same conclusions: that native-like three-dimensional structure forms in large fragments of albumin; moreover it forms faster in fragments than it does in the whole molecule. Johanson et al. (1977, 1981) also found an inverse relationship between peptide size and (a) the rate of peptide —S—S— formation and (b) the rate of formation of secondary structure as judged by CD at 221 nm.

These investigators also compared the folding rates of mercaptalbumin (Cys–SH at position 34 free to react) and mercaptalbumin whose free thiol had been alkylated with iodoacetamide. The folding rates of the alkylated and unalkylated albumins were indistinguishable, whether measured with unfractionated antibodies to albumin or with an anti-1–306 subpopulation. We clearly can conclude that Cys-34 plays no essential role in oxidative renaturation of albumin or its N-terminal one-half.

Since the sites for palmitate binding and bilirubin binding are in different halves of the albumin molecule, it was attractive to compare the return of these two different functions. The results of this comparison are shown in Fig. 9, where bilirubin binding is seen to be more extensively recovered than palmitate binding. Regeneration of the two binding sites is therefore independent. In a third variant of this system, regeneration was carried out in the presence of palmitate (at 3 μM compared with 1 μM albumin). Figure 9 also shows that palmitate at this level (sufficient to bind extensively to the fatty acid site) had no effect on regeneration of the bilirubin site. This adds further evidence in support of the idea of independent folding of the two sites.

Good overall agreement was obtained in the conclusions reached on folding-by-parts in albumin by immunochemical (Teale and Benjamin, 1976, 1977) and small ligand binding measurements (Johanson et al., 1977, 1981). This serves to increase our confidence in the usefulness of immunochemical approaches to this kind of problem.

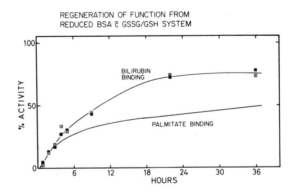

FIG. 9. Nonparallel regeneration of the functions for bilirubin binding and fatty acid binding in bovine plasma albumin. The oxidative regeneration of reduced albumin was carried out at protein concentrations of 1 μM at 25°C in 0.10 M Tris–chloride buffer, pH 8.0, containing 1 mM EDTA and 1 mM reduced and 0.10 mM oxidized glutathione (Johanson et al., 1977, 1981).

G. Tryptophan Synthetase

This enzyme in *E. coli* is an $\alpha_2\beta_2$ tetramer which employs pyridoxal phosphate as coenzyme and L-serine and indoleglycerol phosphate as cosubstrates in the synthesis of tryptophan. The β-subunit can be obtained separately in the form of a dimer, β_2, which shows specific binding sites for pyridoxal phosphate, serine, and indole (Wilson and Crawford, 1965; Goldberg *et al.*, 1968). The β protomer molecular weight is 45,000. Limited proteolysis of β_2 with trypsin yields a "nicked" dimer composed of two nonoverlapping peptide chains whose molecular weights are 12,000 and 29,000 (Högberg-Raibaud and Goldberg, 1977a). Nicked β_2 maintains the specific set of binding and spectral characteristics shown by β_2 before proteolysis. In a subsequent study, Högberg-Raibaud and Goldberg (1977b) separated nicked β_2 into its two constituent peptides by gel permeation chromatography of a 6 M urea solution. Following dialysis to remove the urea, soluble preparations of the peptides were obtained: F_1 (the larger) and F_2 (the smaller). These solutions then underwent a series of experimental tests to determine (*a*) whether they could recombine to generate the properties of nicked β_2, and (*b*) whether F_1 and F_2 have three-dimensional structures similar to those they assume in nicked β_2.

Neither F_1 nor F_2 alone gave the characteristic fluorescence of β_2 and nicked β_2 in the presence of L-serine and pyridoxal phosphate. However, titration of a fixed amount of F_1 with F_2 gave rise to a fluorescence intensity 80–90% that of nicked β_2 at a stoichiometric ratio of F_1 to F_2. Moreover, both the excitation and emission spectra of the stoichiometric mixture were the same as for nicked β_2. In addition, the same specific quenching of this fluorescence was shown in recombined F_1 and F_2 as in nicked β_2. Further, the dissociation constants for L-serine and for indole were determined to be the same within experimental error for recombined F_1 and F_2, as for nicked β_2. No significant differences were found between nicked β_2 and reconstituted $F_1 F_2$ in the intrinsic fluorescence of the aromatic residues, or in the sedimentation coefficients or the 200–250 nm CD spectra. From the foregoing independent lines of evidence, F_1 and F_2 combine to produce a structure very similar to that of nicked β_2.

Studies of the properties of the isolated fragments emphasized the question, "Can F_1 and F_2 acquire their native conformations independently?" A combination of sedimentation velocity and sedimentation equilibrium studies showed frictional ratios f/f_0 of 1.3 and 1.4 for F_1 and F_2, respectively, implying compact globular structures in both cases. The molecular weight of F_1 was determined by sedimentation equilibrium to be 27,000 while that of F_2 was 24,000. From this it is clear that F_1 is monomeric while F_2 is a dimer. This suggests that F_2 may represent those parts of β_2 that are involved in its dimerization.

F_1 contains one tryptophan residue while F_2 contains none. F_2 exhibits only a very weak tyrosine fluorescence, while F_1 gives rise to a strong emission band whose $\lambda_{max} = 327$ nm. The amplitude and λ_{max} of the F_1 fluorescence emission spectrum are characteristic of a tryptophan not exposed to solvent. Further, the emission spectrum of nicked β_2 can be very nearly approximated by the algebraic sum of F_1 and F_2 spectra. This supports the idea that the environment of the single tryptophan in F_1 is very similar to that in nicked β_2. Finally, the CD spectrum in the 250–200 nm range for the F_1 fragment added to that for the F_2 fragment gives a resultant spectrum identical with that of nicked β_2. In this writer's view, the CD is the most convincing, but it is surely strengthed by the consistent results from the frictional ratios and the fluorescence spectra. Overall, the results of this article show that the F_1 and F_2 peptides are capable of independent folding to native-like structures. We agree with Högberg-Raibaud and Goldberg (1977b) in the above conclusion and in their suggestion that structures corresponding to the F_1 and F_2 fragments are likely intermediates in the formation of native β_2 subunits.

H. Thermolysin

Thermolysin is a single-chain protease composed of 316 amino acids; its amino acid sequence and three-dimensional structure (Colman *et al.*, 1972) have been determined. The native protein has no disulfide bonds, but does contain both Ca^{2+} and Zn^{2+}, which strongly stabilize the structure.

Cleavage of the peptide chain with cyanogen bromide leads to peptide fragments 1–120, 121–205, and 206–316. The CD of these fragments in various solvent systems has been determined by Fontana and Vita (1977). Fragment 1–120 was relatively insoluble except at pH $\geq$ 9.6; the CD spectrum obtained on this solution was reported only as "indicative of an essentially disordered conformation." Fragment 121–205 contains those residues of the native enzyme primarily engaged in the binding of the Zn^{2+} and three of the four Ca^{2+} in the native molecule. In the absence of Ca^{2+}, the CD spectrum was characterized by an inflection point, not a true minimum, near 222 nm wavelength. The mean residue ellipticity, $[\theta]_{220} \simeq -4600$, while at 0.9 M Ca^{2+}, $[\theta]_{220} \simeq -9000$. The full spectrum between 205 and 250 nm showed the double minima at 222 and 208 nm of characteristic α-helix both in homopolypeptides and in proteins (Saxena and Wetlaufer,

1971). The CD spectrum of this peptide in 0.2 M NaCl showed no significant difference from that of the Ca^{2+}-free solution above. The results support the conclusion that calcium ions play a specific role in the conformation of this peptide fragment.

Fragment 206–316 exhibits a CD spectrum of even greater negative amplitude ($[\theta]_{220} = -13,600$). As seen in Fig. 10, this fragment shows a temperature-dependent CD spectrum, with a highly cooperative transition evident between 60 and 70°C. The thermal transition appeared to be completely reversible, since the same temperature dependence of $[\theta]_{220}$ was found on cooling as on heating. Reversible solvent-induced transitions were also found from examination of the CD spectrum of fragment 206–316 in various concentrations of urea and guanidinium chloride. The half-transition concentrations were 5 M urea and 2 M guanidinium chloride. As with the thermal transition, reversibility was demonstrated.

Fontana and Vita took the foregoing experiments as evidence not only that native-like structure persists in two fragments of thermolysin, but also that folding to native-like structure can occur with the denatured fragments.

In a report of further work (Vita *et al.*, 1979) the same group has investigated the question immunochemically. They found by immuno-

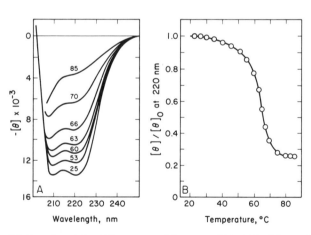

FIG. 10. (A) Far-ultraviolet CD spectra of fragment 206–316 of thermolysin at different temperatures in 20 mM Tris–HCl buffer, pH 7.2. Numbers near the curves indicate temperature in degrees Celsius. The peptide concentration was 0.2 mg/ml. (B) Temperature dependence of $[\theta]/[\theta]_0$ at 220 nm of fragment 206–316, where $[\theta]_0$ is the mean residue ellipticity at 22°C. Reprinted with permission from "Peptides Proceedings of the Fifth American Peptide Symposium," John Wiley & Sons, Inc., 1977. Copyright by John Wiley & Sons, Inc.

diffusion and by radioimmunoassay a significant amount of cross-reactivity between both of the 121–205 and the 206–316 fragment and native thermolysin. Denatured thermolysin was reported to show no cross-reactivity. These results, coupled with the optical evidence of reversible denaturation of thermolysin fragments, give strong support to the conclusion that these fragments have the ability to fold to native-like structures.

I. Elastase

This enzyme is a single-chain protease whose three-dimensional structure strongly resembles that of trypsin and chymotrypsin (Hartley and Shotton, 1971). It is composed of two continuous-chain domains, residues 27–127 and 128–230. Ghelis et al. (1978) briefly reported an investigation of the independent folding of fragment 126–245 of porcine elastase. This fragment, prepared by limited proteolysis, was reduced in strong denaturing solvent to give the theoretical titer of six free SH groups. Following stepwise dialysis to exchange solvent, a presumably oxidized protein was obtained that exhibited several of the characteristics of the native molecule. First, the refolded fragment competitively displaced native (radiolabeled) elastase from the complex with its specific antibody. Evidence was presented that the competition was stoichiometric. Second, putative refolded fragment 126–245 showed a specific substrate-binding affinity (by equilibrium dialysis with solubilized elastin) equal to that obtained for the refolded entire molecule.

There was neither report of yield nor investigation of the kinetics of return of native characteristics. Attempts to prepare intact fragment 16–125 have not yet been successful. The immunochemical and substrate binding evidence support the claim that refolding to native-like structure did indeed occur. It clearly would be desirable to have more evidence on this system.

III. Conclusions

Folding to native-like structure has been demonstrated with fragments of β-galactosidase, lysozyme, serum albumin, penicillinase, and tryptophan synthetase. The capability of protein fragments for independent formation of structure therefore has substantial experimental basis. This generalization also makes plausible the idea that, in general, protein folding occurs "by parts," that is, in a modular fashion.

The structural feature of chain continuity within a compact domain (Wetlaufer, 1973) is realized in the fragments of only two of the systems in which independent folding has been shown. These two are lysozyme and thermolysin. The fragment comprising the C-terminal half of elastase might be added, although that evidence is more limited. Inasmuch as three-dimensional structures are not yet available for β-galactosidase, serum albumin, penicillinase, and tryptophan synthetase, the proposition that peptides capable of independent self-assembly are continuous-chain domains is untestable for these proteins. This idea is supported by two clear examples, which is not enough evidence to decide on its general validity.

Several groups have recently taken initial steps toward a theory of protein folding based on modular principles (Rose *et al.*, 1976; Karplus and Weaver, 1976; Robson and Pain, 1976a; Kanehisa and Tsong, 1978; Finkelstein and Ptitsyn, 1976; Tanaka and Scheraga, 1976; Honig *et al.*, 1976). We believe that the work surveyed in this article provides a sound experimental basis for a folding theory employing ideas of modular assembly.

Since the first attempt to recognize and generalize protein domains (Wetlaufer, 1973), the need for objective criteria has been apparent. The use of distance maps by Liljas and Rossman (1974) was interesting, but did not lead to generally useful criteria. The approach proposed by Rose (1979) appears to be more promising, not only in providing needed criteria, but also for restructuring the domain issue in a hierarchical form. Further discussion of this issue and protein folding theory is beyond the limit that we have set for this article.

Those cases in which protein fragments appeared not to form native structure deserve further consideration. RNase A, staphylococcal nuclease, and proinsulin are all small proteins, but mass (or chain length) alone cannot be a sufficient condition for structure formation. This follows from the fact that some of the fragments that could form native-like structure were smaller than the apparently incompetent fragments of the two nucleases. With the reservation in mind that an appropriate amino acid sequence is a corequisite, it is still of interest to examine the question of minimum size of a peptide that can assume a stable compact structure. The immunogenicity of β-galactosidase peptides containing as few as 15 amino acid residues is suggestive, but the possibility that these are sequence-dependent, conformation-independent immunogens weakens that line of argument. Changing the question slightly to ask what are the smallest known proteins or fragments thereof that show stable compact structure, we find insulin (51 residues, 3 —S—S—), bovine pancreatic trypsin inhibitor (58 residues, 3 —S—S—), and fragment B of *Staphylococcus aureus* protein A (a cell-wall constituent) (45 residues, no —S—S—) (Deisenhofer *et al.*, 1978). Glucagon (29 residues, no

—S—S—) appears to have some specific compact structure, but also gives evidence of structural "softness" (Epand, 1972a,b). The N-terminal tridecapeptide of RNase A appears to contain only a few percent α-helix (Brown and Klee, 1971). From this very small collection of linear peptides, it appears that the lower size limit for a stable compact structure is in the range of 20 and 40 residues. This agrees with estimates that can be made from consideration of surface/volume ratios as a function of peptide size (Rose and Wetlaufer, 1977).

It now appears that eukaryotic DNA is often if not always a kind of linear mosaic, with base sequences coding for a protein (exons) occurring in blocks separated by untranslated sequences (introns). This has led to speculation on the possible functional advantages of such an arrangement. Gilbert (1978) has argued that "genes-in-pieces" provides additional mechanisms for genetic diversification. Blake (1978) has expanded on this argument to suggest that the expressed regions of DNA correspond to "integrally folded protein units— domains or supersecondary structures [subdomains]." This argument appears to be supported (Blake, 1979) by the findings of Sakano *et al.* (1979) on the DNA sequences of a gene specifying a myeloma protein. That work showed that each of the six structional/functional units of the γ-globulin heavy chain corresponds to a separate exon of DNA. Although this is a striking correlation, it is premature to conclude that "genes-in-pieces" implies "proteins-in-pieces" as a general rule. We take issue with Blake's claim (1978) that "no clear physicochemical explanation of their presence [that is; domains] has been proposed." Multiple domains offer large proteins the advantage of rapid self-assembly. A modular folding process offers proteins the advantage of rapid self-assembly, over a broad range of module sizes (Wetlaufer, 1973; Karplus and Weaver, 1976; Rose *et al.*, 1976). We take it as given that biological competition occurs in real time; it then follows that rapid assembly of functional units offers competitive advantage.

We doubt that there will prove to be a generally close linkage between DNA modules (exons) and protein modules (domains or subdomains). In addition to the above arguments, the proposed linkage must also deal with the awkward problem that is posed by prokaryotic DNA, for which there is no evidence of introns. Nonetheless, we will continue to watch with great interest for additional relevant evidence.

Most of the proteins on which fragment folding studies have been carried out are extracellular. Of the nine discussed in this article, only tryptophan synthetase and β-galactosidase are not secreted. Many secreted proteins are synthesized with 20 or so additional amino acid residues at the N-terminus of the peptide chain (Blobel and Dobberstein, 1975). This so-called "leader sequence" appears to be instrumental in "leading" the N-terminus of the nascent peptide chain through the phospholipid membrane to which the ribosome is attached. Penetration and traversal of the membrane appears to occur while biosynthesis of the peptide chain is in progress. The leader sequence is proteolytically cleaved shortly after it has traversed the membrane. Cytoplasmic proteins appear to be synthesized without leader peptides, on ribosomes that are not membrane bound. Wickner (1979) discusses two plausible but unproved mechanisms by which the leader peptide could facilitate passage of the nascent peptide chain through the membrane. Neither of these—the signal hypothesis and the membrane trigger hypothesis—has discernible protein folding consequences. While we recognize the possibility that secretion can impose additional constraints on protein folding, there is no reason at present to believe that folding mechanisms are substantially different for intracellular and secreted proteins.

Although there are many remaining questions about the folding of protein fragments, the weight of present evidence strongly supports the idea that different parts of protein molecules can form native-like structure independently. We note in closing that such folding in early termination mutants of β-galactosidase is, at the very first, *in vivo* folding. Limited evidence on early termination mutants of SV40 virus A protein (Rundell *et al.*, 1977) supports the conclusion that some protein fragments show the ability to form native-like structures *in vivo*. Protein folding "by parts" therefore is a process that goes on in real life as well as *in vitro*.

ACKNOWLEDGMENT

This work was supported by USPHS grant 7R01 GM-23713.

REFERENCES

Adams, B., Burgess, R. J., Carrey, E. A., MacIntosh, I. R., Mitchinson, C., Thomas, R. M., and Pain, R. H. (1980). *In* "Protein Folding" (R. Jaenicke, ed.), pp. 447–467. Elsevier, Amsterdam.

Anderson, W. L., and Wetlaufer, D. B. (1976). *J. Biol. Chem.* **251**, 3147–3153.

Andria, G., and Taniuchi, H. (1978). *J. Biol. Chem.* **253**, 2262–2270.

Anfinsen, C. B. (1956). *J. Biol. Chem.* **221**, 405–412.

Anfinsen, C. B., and Scheraga, H. A. (1975). *Adv. Protein Chem.* **29**, 205–300.

Arnon, R., and Sela, M. (1969). *Proc. Natl. Acad. Sci. U. S. A.* **62**, 163–170.

Arnon, R., Maron, E., Sela, M., and Anfinsen, C. B. (1971). *Proc. Natl. Acad. Sci. U. S. A.* **68**, 1450–1455.

Atassi, M. Z., Lee, C.-L., and Habeeb, A. F. S. A. (1976). *Immunochemistry* **13**, 7–14.

Baldwin, R. L. (1975). *Annu. Rev. Biochem.* **44**, 453–475.

Blake, C. C. F. (1978). *Nature (London)* **273**, 267.

Blake, C. C. F. (1979). Nature (London) **277**, 598.
Blobel, G., and Dobberstein, B. (1975). J. Cell Biol. **67**, 835–851.
Brown, J. E., and Klee, W. A. (1971). Biochemistry **10**, 470–476.
Brown, J. R. (1974). Fed. Proc., Fed. Am. Soc. Exp. Biol. **33**, 1389.
Brown, J. R. (1975). Fed. Proc., Fed. Am. Soc. Exp. Biol. **34**, 591.
Brown, J. R., and Hartley, B. S. (1966). Biochem. J. **101**, 214–216.
Carrey, E. A., and Pain, R. H. (1977). Biochem. Soc. Trans. **5**, 689–692.
Celada, F., Ullmann, A., and Monod, J. (1974). Biochemistry **13**, 5543–5547.
Celada, F., Fowler, A. V., and Zabin, I. (1978). Biochemistry **17**, 5156–5160.
Colman, P. M., Jansonius, J. N., and Matthews, B. W. (1972). J. Mol. Biol. **70**, 701–724.
Creighton, T. E. (1978). Prog. Biophys. Mol. Biol. **33**, 231–297.
Deisenhofer, J., Jones, T. A., Huber, R., Sjödahl, J., and Sjoquist, J. (1978). Hoppe-Seyler's Z. Physiol. Chem. **359**, 975–985.
Dixon, G. A., and Wardlaw, A. C. (1960). Nature (London) **188**, 721.
Du, Y. C., Jiang, R. Q., and Tsou, C. L. (1965). Sci. Sin. **14**, 229–236.
Epand, R. M. (1972a). J. Biol. Chem. **247**, 2132–2138.
Epand, R. M. (1972b). Biochemistry **11**, 3571–3575.
Finkelstein, A. V., and Ptitsyn, O. B. (1976). J. Mol. Biol. **103**, 15–27.
Fontana, A., and Vita, C. (1977). Pept., Proc. Am. Pept. Symp. 5th, 1977 pp. 432–435.
Foster, J. F. (1960). In "The Plasma Proteins" (F. W. Putnam, ed.), 1st ed., Vol. 1, pp. 179–239. Academic Press, New York.
Fowler, A. V., Zabin, I., Sinnott, M. L., and Smith, P. J. (1978). J. Biol. Chem. **253**, 5283–5285.
Friedman, F. K., and Beychok, S. (1979). Annu. Rev. Biochem. **48**, 217–250.
Fujio, H., Imanishi, M., Nishioka, K., and Amano, T. (1968). Biken J. **11**, 219.
Ghelis, C., Tempete-Gaillourdet, M., and Yon, J. M. (1978). Biochem. Biophys. Res. Commun. **84**, 31–36.
Gilbert, W. (1978). Nature (London) **271**, 501.
Givol, D., DeLorenzo, F., Goldberger, R. F., and Anfinsen, C. B. (1965). Proc. Natl. Acad. Sci. U. S. A. **53**, 676–684.
Goldberg, M. E. (1969). J. Mol. Biol. **46**, 441–446.
Goldberg, M. E., York, S., and Stryer, L. (1968). Biochemistry **7**, 3662–3667.
Goldberger, R. F., Epstein, C. J., and Anfinsen, C. B. (1964). J. Biol. Chem. **239**, 1406–1410.
Gutte, B. (1975a). J. Biol. Chem. **250**, 889–904.
Gutte, B. (1975b). Biochem. Soc. Trans. **3**, 897–899.
Gutte, B. (1977). J. Biol. Chem. **252**, 663–670.
Gutte, B. (1978). J. Biol. Chem. **253**, 3837–3842.
Habeeb, A. F. S. A., and Atassi, M. Z. (1976). J. Biol. Chem. **251**, 4616–4621.
Haber, E. (1964). Proc. Natl. Acad. Sci. U. S. A. **52**, 1099–1106.
Hartley, B. S., and Shotton, D. M. (1971). In "The Enzymes" (P. D. Boyer, ed.), 3rd ed., Vol. 3, pp. 323–373. Academic Press, New York.
Högberg-Raibaud, A., and Goldberg, M. E. (1977a). Proc. Natl. Acad. Sci. U. S. A. **74**, 442–446.
Högberg-Raibaud, A., and Goldberg, M. E. (1977b). Biochemistry **16**, 4014–4020.
Honig, B., Ray, A., and Levinthal, C. (1976). Proc. Natl. Acad. Sci. U. S. A. **73**, 1974–1978.
Johanson, K. O., Wetlaufer, D. B., Reed, R. G., and Peters, T., Jr. (1977). 174th Annu. Meet. Am. Chem. Soc. Abstr. Biol. Div. No. 68.

Johanson, K. O., Wetlaufer, D. B., Reed, R. G., and Peters, T., Jr. (1981). *J. Biol. Chem.* **256**, 445–450.

Johnson, E. R. (1974). PhD Thesis, University of Minnesota, Minneapolis.

Johnson, E. R., Anderson, W. L., Wetlaufer, D. B., Lee, C.-L., and Atassi, M. Z. (1975). *Fed. Proc., Fed. Am. Soc. Exp. Biol.* **34**, Abstr. 197.

Johnson, E. R., Anderson, W. L., Wetlaufer, D. B., Lee, C.-L., and Atassi, M. Z. (1978). *J. Biol. Chem.* **253**, 3408–3414.

Kanehisa, M. E., and Tsong, T. Y. (1978). *J. Mol. Biol.* **124**, 177–194.

Karplus, M., and Weaver, D. L. (1976). *Nature (London)* **260**, 404–406.

Kartha, G., Bello, J., and Harker, D. (1967). *Nature (London)* **213**, 862–865.

Kato, I., and Anfinsen, C. B. (1969). *J. Biol. Chem.* **244**, 1004–1007.

Lee, C.-L., and Atassi, M. Z. (1975). *Biochim. Biophys. Acta* **405**, 464–474.

Liljas, A., and Rossman, M. G. (1974). *Annu. Rev. Biochem.* **43**, 475–507.

Mathyssens, G. E., Giebens, G., and Kanarek, L. (1974). *Eur. J. Biochem.* **43**, 353–362.

Nemethy, G., and Scheraga, H. A. (1977). *Q. Rev. Biophys.* **10**, 239–352.

Reed, R. G., Feldhoff, R. C., Clute, O., and Peters, T., Jr. (1975a). *Biochemistry* **14**, 4578–4583.

Reed, R. G., Gates, T., and Peters, T., Jr. (1975b). *Anal. Biochem.* **69**, 361–371.

Richards, F. M., and Vithayathil, P. J. (1959). *J. Biol. Chem.* **234**, 1459–1464.

Robson, B., and Pain, R. H. (1973). *In* "Conformation of Biological Molecules and Polymers" (E. D. Bergman and A. Pullman, eds.), pp. 161–172. Academic Press, New York.

Robson, B., and Pain, R. H. (1976a). *Biochem. J.* **155**, 325–330.

Robson, B., and Pain, R. H. (1976b). *Biochem. J.* **155**, 331–344.

Rose, G. D. (1979). *J. Mol. Biol.* **134**, 447–470.

Rose, G. D., and Wetlaufer, D. B. (1977). *Nature (London)* **268**, 769–770.

Rose, G. D., Winters, R. H., and Wetlaufer, D. B. (1976). *FEBS Lett.* **63**, 10–16.

Rundell, K., Collins, J. K., Tegtmeyer, P., Ozer, H. L., Lai, C.-J., and Nathans, D. (1977). *J. Virol.* **21**, 636–646.

Sakano, H., Rogers, J. H., Kiippi, K., Brack, C., Traunecker, A., Maki, R., Wall, R., and Tonegawa, S. (1979). *Nature (London)* **277**, 627–633.

Saxena, V. P., and Wetlaufer, D. B. (1970). *Biochemistry* **9**, 5015–5023.

Saxena, V. P., and Wetlaufer, D. B. (1971). *Proc. Natl. Acad. Sci. U. S. A.* **68**, 969–972.

Steiner, D. B., and Clark, J. L. (1968). *Proc. Natl. Acad. Sci. U. S. A.* **60**, 622–629.

Tanaka, S., and Scheraga, H. A. (1976). *Macromolecules* **9**, 945–953.

Taniuchi, H. (1970). *J. Biol. Chem.* **245**, 5459–5468.

Taniuchi, H., and Anfinsen, C. B. (1969). *J. Biol. Chem.* **244**, 3864–3875.

Taniuchi, H., Parker, D. S., and Bohnert, J. L. (1977). *J. Biol. Chem.* **252**, 125–140.

Taylor, R. P., and Silver, A. (1976). *J. Am. Chem. Soc.* **98**, 4650–4651.

Teale, J. M., and Benjamin, D. C. (1976). *J. Biol. Chem.* **251**, 4609–4615.

Teale, J. M., and Benjamin, D. C. (1977). *J. Biol. Chem.* **252**, 4521–4526.

Ullmann, A., and Monod, J. (1970). *In* "The Lactose Operon" (J. R. Beckwith and D. Zipser, eds.), pp. 265–272. Cold Spring Harbor Lab., Cold Spring Harbor, New York.

Ullmann A., and Perrin, D. (1970). *In* "The Lactose Operon" (J. R. Beckwith and D. Zipser, eds.), pp. 143–172. Cold Spring Harbor Lab., Cold Spring Harbor, New York.

Ullmann, A., Perrin, D., Jacob, F., and Monod, J. (1965). *J. Mol. Biol.* **12**, 918–923.

Ullmann, A., Jacob, F., and Monod, J. (1967). *J. Mol. Biol.* **24**, 339–343.
Ullmann, A., Jacob, F., and Monod, J. (1968). *J. Mol. Biol.* **32**, 1.
Villarejo, M. R., and Zabin, I. (1973). *Nature (London), New Biol.* **242**, 50–52.
Vita, C., Fontana, A., Seeman, J. R., and Chaiken, I. M. (1979). *Biochemistry* **18**, 3023–3031.
Wetlaufer, D. B. (1973). *Proc. Natl. Acad. Sci. U. S. A.* **70**, 697–701.
Wetlaufer, D. B., and Ristow, S. S. (1973). *Annu. Rev. Biochem.* **42**, 135–158.
Whitney, P. L., and Tanford, C. (1965). *Proc. Natl. Acad. Sci. U. S. A.* **53**, 524–532.
Wickner, W. (1979). *Annu. Rev. Biochem.* **48**, 23–45.
Wilson, D. A., and Crawford, I. P. (1965). *J. Biol. Chem.* **240**, 4801–4808.
Wyckoff, H. W., Tsernoglou, D., Hanson, A. W., Knox, J. R., Lee, R., and Richards, F. M. (1970). *J. Biol. Chem.* **245**, 305–328.
Zabin, I., and Villarejo, M. (1975). *Annu. Rev. Biochem.* **44**, 295–313.
Zahn, H., Gutte, B., Pfeffer, E. F., and Ammon, J. (1966). *Justus Liebigs Ann. Chem.* **691**, 225–231.

THE THEORY OF PRESSURE EFFECTS ON ENZYMES

By EDDIE MORILD[1]

Norwegian Underwater Institute, Ytre Laksevåg/Bergen, Norway

I. LIST OF SYMBOLS

A	Reaction species, constant
a	Thermodynamic activity
$\alpha = \dfrac{1}{V}\left(\dfrac{\partial V}{\partial T}\right)_p$	Expansibility, relative
α	Fraction of reaction

[1] Also at the Department of Chemistry, University of Bergen, Bergen, Norway.

ADVANCES IN
PROTEIN CHEMISTRY, Vol. 34

B	Reaction species, constant
$\beta = -\dfrac{1}{V}\left(\dfrac{\partial V}{\partial p}\right)_T$	Compressibility, relative
C	Reaction species
C_p	Heat capacity
c	General concentration (molarity)
D	Reaction species
[E]	Enzyme concentration
E_x	Experimental (Arrhenius) activation energy
ΔE	Internal energy change
$\Delta E\ddagger$	Transition state activation energy
ES, EP	Enzyme–ligand complexes
ϵ	Extinction coefficient
f	Fugacity
ΔG	Gibbs free energy change
$\Delta G\ddagger$	Transition state activation free energy
H_y	Henry's constant
ΔH	Enthalpy change
$\Delta H\ddagger$	Transition state activation enthalpy
h	Planck's constant
I	Ionic strength
i	Subscript for reaction species, solutes, etc.
K	General equilibrium constant
K_A, K_B	General kinetic constants
K_c	Equilibrium constant on the molar concentration scale
K_m	Michaelis constant
K_w	Ionization constant of water
K_x	Equilibrium constant on the mole fraction scale
k	General rate constant
k_B	Boltzmann's constant
k_i	Rate constant of step i
$\kappa = -\left(\dfrac{\partial V}{\partial p}\right)_T$	Compressibility, absolute
μ	Chemical potential
ν	Stoichiometric coefficient
p	Pressure
r	Transmission coefficient
[S]	Substrate concentration
ΔS	Entropy change

$\Delta S\ddagger$	Transition state activation entropy
T	Temperature
V_A, V_B	Volume of reaction species
ΔV	General volume change
ΔV_b	Volume change of the binding step
$\Delta V_c\ddagger$	Volume change of the catalytic step
ΔV_D	Volume change of denaturation
ΔV_w	Volume change of ionization of water
ΔV_z	Volume change of ionization
$\Delta V\ddagger$	Transition state activation volume
v	Reaction rate
W	Maximum reaction rate
$X\ddagger$	Transition state activated complex
x	Mole fraction
$\zeta = \left(\dfrac{\partial V}{\partial T}\right)_p$	Expansibility, absolute

II. Introduction

The study of the effects of pressure upon biochemical systems is important for many reasons. An understanding of how temperature, pressure, and the combination of both affect physiological processes is basic to an understanding of the biology of the deep sea. Pressure affects a wide variety of processes in biological systems and is therefore a powerful tool in the investigation of the mechanisms in these processes. There is also a large gap in our knowledge about the response of man to high pressures, as experienced during deep sea diving.

There are various experimental difficulties involved in measuring biochemical systems at high pressure. Nevertheless, almost every method used routinely at atmospheric pressure today seems to have been carried out successfully at high pressure. Expansion in the field of high-pressure biochemistry and biophysics has been slowed by the lack of a solid theoretical basis for the interpretation of observations and by the large number of ambiguous interpretations of pressure effects. In many ways, the volume changes deduced from pressure experiments may be more intuitively interpreted and more easily visualized than other thermodynamic quantities. Sometimes interpretations are too superficial due to a lack of understanding of how pressure really influences molecules in dynamic systems. From pressure measurements of equilibria one is able to calculate various kinds of volume changes that occur in going from reactants to products. These may include ordinary chemical and biochemical equilibria, ionization equilibria, redox equilibria, and dissociation–association equilibria, among others. Mixing and solubility properties of substances from different phases may also be obtained from pressure experiments. In organic and inorganic reaction kinetics, pressure has been used to clarify reaction mechanisms. By running the reactions in different media and under different conditions, high-pressure data can give information about transition states, size, bond cleavage and formation, polarity and charge development, and hydration. In complicated reactions such as enzyme reactions, information may be obtained about volume changes not only for the overall reaction, but also for intermediate steps in the mechanism. Because rate may increase or decrease with pressure, pressure may be used either to increase or decrease a certain reaction product, or to change the relative amount of, for example, isomeric products. With the introduction of optical high-pressure cells a wide variety of spectrophotometric techniques can be used; and, in recent years, we have also seen the development of high-pressure electron spin resonance (ESR) and nuclear magnetic resonance (NMR). In the future, more detailed information will certainly be obtained. Especially interesting will be the development of methods to investigate heterogeneous systems, membranes, cell organelles, nerves, and others.

For the time being, our basic understanding of pressure effects is far from complete. However, some new developments concerning theory and application have occurred over the years. A short theoretical treatment of pressure effects was presented almost 30 years ago (Laidler, 1951). In this article we will present an extensive treatment of the present theoretical basis for pressure effects, incorporating contemporary knowledge of enzyme kinetics, physical biochemistry, and high-pressure theory. The theoretical level in this field is still not very sophisticated, but it is important enough so that theoretical considerations should be applied when future experiments are planned.

III. REACTION RATE THEORY AND KINETICS

A. *The Transition State Theory*

In theoretical kinetics today there are still no serious competitors to the transition state theory of Eyring and co-workers (Glasstone *et al.*, 1941). In its most stringent sense it applies only to simple homogeneous gas reactions. The treatment of simple reactions in solution requires additional knowledge of the properties of liquids, and the theory becomes less rigorous and less fundamental. In the extension of the theory to complex reactions between complex molecules in a complex solvent like water, approximations and averaged quantities must be introduced. Clearly, at this level the theory loses nearly all its quantum and statistical mechanical foundation and formalism and must be oriented toward phenomenological thermodynamics. In this sense the theory is largely descriptive, with its formal basis in the quasi-equilibrium thermodynamics of transition from reactants to activated states. The precise meaning of the "activated state," the "activated complex," or the "transition state" is understandable only with reference to quantum and statistical mechanics. One should therefore be aware that there are many possible activated states, according to the energy levels occupied by the complex, due to the number of variable coordinates. The activated state corresponding to the top of the energy barrier is therefore not necessarily identical to that corresponding to the largest (or smallest) volume of the complex.

In applying this theory to enzyme kinetics, one should be careful with "explanations" of the activation quantities found. Generally, these will be averaged quantities which in most cases have lost their character as single-step quantities. These aspects will be treated more fully in later sections.

However, there is no doubt that the activation quantities found are important for classifying purposes and as measures of temperature and pressure effects.

According to the theory, a chemical reaction proceeds via a transition state $X\ddagger$. A general, simple reaction is

$$A + B \rightleftharpoons X\ddagger \xrightarrow{k} C + D \tag{1}$$

The quasi-thermodynamic assumption is that $X\ddagger$ is in equilibrium with the reactants, despite the continuous leakage of $X\ddagger$ to products. The quasi-equilibrium constant $K\ddagger$ is then defined in terms of concentration by

$$K\ddagger = [X]\ddagger/([A][B]) \tag{2}$$

We do not introduce activity coefficients in the present discussion (see Section V,B). The next assumption in the theory is that the formation of products is proportional to $[X]\ddagger$ and therefore to $K\ddagger$. After an intricate mathematical derivation, the rate constant k of the reaction is obtained in the form

$$k = r(k_B T/h)K\ddagger \tag{3}$$

Here, k_B is Boltzmann's constant and h is Planck's constant. One is left a little uncomfortable by the statement that r is "usually assumed equal to unity." In any case, as a quantum mechanical parameter, the transmission coefficient r may be assumed to be temperature- and pressure-independent or just absorbed into the rate constant itself. Then the connection between the quasi-thermodynamic equilibrium constant $K\ddagger$ and the activation free energy $\Delta G\ddagger$ is made through the standard relation from equilibrium thermodynamics

$$\Delta G\ddagger = -RT \ln K\ddagger \tag{4}$$

together with

$$\Delta G\ddagger = \Delta H\ddagger - T\Delta S\ddagger \tag{5}$$

Elimination of $K\ddagger$ gives the transition state expression for the rate constant k as

$$k = (k_B T/h) \exp(\Delta S\ddagger/R) \exp(-\Delta H\ddagger/RT) \tag{6}$$

This equation has a long tradition in chemical kinetics. The experimentally derived activation quantities are generally both consistent and useful, and seem to bear the highest information content about rate processes that one can wish.

B. Interpretation of Volume Changes

Let us consider a general chemical equilibrium

$$\sum_i \nu_i S_i = 0 \tag{7}$$

where S_i are the participating species (reactants and products) and ν_i are the stoichiometric coefficients, negative for reactants and positive for products. The thermodynamic equilibrium condition is

$$\sum_i \nu_i \mu_i = 0 \tag{8}$$

with the chemical potential of species expressed by

$$\mu_i = \mu_i^0 + RT \ln a_i \tag{9}$$

where μ_i^0 is the chemical potential in a chosen standard state, and a_i is the activity of i. The combination of Eqs. (8) and (9) gives us the set of relations

$$\Delta G^0 = \sum_i \nu_i \mu_i^0 = -RT \ln \prod_i (a_i)\nu_i = -RT \ln K \tag{10}$$

where K is the equilibrium constant and ΔG^0 is the change in the Gibbs energy in the standard state. The molar volume of component i in the standard state is given by Eq. (11):

$$\left(\frac{\partial \mu_i^0}{\partial p}\right)_T = V_i^0 \tag{11}$$

Differentiation of Eq. (10) with respect to pressure yields

$$-RT\left(\frac{\partial \ln K}{\partial p}\right)_T = \sum_i v_i \left(\frac{\partial \mu_i^0}{\partial p}\right)_T = \sum_i v_i V_i^0 = \Delta V^0 \tag{12}$$

ΔV^0 is then the excess molar volume of products over that of reactants, in their standard states. For dilute solutions, where activity corrections may be neglected, and where K_x is expressed in mole fraction units

$$-RT\left(\frac{\partial \ln K_x}{\partial p}\right)_T = \Delta V \tag{13}$$

where ΔV now means the excess of the partial molar volumes at high dilution. Equation (13) will also be valid when K is expressed in molalities. For the molarity scale, however, where the concentration c_i is related to the mole fraction x_i by

$$c_i = x_i/V_0 \tag{14}$$

V_0 being the molar volume of the solvent, we have

$$K_c = K_x V_0^{-\sum_i v_i} \tag{15}$$

and V_0 varies with pressure.

This leads to Eq. (16):

$$\left(\frac{\partial \ln K_c}{\partial p}\right)_T = \left(\frac{\partial \ln K_x}{\partial p}\right)_T - \sum v_i \left(\frac{\partial \ln V_0}{\partial p}\right)_T = -\frac{\Delta V}{RT} + \sum v_i \beta_0 \tag{16}$$

β_0 is the compressibility coefficient of the solvent. In water at 25°C and 1 bar the factor $RT\beta_0$ is about 1 cm³ mol⁻¹ and sometimes $\sum v_i = 0$. Often there will be no great error from neglecting this correction term (Weale, 1967). In kinetics, the situation is not quite analogous, and it is necessary to be very careful with the interpretation of the measured volume changes. A number of similar interpretations of volumes of activation have been put forward. Some reviews have been given by Weale (1967), le Noble (1967), Kohnstam (1970), Eckert (1972), Stranks (1974), Jenner (1975), and Asano and le Noble (1978) on organic and inorganic reactions. Activation volumes of enzyme reactions are briefly discussed by Laidler (1951), Laidler and Bunting (1973), Johnson *et al.* (1974), Low and Somero (1975a), and Mac-Donald (1975). Recently, two reviews of high-pressure biochemistry have also been presented by Heremans (1978, 1979).

An unfortunate misprint in the Zimmerman book (Johnson and Eyring, 1970, p. 18) should be noted, since several workers have used this article as a theoretical reference. It is there stated that $\Delta V\ddagger$ is a *ratio* between two volumes, when it is actually a *difference*. The conclusions reached in another paper (Penniston, 1971) have hitherto been used as a working basis in several papers, but do not at present seem to generate strong support. This is discussed in Section VI,B.

Many reports about activation volume determinations of enzyme reactions do not state the experimental conditions clearly. In such cases it is difficult to interpret the $\Delta V\ddagger$s. As $\Delta V\ddagger$s nearly always are dependent upon substrate concentration, it is also difficult to interpret effects *in vivo* from *in vitro* experiments when substrate concentration is not specified. With only one more determination of $\Delta V\ddagger$, e.g., one at high and one at low substrate concentration (with respect to K_m), much more information can be obtained than from just a single determination.

In the original transition state theory, the activation volume is defined as the difference between the volume of the activated state $V_X\ddagger$ and the volumes of the initial reactants, V_A and V_B,

$$\Delta V\ddagger = V_X\ddagger - (V_A + V_B) \tag{17}$$

The assumption was that the activation process took place in one single step, the catalytic step. It is almost impossible to determine the absolute value of $V_X\ddagger$ directly. Accordingly, one cannot measure $\Delta V\ddagger$ directly. All determinations of $\Delta V\ddagger$s are indirect, and are obtained only through measurements of the rate of reaction at different pressures.

Therefore, our concept of the activation volume is very vague, but it certainly improves if defined in relation to a specified model of the reaction. At this point the use of activation quantities in enzyme kinetics departs from common usage in less complex mechanisms. For the simplest one-substrate enzyme reaction, it can easily be imagined that the process occurs in at least two steps:

1. The substrate binding step, where the substrate and the enzyme form a complex:

$$E + S = ES \tag{18}$$

2. The catalytic step, where the enzyme–substrate complex is activated:

$$ES = ES\ddagger \tag{19}$$

The experimental activation volume is unambiguously determined from the slope of ln k vs. p; but its interpretation is not necessarily that of the original definition, the difference between the volumes of the activated complex and the reactants. In most cases the activation volume is a linear combination of volume changes in steps 1 and 2. In other cases one may have linear combinations of 3 and 4 volume contributions to $\Delta V\ddagger$. This will be treated in detail in the next section.

Formally, the general definition of the activation volume for all reactions is

$$\Delta V\ddagger = V\ddagger - \sum \nu_i V_i \tag{20}$$

where $V\ddagger$ is the partial molar volume of the activated complex, V_i is the partial molar volume of the reactants, and ν_i is the stoichiometric coefficient of reactant i. $\Delta V\ddagger$ is the excess of the partial molar volume of the transition state over the sum of the partial molar volumes of the reactants, all at the composition of the mixture.

From equilibrium thermodynamics we have the relation

$$\left(\frac{\partial \Delta G\ddagger}{\partial p}\right)_T = \Delta V\ddagger \tag{21}$$

and then, from Eqs. (4) and (21), we obtain the equation

$$\left(\frac{\partial \ln k}{\partial p}\right)_T = -\frac{\Delta V\ddagger}{RT} + \sum \nu_i(RT\beta_0) \tag{22}$$

The complications of compressibility and nonlinearity of the plots of ln k vs. p are discussed in Section III,C.

It is reasonable to believe that all kinds of volume changes observed in chemical processes in solution are the result of at least two contributions. First, there is an intrinsic, structural volume change in the molecules themselves due to the breaking and formation of chemical bonds (and interactions). Much work has been concerned with estimation of the magnitude of volume changes accompanying the breaking and formation of single bonds. These studies indicate that the contraction or expansion following changes of interatomic distances in substrate molecules usually falls within 5 to 15 cm^2 mol^{-1} per bond (Weale, 1967; Le Noble, 1967). Second, there is a volume change resulting from the rearrangement of surrounding solvent molecules during reaction. This solvation term is most pronounced when changes in electric charges and dipoles are made in the reacting molecules. This effect is especially prominent when water is the solvent. The water molecules, being strong dipoles, will reorient toward developing charges and thereby lose the open structure characteristic of bulk water. The volume change introduced with the development of a full electronic charge is between -10 and -30 cm^3 mol^{-1} (Weale, 1967). Usually, these two contributions cannot be separated completely.

Thus, even if the volume quantity is easier to interpret than other thermodynamic quantities such as ΔH, ΔS, or ΔC_p, there are still difficulties in giving the activation volume an unambiguous interpretation.

In reactions involving proteins, the interpretation of such volume changes becomes extremely complicated because so many processes take place almost simultaneously. Theoretically, we can split the overall activation process into the following phases:

1. In the initial phase, prior to the binding of the substrate to the enzyme, the enzyme is surrounded by water with some ions of various species (Kuntz and Kauzmann, 1974). The enzyme is solvated, or wrapped in a more or less continuous cover of water molecules interacting with the different groups in the enzyme surface. Some of these groups may be easily accessible to the water molecules, while other groups may be more deeply buried. The interactions are of three different kinds. Adjacent to a charged or polar group, the water dipoles will be oriented in the electrostatic field of the group, forming a rather dense aggregate. Adjacent to a nonpolar group, the water molecules will be arranged in a more structured fashion than in bulk water, dependent on the size of the group and nature of the immediate nearby groups. The so-called hydrophobic hydration, being a cooperative phenomenon, may be sensitive to local disturbances. The water molecules will also, to varying degrees, be hydrogen-bonded to each other and to the enzyme surface. Evidently, this hydration sheet will, to a large extent, determine the energy level of the enzyme molecule.

2. The binding phase can be divided into the following processes:

a. The binding of the substrate to the enzyme surface at the active site often involves all three interaction types. For some enzymes it has been possible to localize a hydrophobic or a polar interaction site. During the interaction process, minor intrinsic structural changes can occur in the substrate and in the active site of the enzyme. These changes may lead to detectable volume changes.

b. In the overall interaction, which is of an attractive nature, the substrate will closely approach the enzyme, corresponding to a low energy level. Most probably, several water molecules will have to be displaced during this process, going from one structured arrangement to a differently structured bulk water phase. New arrangements of water molecules may also be possible. These changes may lead to detectable volume changes.

c. For accommodation of the substrate, the enzyme molecule may have to rearrange amino acid residues either at or remote from the substrate binding site (conformational changes). These structural changes in the enzyme accordingly introduce further rearrangements in the close water layer around the enzyme surface, all of which may lead to detectable volume changes.

3. The catalytic phase can again be divided into the following processes:

a. Further conformational changes in the tertiary structure of the enzyme may occur during the formation of the transition state.

b. Further rearrangements of water molecules may occur.

c. Changes in the interactions between substrate and enzyme may follow.

d. The catalytic process itself is now going on, due to the enzyme action. This usually involves the breaking or formation of chemical bonds in the substrate(s), and thereby introduces new structural arrangements.

e. This will again be followed by new water arrangements, until the enzyme–substrate complex reaches the transition state.

All these more or less simultaneous processes will certainly lead to detectable volume changes.

Under specific conditions one may be able to distinguish the last two phases and separate their volume contributions, whereas under other conditions one may not. By means of special techniques and with very simple enzymes it may even be possible to separate volume changes within one single phase. However, with ordinary techniques and especially without a thorough theoretical treatment, the measured volume changes will be mixtures of all the above-mentioned contributions.

C. Compressibility

In kinetics, the use of high pressure introduces various aspects of the compressibility that may sometimes seem confusing. This is partly related to the fact that our rate constants are phenomenological quantities determined by arbitrary sets of experimental conditions at the same time as they are described by one single model theory.

For the general reaction scheme

$$A \rightleftharpoons X\ddagger \rightarrow P \tag{23}$$

the usual first-order reaction rate equation is given by

$$v = -(d[A]/dt) = k[A] \tag{24}$$

A logarithmic differentiation of this equation with respect to pressure gives us

$$\left(\frac{\partial \ln v}{\partial p}\right)_T = \left(\frac{\partial \ln k}{\partial p}\right)_T + \left(\frac{\partial \ln[A]}{\partial p}\right)_T \tag{25}$$

It has repeatedly been stressed that, in this case, the quantities involved may be favorably expressed in pressure-independent units such as molality or mole fractions (Asano and Le Noble, 1978). Then the last term disappears, and we can write

$$\left(\frac{\partial \ln v}{\partial p}\right)_{T,c} = \left(\frac{\partial \ln k}{\partial p}\right)_{T,c} \tag{26}$$

where subscript c indicates that the concentrations are held constant. If molar units were used, however, we could have

$$\left(\frac{\partial \ln v}{\partial p}\right)_T = \left(\frac{\partial \ln k}{\partial p}\right)_T + \beta \tag{27}$$

where β is the compressibility of the solvent.[2] So far we have used only the phenomenological Eq. (24). Introducing the thermodynamic equilibrium constant of the quasi-equilibrium in Eq. (23),

$$K\ddagger = [X\ddagger]/[A] \tag{28}$$

the transition state theory requires that

$$\left(\frac{\partial \ln k}{\partial p}\right)_T = \left(\frac{\partial \ln K\ddagger}{\partial p}\right)_T \tag{29}$$

and in this case,

$$\left(\frac{\partial \ln K\ddagger}{\partial p}\right)_T = -\frac{\Delta V\ddagger}{RT} \tag{30}$$

---

[2] Note that in this case the rate constant will have to be expressed in pressure-dependent units. This is also true for Eq. (22).

For first-order reactions then, there is no compressibility term in the expression for ln k, no matter what concentration scale is used. For higher order reactions involving molar concentrations, Eq. (22) could be applied when accurate rate data are available. Whether Eq. (27) should be applied depends on the method used for obtaining the data. If a spectrophotometric determination of the relative decrease in [A] is used, a relative measure of $(\partial \ln k/\partial p)_T$ is obtained from Eq. (27). If an absolute determination of [A] can be made at various times, Eq. (24) can be used directly, and k and $(\partial \ln k/\partial p)_T$ can be immediately obtained. The situation is easily generalized to higher order kinetics. In some cases, where $\Delta V\ddagger < 0$ and the method of measurement detects [A] but not [X$\ddagger$], there may be a slight displacement of the quasi-equilibrium with pressure which leads to different initial concentrations of A. When $\Delta V\ddagger$ can be determined from Eq. (22), it may appear pressure-dependent, i.e.,

$$\Delta V_p\ddagger = \Delta V_0\ddagger - \Delta\kappa\ddagger p \tag{31}$$

Here $\Delta V_0\ddagger$ and $\Delta V_p\ddagger$ are the activation volumes at atmospheric and high pressure, respectively, and $\Delta\kappa\ddagger$ is the activation absolute compressibility. The latter quantity is defined by the difference in absolute compressibility between the activated state and the reactants,

$$\Delta\kappa\ddagger = \kappa\ddagger - \sum \kappa_i = -\left(\frac{\partial V\ddagger}{\partial p}\right)_T + \sum_i \left(\frac{\partial V_i}{\partial p}\right)_T \tag{32}$$

Using Eqs. (29), (30), (31), and (32), the integrated equation for ln k_p at pressure p becomes

$$\ln k_p = \ln k_0 - \frac{\Delta V_0\ddagger}{RT} \cdot p + \frac{1}{2}\frac{\Delta\kappa\ddagger}{RT} \cdot p^2 \tag{33}$$

This equation is useful for extrapolation purposes, e.g., to find $\Delta V_0\ddagger$ from data at pressures higher than one atmosphere.

D. One-Substrate Reactions

The usual starting point in enzyme kinetics is the Michaelis–Menten equation for the reaction rate v. This also seems a convenient starting point for interpretation of pressure effects on enzyme mechanisms. It will be shown that this formalism may be deceptive if the definitions and interpretations have not been made clear from the beginning. For the mechanism

$$E + S \xrightleftharpoons[k_{-1}]{k_1} ES \xrightarrow{k_2} E + P \tag{34}$$

the rate equation is

$$v = \frac{W}{1 + K_m/[S]} \tag{35}$$

where W is the maximum rate.

Analysis of rate data can be made according to this equation after a differentiation with respect to pressure

$$\ln v = \ln W - \ln(1 + K_m/[S])$$

$$\left(\frac{\partial \ln v}{\partial p}\right)_T = \left(\frac{\partial \ln W}{\partial p}\right)_T - \frac{K_m/[S]}{1 + K_m/[S]}\left(\frac{\partial \ln K_m}{\partial p}\right)_T$$

$$-\frac{\Delta V\ddagger}{RT} = -\frac{\Delta V_c\ddagger}{RT} + \frac{K_m}{K_m + [S]}\frac{\Delta V_b}{RT}$$

$$\Delta V\ddagger = \Delta V_c\ddagger - \frac{K_m}{K_m + [S]}\Delta V_b \tag{36}$$

Note that this derivation leaves a pressure-dependent factor in the ΔV_b term.

Formally, Eq. (36) separates the contribution from the catalytic and the binding step treated in Section III,B. These formal volume changes can be visualized as in Fig. 1. First, it must be clear that the apparent contradiction between the minus sign in Eq. (36) and the volume change ΔV_b in the figure stems from the definition of the Mi-

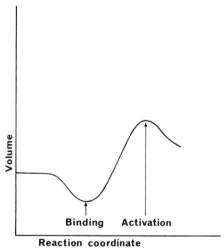

FIG. 1. A possible reaction time course of the volume of reactant through a binding step and an activation step.

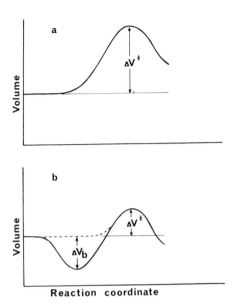

FIG. 2. Two possible interpretations of a measured activation volume $\Delta V\ddagger$, according to Eq. (36). (a) At high substrate concentration, the measured activation volume is dominated by the volume change in the catalytic step. (b) At low substrate concentration, an apparent lower activation volume is due to a contribution from a negative binding volume.

chaelis constant as a dissociation constant, while the binding volume is seen as an association volume, with a definition as, e.g.,

$$\Delta V_b = V_{ES} - (V_E + V_S) \tag{37}$$

Using now the original definition of activation volume, this can be visualized as in Fig. 2a. At high substrate concentrations, this last interpretation should be correct. In this case the binding volume, ΔV_b, makes a negligible contribution to the measured volume and the volume change of the catalytic step, $\Delta V_c\ddagger$, dominates. At low substrate concentrations, however, the binding contribution becomes larger, until in the limit,

$$\Delta V\ddagger = \Delta V_c\ddagger - \Delta V_b \tag{38}$$

The interpretation according to the original definition is indicated in Fig. 2b, as if the catalytic volume change had become smaller, without noticing that there is a significant contribution from the binding volume. It is clear that there is much more information to extract when an appropriate model is chosen for the system and when measurements are done at various substrate concentrations.

However, the situation is more complicated than this. Although there now seems to be consistency between interpretation and definitions, there is still a problem arising from the definition of K_m. From the reaction mechanism (34)

$$K_m \neq k_{-1}/k_1 \tag{39}$$

but

$$K_m = (k_{-1} + k_2)/k_1 \tag{40}$$

From this definition it appears that K_m, and also the binding volume appropriately derived from K_m, includes contributions from the catalytic step. From Eq. (40)

$$\ln K_m = \ln(k_{-1} + k_2) - \ln k_1$$

Differentiation with respect to pressure gives

$$\left(\frac{\partial \ln K_m}{\partial p}\right)_T = \left(\frac{\partial \ln(k_{-1} + k_2)}{\partial p}\right)_T - \left(\frac{\partial \ln k_1}{\partial p}\right)_T$$

$$= \frac{k_{-1}}{k_{-1} + k_2}\left(\frac{\partial \ln k_{-1}}{\partial p}\right)_T + \frac{k_2}{k_{-1} + k_2}\left(\frac{\partial \ln k_2}{\partial p}\right)_T - \left(\frac{\partial \ln k_1}{\partial p}\right)_T$$

Using the definition of activation volume as given by Eq. (21) for every individual rate constant, the binding volume derived from K_m becomes

$$\left(\frac{\partial \ln K_m}{\partial p}\right)_T = -\frac{\Delta V_b}{RT} = -\frac{k_{-1}}{k_{-1} + k_2}\frac{\Delta V_{-1}\ddagger}{RT} - \frac{k_2}{k_{-1} + k_2}\frac{\Delta V_2\ddagger}{RT} + \frac{\Delta V_1\ddagger}{RT}$$

$$-\Delta V_b = \frac{k_{-1}\Delta V_{-1}\ddagger + k_2\Delta V_2\ddagger}{k_{-1} + k_2} - \Delta V_1\ddagger \tag{41}$$

Here, $\Delta V_2\ddagger$ is the same as our former $\Delta V_c\ddagger$. However, another problem now becomes apparent, namely, that we must include two activated states in our model, one for the binding step and one for the catalytic step. Denoting the volumes of these two activated states by $V_{ES}\ddagger$ and $V_{EP}\ddagger$, we can define the previously introduced volume changes as

$$\Delta V_1\ddagger = V_{ES}\ddagger - (V_E + V_S) \tag{42}$$

$$\Delta V_{-1}\ddagger = V_{ES}\ddagger - V_{ES} \tag{43}$$

$$\Delta V_2\ddagger = V_{EP}\ddagger - V_{ES} \tag{44}$$

The volume changes in this system as functions of the reaction coordinate are the same as those shown in Fig. 5. There is nothing more to be done with Eq. (41) if the volume changes $\Delta V_{-1}\ddagger$ and $\Delta V_2\ddagger$ and the rate constants k_{-1} and k_2 are of comparable order of magnitude. The binding volume is then a really complicated quantity to interpret.

Fortunately, a usual assumption for applying the steady-state condition to mechanism (34) is that $k_2 \ll k_{-1}$. This resolves the problem, because then the Michaelis constant can be interpreted as an ordinary single-step equilibrium constant and the binding volume as a simple association volume.

Although a treatment like this can be considered straightforward and even trivial, very few workers have applied similar models for extracting maximum information from their data. We shall now treat a slightly more complicated model and look at the various special cases that occur.

In this model two intermediate metastable states are assumed to exist, one for the enzyme–substrate complex and one for the enzyme–product complex. Associated with every rate constant there is assumed to exist an activated state, and we apply the same notation as before.

$$E + S \underset{k_{-1}}{\overset{k_1}{\rightleftharpoons}} ES \underset{k_{-2}}{\overset{k_2}{\rightleftharpoons}} EP \underset{k_{-3}}{\overset{k_3}{\rightleftharpoons}} E + P \tag{45}$$

Here E is the enzyme, S is the substrate, P is the product, and ES and EP are intermediate complexes. For simplicity we now assume that k_{-2} and k_{-3} are so small that they can be neglected, and the scheme is

$$E + S \underset{k_{-1}}{\overset{k_1}{\rightleftharpoons}} ES \overset{k_2}{\longrightarrow} EP \overset{k_3}{\longrightarrow} E + P \tag{46}$$

The general rate equation resulting from the steady-state hypothesis is

$$v = \frac{k_2[E]_0[S]}{[(k_{-1} + k_2)/k_1] + [(k_2 + k_3)/k_3][S]} \tag{47}$$

Two distinctly different cases may be considered, one at high substrate concentration and one at low substrate concentration. In the following, compressibilities are neglected.

1. High [S]

Equation (47) now rearranges to

$$v = \frac{k_2 k_3}{k_2 + k_3}[E]_0 = k_0[E]_0 \tag{48}$$

If we insert the integrated form of Eq. (21) for the rate constants

$$k(p) = k(0) \exp(-p\, \Delta V\ddagger/RT) \qquad (49)$$

[$k(p)$ and $k(0)$ are the rate constant at the pressures p and 1 bar, respectively], we obtain

$$v(p) = \frac{k_2(0)k_3(0)[\mathrm{E}]_0 \exp\{-p(\Delta V_2\ddagger + \Delta V_3\ddagger)/RT\}}{k_2(0) \exp(-p\Delta V_2\ddagger/RT) + k_3(0) \exp(-p\Delta V_3\ddagger/RT)} \qquad (50)$$

From a study of the pressure dependence of $v(p)$ we observe an overall volume change $\Delta V_0\ddagger$, given by

$$\left(\frac{\partial \ln v}{\partial p}\right)_T = \left(\frac{\partial \ln k_0}{\partial p}\right)_T = -\frac{\Delta V_0\ddagger}{RT} \qquad (51)$$

Equation (48) yields

$$\ln k_0 = \ln k_2 + \ln k_3 - \ln (k_2 + k_3)$$

$$\left(\frac{\partial \ln k_0}{\partial p}\right)_T = \left(\frac{\partial \ln k_2}{\partial p}\right)_T + \left(\frac{\partial \ln k_3}{\partial p}\right)_T - \frac{k_2}{k_2 + k_3}\left(\frac{\partial \ln k_2}{\partial p}\right)_T$$

$$- \frac{k_3}{k_2 + k_3}\left(\frac{\partial \ln k_3}{\partial p}\right)_T$$

$$\left(\frac{\partial \ln k_0}{\partial p}\right)_T = \frac{k_3}{k_2 + k_3}\left(\frac{\partial \ln k_2}{\partial p}\right)_T + \frac{k_2}{k_2 + k_3}\left(\frac{\partial \ln k_3}{\partial p}\right)_T$$

Using Eq. (21), this now leads to

$$\Delta V_0\ddagger = \frac{k_3\Delta V_2\ddagger + k_2\Delta V_3\ddagger}{k_2 + k_3} \qquad (52)$$

i.e., the observed volume change is a weighted average of the single volume changes, with the rate constants as the weighting factors. A possible picture of the volume change as a function of the reaction coordinate can be seen in Fig. 3.

Let us now assume that step 3 is rate-determining, $k_2 \gg k_3$. From Eq. (50) we obtain

$$v(p) = k_3(0)[\mathrm{E}]_0 \exp(-p\Delta V_3\ddagger/RT) \qquad (53)$$

and also from Eq. (52) we see that

$$\Delta V_0\ddagger \simeq \Delta V_3\ddagger \qquad (54)$$

At high substrate concentration then, the observed overall volume change will be more or less equal to the volume change of the rate-determining step. If $\Delta V_3\ddagger > 0$, this step will be even more "rate-determining" at higher pressure, as seen from Eq. (53). If $\Delta V_3\ddagger < 0$, the reaction rate will increase with pressure in a certain pressure in-

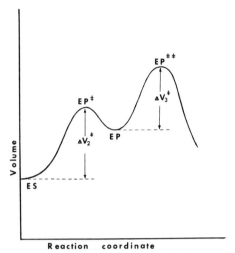

FIG. 3. At saturating conditions (high |S|), the initial volume can be considered to be the volume of the ES complex. Either step 2 or step 3 may be rate-determining, each characterized by the activation volumes $\Delta V_2\ddagger$ and $\Delta V_3\ddagger$, respectively.

terval (see below). When the rate-determining step is step 2, $k_3 \gg k_2$.

As before, the rate will now be

$$v(p) = k_2(0)[E]_0 \exp(-p\Delta V_2\ddagger/RT) \qquad (55)$$

and

$$\Delta V_0\ddagger \simeq \Delta V_2\ddagger \qquad (56)$$

An interesting situation will occur if $\Delta V_2 < 0$. This is illustrated in Fig. 4. The intramolecular rearrangement from ES to EP via the activated state EP‡ may well be accompanied by a negative volume change. The rate of this single step will increase with pressure, and the rate-determining character of the step disappears. However, if $\Delta V_3\ddagger > 0$, the constant $k_3(p)$ decreases with pressure. *At some specific value of* p, *the rate will reach a maximum where step 2 is no longer rate-determining, but step 3 has become rate-determining.* Then the rate will steadily decrease again with pressure. This situation is more thoroughly treated in Section III,F.

2. Low [S]

Equation (47) now rearranges to

$$v = \frac{k_1 k_2}{k_1 + k_2} [E]_0[S] = k_0[E]_0[S]_0 \qquad (57)$$

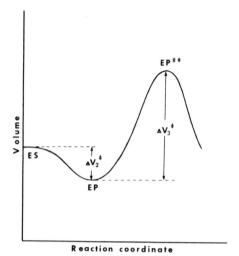

FIG. 4. If the rate-determining step is associated with a negative volume change $(\Delta V_2\ddagger)$, the rate will increase with pressure until another step (step 3) becomes rate-determining.

The pressure dependence is given by

$$v(p) = \frac{k_1(0)k_2(0)[E]_0[S] \exp\{-p(\Delta V_1\ddagger + \Delta V_2\ddagger/RT\}}{k_1(0) \exp(-p\Delta V_{-1}\ddagger/RT) + k_2(0) \exp(-p\Delta V_2\ddagger/RT)} \qquad (58)$$

To find the volume change, we start with

$$\ln k_0 = \ln k_1 + \ln k_2 - \ln (k_1 + k_2)$$

$$\left(\frac{\partial \ln k_0}{\partial p}\right)_T = \left(\frac{\partial \ln k_1}{\partial p}\right)_T + \left(\frac{\partial \ln k_2}{\partial p}\right)_T - \frac{k_{-1}}{k_{-1} + k_2}\left(\frac{\partial \ln k_{-1}}{\partial p}\right)_T$$

$$\qquad - \frac{k_2}{k_{-1} + k_2}\left(\frac{\partial \ln k_2}{\partial p}\right)_T$$

$$\left(\frac{\partial \ln k_0}{\partial p}\right)_T = \left(\frac{\partial \ln k_1}{\partial p}\right)_T + \frac{k_{-1}}{k_{-1} + k_2}\left(\frac{\partial \ln k_2}{\partial p}\right)_T - \frac{k_{-1}}{k_{-1} + k_2}\left(\frac{\partial \ln k_{-1}}{\partial p}\right)_T$$

and the result is

$$\Delta V_0\ddagger = \frac{k_{-1}(\Delta V_1\ddagger - \Delta V_{-1}\ddagger + \Delta V_2\ddagger) + k_2\Delta V_1\ddagger}{k_{-1} + k_2} \qquad (59)$$

This situation is illustrated in Fig. 5.

If $k_2 \gg k_{-1}$, then the pressure effect is determined by

$$\Delta V_0\ddagger \simeq \Delta V_1\ddagger \qquad (60)$$

and

$$v(p) = k_1(0)[E]_0[S] \exp(-p\Delta V_1^{\ddagger}/RT) \qquad (61)$$

as seen from Eq. (58).

In this case, the rate of disappearance of ES to EP is so large that the reverse formation of ES is unimportant and step 1 becomes rate-determining. The last case is the one where $k_{-1} \gg k_2$.

The rate expression becomes

$$v(p) = \frac{k_1(0)k_2(0)[E]_0[S] \exp\{-p(\Delta V_1^{\ddagger} - \Delta V_{-1}^{\ddagger} + \Delta V_2^{\ddagger})/RT\}}{k_{-1}(0)} \qquad (62)$$

and the overall volume change, as is also seen from Eq. (59) is

$$\Delta V_0^{\ddagger} = \Delta V_1^{\ddagger} - \Delta V_{-1}^{\ddagger} + \Delta V_2^{\ddagger} \qquad (63)$$

If we now use the fact that the equilibrium constant of step 1 is defined by

$$K = k_1/k_{-1} \qquad (64)$$

then its pressure dependence follows from

$$K(p) = \frac{k_1(0) \exp(-p\Delta V_1^{\ddagger}/RT)}{k_{-1}(0) \exp(-p\Delta V_{-1}^{\ddagger}/RT)} = K(0) \exp(-p\Delta V/RT) \qquad (65)$$

where

$$\Delta V = \Delta V_1^{\ddagger} - \Delta V_{-1}^{\ddagger} \qquad (66)$$

is the dissociation (equilibrium) volume.

Equation (62) then reduces to

$$v(p) = k_2(0)K(0)[E]_0[S] \exp\{-p(\Delta V + \Delta V_2^{\ddagger})/RT\} \qquad (67)$$

with k_2 as the rate-determining constant, and with the effect of pressure dependent on the composite volume change

$$\Delta V_0^{\ddagger} = \Delta V + \Delta V_2^{\ddagger} = V_{EP}^{\ddagger} - V_E - V_S \qquad (68)$$

(see Fig. 5).

One may prefer to treat Eq. (47) in the simpler form

$$v = \frac{W}{1 + K_m/[S]}$$

The interpretation of the volume change of the catalytic step will then be the same as of $\Delta V_0^{\ddagger}$ in Eq. (42), and the volume change of the "binding" step will be derived from the Michaelis constant

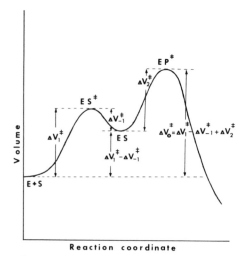

FIG. 5. At low substrate concentration step 3 becomes unimportant. The figure shows the relations between the volume changes involved in Eqs. (59) and (66).

$$K_m = \frac{k_3(k_1 + k_2)}{k_1(k_2 + k_3)}$$

$$\ln K_m = \ln k_3 - \ln k_1 + \ln(k_{-1} + k_2) - \ln(k_2 + k_3)$$

After pressure differentiation we obtain

$$\Delta V_b = \Delta V_3\ddagger - \Delta V_1\ddagger + \frac{k_{-1}\Delta V_{-1}\ddagger + k_2\Delta V_2\ddagger}{k_{-1} + k_2} - \frac{k_2\Delta V_2\ddagger + k_3\Delta V_3\ddagger}{k_2 + k_3} \qquad (69)$$

which is not easy to deal with. In this case it would be wiser to use the procedure introduced above, at extremes of substrate concentrations.

E. Two-Substrate Reactions

Several two-substrate mechanisms are homeomorphous, and lead to the general rate equation

$$v = \frac{W[A][B]}{K'_A K_B + K_B[A] + K_A[B] + [A][B]} \qquad (70)$$

where [A] and [B] are the substrate concentrations, W is the maximum rate, and the K_is are functions of single-step rate constants. The equation may also be written

$$v = W \left[1 + \frac{K_A}{[A]} + \frac{K_B}{[B]} + \frac{K'_A K_B}{[A][B]} \right]^{-1} \tag{71}$$

or

$$\ln v = \ln W - \ln \left[1 + \frac{K_A}{[A]} + \frac{K_B}{[B]} + \frac{K'_A K_B}{[A][B]} \right] \tag{72}$$

Differentiation with respect to pressure gives

$$\left(\frac{\partial \ln v}{\partial p} \right)_T = \left(\frac{\partial \ln W}{\partial p} \right)_T - \frac{v}{W} \left[\frac{1}{[A]} \left(\frac{\partial K_A}{\partial p} \right)_T + \frac{1}{[B]} \left(\frac{\partial K_B}{\partial p} \right)_T \right.$$
$$\left. + \frac{1}{[A][B]} \left(\frac{\partial K'_A K_B}{\partial p} \right)_T \right]$$

which can be rearranged to

$$\left(\frac{\partial \ln v}{\partial p} \right)_T = \left(\frac{\partial \ln W}{\partial p} \right) - \frac{v}{W} \left[\frac{K_A}{[A]} \left(\frac{\partial \ln K_A}{\partial p} \right)_T + \frac{K_B}{[B]} \left(\frac{\partial \ln K_B}{\partial p} \right)_T \right.$$
$$\left. + \frac{K'_A K_B}{[A][B]} \left(\frac{\partial \ln K'_A K_B}{\partial p} \right)_T \right] \tag{73}$$

With definition (21) we now get the overall volume change

$$\Delta V\ddagger = \Delta V_c\ddagger - \frac{v}{W} \left[\frac{K_A}{[A]} \Delta V_A\ddagger + \frac{K_B}{[B]} \Delta V_B\ddagger \right.$$
$$\left. + \frac{K'_A K_B}{[A][B]} (\Delta V_{A'}\ddagger + \Delta V_B\ddagger) \right] \tag{74}$$

All these volume changes, $\Delta V_i\ddagger$, are associated with the observed variation of the kinetic constants K_i with pressure. They cannot always be interpreted in terms of single reactions, but must be analyzed according to the explicit expressions of the K_is for the mechanism in question. In Eq. (74), v, W, and the K_is must be evaluated at the same pressure as the derivatives, $\Delta V_i\ddagger$.

The pressure dependence of Eq. (71) is now given by

$$v(p) = W(0) \exp(-p\Delta V_c\ddagger/RT) \left[1 + \frac{K_A(0) \exp(-p\Delta V_A\ddagger/RT)}{[A]} \right.$$
$$+ \frac{K_B(0) \exp(-p\Delta V_B\ddagger/RT)}{[B]}$$
$$\left. + \frac{K'_A(0)K_B(0) \exp(-p(\Delta V'_A\ddagger + \Delta V_B\ddagger)/RT)}{[A][B]} \right]^{-1} \tag{75}$$

There are obvious special cases of this equation, e.g., at either high [A] or [B], whereby it reduces to pseudo-one-substrate reactions as treated above. Let us now consider some typical mechanisms as examples.

1. Random Ternary-Complex Mechanism

The exact steady-state solution for this mechanism is too complicated to be of any interest in this context. If, however, the rate constant k_3 for disappearance of the ternary complex EAB is so small that there is practically equilibrium as far as EA, EB, and EAB are concerned, the equation reduces to a simple form. In this case, the scheme (76) leads to the kinetic constants $W = k_3[E]_0$, $K_A = k_{-2}/k_2$, $K_B = k'_{-2}/k'_2$, and $K'_A = k_{-1}/k_1$.

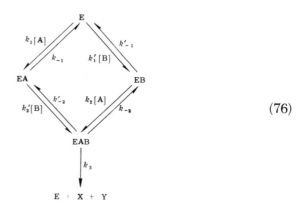

$$(76)$$

The interpretation of the volume changes will then be as follows:

$$\Delta V_c\ddagger = V_{EAB}\ddagger - V_{EAB} = \Delta V_3\ddagger$$
$$\Delta V_A = V_A + V_{EB} - V_{EAB}$$
$$\Delta V_B = V_B + V_{EA} - V_{EAB}$$
$$\Delta V_{A'} = V_E + V_A - V_{EA} \tag{77}$$

The last three volume changes are not activation volumes, but dissociation (equilibrium) volumes. Figure 6 shows the volume as a function of the reaction coordinate for this reaction. Note that the activation volumes are not known, and therefore we use dashed curves.

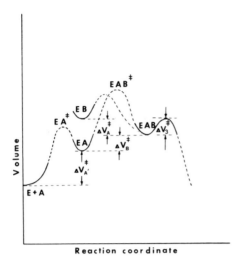

FIG. 6. A set of possible relations between volume changes in a two-substrate reaction with a random ternary-complex mechanism. Note that these volume changes as defined by Eq. (77) are dissociation volumes.

2. Ordered Ternary-Complex Mechanism

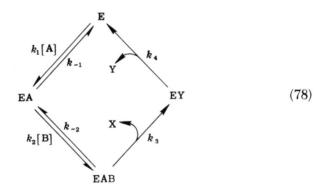

(78)

This mechanism can be considered as a special case of the foregoing random mechanism, where the complex EB cannot be formed. If the ternary complex is very short-lived, i.e., $k_3 \gg k_{-2}$, we can interpret the kinetic constants as $W = k_3[E]_0$, $K_A = k_3/k_2$, $K_B = k_3/k_1$, and $K'_A = k_{-1}/k_2$. Accordingly, the volume changes will be

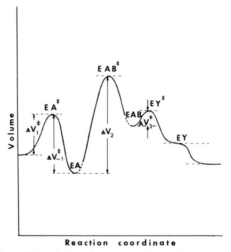

FIG. 7. A set of possible relations between volume changes in a two-substrate reaction with an ordered ternary-complex mechanism. The activation volumes are defined by Eq. (79).

$$\Delta V_A\ddagger = \Delta V_3\ddagger - \Delta V_2\ddagger$$
$$\Delta V_B\ddagger = \Delta V_3\ddagger - \Delta V_1\ddagger$$
$$\Delta V_A'\ddagger = \Delta V_{-1}\ddagger - \Delta V_2\ddagger$$
$$\Delta V_c\ddagger = \Delta V_3\ddagger \qquad (79)$$

In this case there are no equilibria, and all volumes are activation volumes (see Fig. 7).

Both sets of volume changes (77) and (79) can be found experimentally if all kinetic constants are evaluated at each of a series of pressures (Morild, 1977a,b; Morild and Tvedt, 1978).

F. Pressure Change of the Rate-Determining Step

As mentioned in Section III,D, there may be a problem if the reaction undergoes a change in the rate-determining step as pressure increases. In effect, this corresponds to the change from one reaction mechanism to another, and the question is whether the rate equation based on the first mechanism will still be valid for the second mechanism. The question may be of an algebraic nature: Does the change of rate constants with pressure imply an algebraic form of the rate equation that can remain unchanged if only the rate constants are allowed to change, or does it imply a discontinuous transition from one algebraic form to another?

To answer this question one has to look at the way the rate equation is derived. A rate equation based on a certain reaction mechanism may have been derived after the introduction of some approximations valid at atmospheric pressure. If, at higher pressure these approximations are no longer valid, a continuous use of the rate equation may lead to erroneous results. As approximations usually are introduced to reduce the number of parameters, it should be evident that equations with differing numbers of parameters most probably have different algebraic forms. The omission of a critical, initially small, but increasingly more important rate constant with increasing pressure will unavoidably lead to suspect interpretations.

Let us look at the example given in Section III,D, and let Eq. (52) be valid at atmospheric pressure:

$$\Delta V_0\ddagger = \frac{k_3\Delta V_2\ddagger + k_2\Delta V_3\ddagger}{k_2 + k_3}$$

If now $\Delta V_2\ddagger \simeq |\Delta V_3\ddagger|$ but $\Delta V_3\ddagger < 0$ at the same time as $k_2 \gg k_3$, we will have the following situation: At low pressure, $\Delta V_0\ddagger \simeq \Delta V_3\ddagger$ and the rate increases with pressure. As pressure increases, k_2 will decrease and k_3 will increase until at a certain pressure $k_2\Delta V_3\ddagger = -k_3\Delta V_2\ddagger$. Then $\Delta V_0\ddagger = 0$ and the curve of ln k versus pressure will go through a maximum. The approximation that $k_2 \gg k_3$ is no longer valid, and k_3 cannot be neglected anymore. From this point the roles are changed, and at increasing pressure k_2 will be the smaller constant and step 2 will be the rate-determining step. We can also have the case where $\Delta V_2\ddagger < 0$, and the rate will go to a minimum. These cases are symmetrical with respect to the other initial case where $k_2 \ll k_3$. In both cases one may have a change of the rate-determining step if the activation volumes have different signs. Let us also look at Eq. (47)

$$v = \frac{k_2[E]_0[S]}{[(k_{-1} + k_2)/k_1] + [(k_2 + k_3)/k_3] [S]}$$

and assume that step 2 is rate-determining, with $k_2 \ll \{k_1, k_{-1}, k_3\}$. Assume that $k_1 \simeq k_{-1}$ and look at the case at a high substrate concentration. Equation (47) may now be reduced to

$$v \simeq \frac{k_2[E]_0[S]}{1 + [S]} \simeq k_2[E]_0 \tag{80}$$

which is valid at atmospheric pressure. Now let k_2 and k_3 be nearly unaffected by pressure, but let k_1 decrease and k_{-1} increase. At high pressure, then, the valid rate equation would be more like

$$v \simeq \frac{k_2[E]_0[S]}{k_{-1}/k_1 + [S]} \simeq \frac{k_1 k_2}{k_{-1}} [E]_0[S] \qquad (81)$$

This equation shows that the rate will be reduced with pressure, but according to Eq. (80) this reduction will be absorbed into k_2, which is really constant. The rate constants k_1 and k_{-1} have been removed in the initial approximation, and nothing can be said about the pressure dependences of the steps 1 and -1. The interpretation will be that the rate-determining step 2 becomes slower with pressure, while in fact the rate determination has been displaced to step 1. It is immediately clear that such an interpretation would be disastrous for the clarification of the high-pressure mechanism. The condition for a relatively simple rate equation of the random ternary-complex two-substrate mechanism was a small k_3. This constant k_3 may not be as small at high pressures, and the whole rate equation breaks down.

It is difficult to generalize this matter much further. There are so many circumstances, depending on the actual rate equation, the number, location, and relative magnitude of the rate constants and their volume changes, and—not least—the approximations used. The problem is that it is not easy to know in advance which constants are going to change and which are not. The introduction of more constants may only give alternative interpretations, all equally acceptable. The solution of this problem can best be achieved by ensuring good data and an iterative approach starting with the simplest equation that can gradually be furnished with additional constants until an optimal fit is reached. An important thing to remember is that even if $k_1 \ll k_2$, we can still have

$$\left(\frac{\partial \ln k_1}{\partial p}\right)_T \gg \left(\frac{\partial \ln k_2}{\partial p}\right)_T \qquad (82)$$

G. Time Delays in the Experimental Measurements

A problem is encountered when the pressure cell is assembled, and before the pressure is applied. The reaction mixture reacts at atmospheric pressure for some time until the pressure is raised to its final value. The kinetics under pressure will then be dependent on the amount of reactants that previously reacted at atmospheric pressure and during the pressure increase. The same problem appears when the cell is disassembled if measurements can be done after pressure release only. It is, of course, of interest to make this time lapse as small as possible, but usually it may be of the order of tens of seconds. The importance of this error also depends on whether pressure increases or decreases the reaction rate. Certainly, the error will be greater in the latter case.

Usually, corrections can be made from a simple calibration at atmospheric pressure, subtracting the concentration of products from the concentration of substrates, regarding the new concentration as the initial at the pressure p.

Sometimes, however, initial velocities are needed. If measurements of the rate v_t are made at the time t, simple division of Michaelis–Menten expressions yields for the initial velocity v_0

$$v_0 = \frac{1 + K_p/(a_0 - c_t)}{1 + K_p/a_0} \, v_t \qquad (83)$$

where a_0 is initial concentration, c_t is product concentration at time t, and K_p is the Michaelis constant at pressure p. K_p may be found from a Lineweaver–Burk plot at very high a_0.

But there is still a small uncertainty due to the reaction during raising and lowering of the pressure. However, the time lapse should not be more than a few seconds here, and a pressure experiment should last for more than a few minutes. Of course, some integration factor may be applied to correct for this, but the error is probably not more than of the order of a few percent.

Even if the pressure apparatus can be pressurized in a very short time, the generated heat will introduce transients that may last for a minute, maybe two, depending on pressure. Such transients appear as "humps" on a recorded graph, and sometimes a few oscillations may also result. Altogether, steady state may be attained within 2 or 3 minutes after mixing. Further discussion of such temperature effects is found in the end of Section V,A.

IV. Volume Changes in Intermolecular Interactions

A. Electrostatic Interactions

When an electrically charged ion enters an aqueous solution, a phenomenon known as electrostriction occurs (Distéche, 1972). The coulombic field of the ion aligns the nearby water dipoles radially, binding them compactly so that the reaction

$$M^z + nH_2O \rightarrow M(H_2O)_n^z \text{ (solution)} \qquad (84)$$

is accompanied by an overall volume decrease. The superscript z is the (positive or negative) charge of ion M, and n is the number of nearby water molecules. By assuming that ions can be treated as charged spheres in a continuous dielectric medium (Drude and Nernst, 1894), a theory of electrostriction was developed, where the decrease in volume could be calculated from the equation

$$\Delta V_e = - \frac{N_0 z^2 e^2}{8\pi\epsilon_0 Dr} \left(\frac{\partial \ln D}{\partial p}\right)_T = -B \frac{z^2}{r} \tag{85}$$

where e is the electronic charge, ϵ_0 is the vacuum permittivity, D is the dielectric constant of the medium, r is the radius of the ion, and p is the pressure. At 25°C the constant B is found to be 4.175 cm³ mol⁻¹. Measurements of ΔV_e of various substances in water and organic solvents agreed fairly well with calculations from this equation. Later, an additional term was introduced to allow for the change of r with pressure (Buchanan and Hamann, 1953), that is, to allow for the compressibility of the ions,

$$\Delta V_e = - \frac{N_0 z^2 e^2}{8\pi\epsilon_0 Dr} \left(\frac{\partial \ln D}{\partial p}\right)_T + \frac{N_0 z^2 e^2}{8\pi\epsilon_0 r} \left(1 - \frac{1}{\epsilon_0 D}\right) \left(\frac{\partial \ln r}{\partial p}\right)_T \tag{86}$$

Applied to singly charged ions ($z = 1$), this formula yields $\Delta V_e \simeq -10$ cm³ mol⁻¹ in water at atmospheric pressure and 25°C. The dissociation of a neutral molecule into two ions should then induce a contraction of about -20 cm³ mol⁻¹. Some reviews on the volumes of electrolytes have been given (Millero, 1971, 1972; Hamann, 1974). Experimental results show that, in every case, the ionization of a neutral acid or base

$$HA \rightarrow H^+ + A^-$$
$$BOH \rightarrow B^+ + OH^- \tag{87}$$

involves a contraction, and these processes are accordingly enhanced by pressure increase. The ionization of monobasic acids such as the homologous series of organic acids involves volume changes of about -13 cm³ mol⁻¹. Some ionization reactions involve specific chemical hydration where a water molecule becomes bound into the anion covalently and not just electrostatically. This type of hydration evidently causes a larger volume contraction because the first ionization step of some acids involves volume changes of about -30 cm³ mol⁻¹. This is the case, for example, for carbonic and boric acids. The second ionization step of di- and tribasic acids involves volume changes larger than the first ionization step, as is evident from the factor z^2 in Eq. (85). However, this also depends on structural characteristics, but the volume change ordinarily is about -24 cm³ mol⁻¹.

For zwitterionic amino acids the fields of the two charges cancel at a distance from the ion and z is effectively equal to zero. In the second ionization, to give zwitterions, the electrostatic free energy change is reversed and the dissociation volume becomes less than it would be if the first charge were not there. For instance, the ionization of the glycine cation

$$NH_3^+CH_2CO_2H \rightarrow NH_3^+CH_2CO_2^- + H^+$$

gives a volume change $\Delta V = -6.8$ cm^3 mol^{-1}, whereas one would expect a ΔV of about -13 cm^3 mol^{-1}. However, the loss of a proton

$$NH_3^+CH_2CO_2^- \rightarrow NH_2CH_2CO_2^- + H^+$$

may involve small and positive changes, $\Delta V = +1.3$ cm^3 mol^{-1}. The electrostriction effects of both the COO$^-$ and the NH$_3^+$ group upon water have been extensively studied and it is suggested that these groups cause small and large volume changes, respectively, when they act alone. The electrostrictive effects of charges in $\alpha,1$-amino acids have been calculated to give volume changes of about -15 cm^3 mol^{-1}. From these measurements on model systems it should be evident that the volume associated with the development or neutralization of electric charges on an enzyme surface cannot be easily predicted. The interactions from surrounding ionic, polar, and hydrophobic groups will certainly make the volume change deviate significantly from that of an isolated charge. However, the sign of the volume changes will probably be consistent with those measured on model systems. That is, the temporary neutralization of a cationic substrate with an anionic enzyme active site to make a neutral transition state is expected to result in a positive volume change of about 10 ± 10 cm^3 mol^{-1}. Accordingly, the reaction rate of this process decreases with pressure. The opposite process, the development of charges in an enzyme–substrate transition state, is expected to result in a negative volume change of similar order, with pressure increasing the reaction rate.

Related to this is the volume change associated with dipole development in transition states. This has been investigated theoretically for a model substance of molecules with a size similar to that of water (Morild and Larsen, 1978). The calculations show that the volume changes are very pressure-dependent in this case. A change in dipole moment from 0 to 1×10^{-30} C·m gives a volume decrease of about 30 cm^3 mol^{-1} at 350 bar and about 20 cm^3 mol^{-1} at 750 bar. However, this may not be typical for molecules as large as enzyme–substrate complexes.

Studies of organic fragmentation and decomposition reactions (le

Noble *et al.*, 1976; Neuman and Pankratz, 1973) have shown that the activation volumes in fact are rather small, i.e., about 0 to -10 ml mol^{-1}. This can probably be explained by the simultaneous extension of the bond to be broken when a dipole is formed. These two processes are accompanied by volume changes that partly compensate each other.

B. Hydrophobic Interactions

There has been a noticeable increase in interest in hydrophobic interactions during the last decade. This interest is mostly related to the supposed importance of the hydrophobic interactions as determinants of stability in and interactions between biological macromolecules. A relatively small part of this interest has been focused on the volume changes associated with the interactions. Most of our "knowledge" about these volume change comes from studies on small molecules serving as model systems for the much larger macromolecules.

Ben-Naim (1980) has recently reviewed the work in this field. The results obtained so far are not very satisfying because there does not seem to be a single method that can provide direct information on the properties of the hydrophobic interactions between two simple solutes in water at a realistic interparticle distance. Most studies have been concerned with the volume difference between a "dimer" 2A and a "monomer" A, for example, ethane and methane. However, such comparisons are not realistic because there is a covalent bond between the two monomers in the "dimer," and also two hydrogen atoms are missing in the dimer. The volume of these two hydrogens seems to account for the resulting volume difference.

It is well known that the transfer of nonpolar molecules from nonpolar to polar surroundings results in a decrease in the partial molar volume of the solute. The dimerization studies also show that there is a similar volume decrease when two monomers form a "dimer." This volume decrease is of the order of 20 cm^3 mol^{-1}. It is difficult to understand how there can be first a volume decrease when the nonpolar molecules are transferred from the nonpolar to the polar environment and then a further volume decrease when two molecules come together and "partly reverse" the first transfer. It is a little dangerous to speak of the partial reversal of a process we know so little about. It is believed that the hydrophobic hydration is a cooperative phenomenon, in which the exact microstructure of water is very important for the occupied volume. How this microstructure changes when two molecules associate in a hydrophobic interaction is not particularly well understood. It is possible that the interactions between pairs of molecules differ from the interactions between several molecules. There may be a volume decrease as long as only pairs are allowed to interact, but a volume increase when several molecules interact.

During the last years there have been some computer calculations, simulating interactions between a nonpolar molecule, such as methane, and water molecules. These reports conclude that there seem to be more open arrangements of water molecules around the nonpolar solute, with a slightly higher average number of hydrogen bonds between each water molecule and its neighbors. So far, this is only what would be expected from what is already known.

The measure of the volume change in hydrophobic interactions used in the monomer–dimer studies is defined by the volume difference between a dimer 2A and two monomers A

$$\delta V^{\text{HI}}(\sigma) = \Delta V_{2_A}^0 - 2\Delta V_A^0 \tag{88}$$

where σ indicates the intermolecular distance and ΔV_i^0 is the so-called "local volume" of transfer of a molecule i from the gas phase to the liquid phase, given by

$$\Delta V_i^0 = V_i^0 - RT\beta_0 \tag{89}$$

and V_i^0 is the conventional partial molar volume of the molecule at infinite dilution (for more details, see Ben-Naim, 1980). β_0 is the isothermal compressibility coefficient. This indirect quantity is the only measure of the effect of pressure on hydrophobic interactions in equilibrium processes that has been used hitherto. However, by using isomeric dialkyl aromatics, as suggested by Ben-Naim (1980), Morild (1981) has found a negative volume change of the order of a few cubic centimeters per mole.

On the other hand, some direct measurements of the volume change in hydrophobic interactions have been performed (Oakenfull and Fenwick, 1977) by means of ion-pair formation of long-chain molecules. These measurements indicate that there is a volume increase of about 10 cm^3 mol^{-1} per pair of interacting CH$_2$ groups. It is difficult to judge the significance of this figure, because it is accompanied by the unexpected volume decrease of about -60 cm^3 mol^{-1} associated with the neutralization of the electric charges.

In a measurement of a hydrophobic activation volume, Kettman *et al.* (1966) found a large positive volume change when ribonuclease molecules with attached polyvalyl chains were made to aggregate under pressure. Work is also in progress on the interaction kinetics between long-chain molecules under pressure (Morild and Aksnes, 1981).

Reasoning on the basis of the short relaxation time of water molecules compared to the time of hydrophobic interaction, there may perhaps be small differences between the volume changes involved in equilibrium and kinetic processes. However, so far no realistic comparison has been made.

C. Hydrogen Bonds

The contribution of hydrogen bonds to the partial molar volume of nonpolar solutes in water seems to be unanimously agreed upon (Terasawa et al., 1975; Suzuki and Taniguchi, 1972). More than 20 years ago, investigations of changes in infared spectra with pressure indicated that hydrogen-bond formation was accompanied by negative volume changes. Other results, obtained with different methods, point in the same direction, and it is well known that hydrogen-bond formation results in a decrease in molar volume due to shortening of the interatomic distance. Sometimes, however, the bonds are strongly directed and may create a fairly open structure partly compensating the volume decrease. Calculations of the volume change on formation of one hydrogen bond range from -1 to -3 cm^3 mol^{-1}.

V. Important Parameters

A. Temperature

For more than 40 years there have been reports about interesting relationships between pressure and temperature effects (Johnson et al., 1974), but so far there have been no significant attempts to try to understand these phenomena theoretically. This is not at all strange, since most of the reports of interest involved effects on complex living organisms. This section will not include treatment of any particular phenomena, but rather give a general thermodynamic basis for understanding pressure–temperature relations.

Unless specifically noted, the equations below are assumed to apply to transition state quantities ($\ddagger$) as well as to equilibrium quantities. All Δ quantities are then understood either as differences between activated complex quantities and reactant quantities, or between product and reactant quantities.

Quantum mechanically, there is a clear distinction between the energy absorbed by a system in the form of heat and in the form of work ($p \times \Delta V$). Heat absorption increases the population of the higher energy levels of the system, while work absorption increases the spacing between the energy levels. The basic equation governing the behavior of a system during a pressure and temperature change is

$$d(\Delta G) = -\Delta S dT + \Delta V dp \qquad (90)$$

ΔG is the free energy change of a process with ΔS and ΔV as the accompanying changes in entropy and volume. As a first approximation, one can assume ΔS and ΔV to be nearly constant within a limited p and T range, and use the integrated equation

$$\Delta G = \Delta G_0 - \Delta S(T - T_0) + \Delta V(p - p_0) \qquad (91)$$

for ΔG in the state (p, T), if ΔG_0 was found in the state (p_0, T_0). More stringently, ΔS and ΔV vary both with p and T, and their total differentials are, respectively,

$$d(\Delta S) = \left(\frac{\partial \Delta S}{\partial T}\right)_p dT + \left(\frac{\partial \Delta S}{\partial p}\right)_T dp \qquad (92)$$

and

$$d(\Delta V) = \left(\frac{\partial \Delta V}{\partial T}\right)_p dT + \left(\frac{\partial \Delta V}{\partial p}\right)_T dp \qquad (93)$$

Inserting appropriately defined quantities, these equations may be rewritten as

$$d(\Delta S) = \left(\frac{\Delta C_p}{T}\right)_p dT - \Delta \zeta dp \qquad (94)$$

$$d(\Delta V) = \Delta \zeta dT + \Delta \kappa dp \qquad (95)$$

By integration of these equations between the states (p_0, T_0) and (p, T) we obtain their functional dependence on p and T. Then the results of these integrations can be inserted into Eq. (90) and this equation can be properly integrated.

Assuming that ΔC_p, $\Delta \zeta$, and $\Delta \kappa$ are constant in the integration interval (which is fairly reasonable), the result is

$$\Delta G = \Delta G_0 + \Delta V_0(p - p_0) - \Delta S_0(T - T_0) + (\tfrac{1}{2})\Delta \zeta(p - p_0)^2$$
$$+ 2\Delta \zeta(p - p_0)(T - T_0) - \Delta C_p[T(\ln \frac{T}{T_0} - 1) + T_0] \qquad (96)$$

This formidable equation should, in principle, be able to describe most phenomena with regard to their pressure and temperature behavior. However, its practical use may be limited due to its many parameters. Evaluation of the unknown parameters by means of computerized curve-fitting would require data from rather large pressure

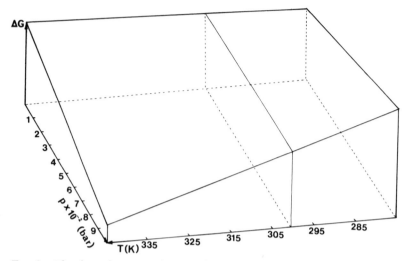

FIG. 8. The three-dimensional ΔG surface defined by Eq. (96) with the following choice of parameters: $\Delta V_0 = 50$ cm^3 mol^{-1}, $\Delta \kappa = -0.001$ cm^3 bar^{-1}, $\Delta \alpha = -1$ cm^3 mol^{-1} K^{-1}, $\Delta C_p = 25$ J mol^{-1} K^{-1}, $\Delta S_0 = -25$ J mol^{-1} K^{-1}, $\Delta G_0 = 50$ kJ mol^{-1}. The surface is located between the ΔG values 42.5 and 55.0 kJ.

and temperature intervals, of the order of 5000 bar and 100K. In Fig. 8, a diagram of the three-dimensional ΔG surface is shown as calculated from Eq. (96). The ranges are $1 < p < 1000$ bar and $275 < T < 345$K, which are the most interesting ones for biological systems. The parameters are chosen more or less arbitrarily, but are believed to be representative for biological systems, considering the limited pool of literature data on these quantities. By changing signs and magnitudes of the parameters, the form of the surface may be altered in many ways. With the present choice, it may be seen how a pressure effect (as represented by ΔV) changes sign as temperature is varied around $T = 300$K. This is also a very common observation in combined pressure and temperature experiments [see the review of these effects in Johnson *et al.* (1974)].

All processes involving interesting pressure and temperature relationships can probably in effect be described by Eq. (96), although the origin of the effect in a complex system may need an elaborate explanation. The parameters of the equation then play the same averaged role as in the transition state theory; i.e., they are only effective or apparent quantities.

The most important single parameter for a discussion of pressure–temperature relationships is no doubt ΔV. If ΔV is temperature-dependent (and it nearly always is), it also controls the pressure dependence of ΔH and ΔS, through

$$\left(\frac{\partial \Delta S}{\partial p}\right)_T = -\left(\frac{\partial \Delta V}{\partial T}\right)_p \tag{97}$$

which is a Maxwell relation, and

$$\left(\frac{\partial \Delta H}{\partial p}\right)_T = \Delta V - T\left(\frac{\partial \Delta V}{\partial T}\right)_p \tag{98}$$

which follows from Eqs. (5) and (97).

It is not easy to find many reliable values of $(\partial \Delta V\ddagger/\partial T)$ in the literature of enzyme reactions, but the values $\Delta \zeta\ddagger = 0.45$ and 1.4 cm^3 mol^{-1} K^{-1} have been observed (Greaney and Somero, 1980; E. Morild and H. Kryvi, unpublished). Results from organic reactions have shown that $\Delta V\ddagger$ and $(\partial \Delta V\ddagger/\partial T)$ almost invariably have the same sign, suggesting that the state of largest volume also has the largest expansibility. It is also found that $\Delta \zeta\ddagger$ in general tends monotonically toward zero as the pressure is increased, like $\Delta V\ddagger$ itself. Regarding the contributions of $\Delta H\ddagger$ and $\Delta S\ddagger$ to $\Delta G\ddagger$ (Eq. (5)), the change in $\Delta G\ddagger$ with pressure results mainly from the $\Delta S\ddagger$ contribution if $\Delta V\ddagger$ and $\Delta \zeta\ddagger$ have the same sign. Then, if $|\Delta \xi\ddagger| > |\Delta V\ddagger/T|$, the enthalpy change with pressure will tend to oppose the kinetic pressure effect. If $\Delta V\ddagger$ and $\Delta \zeta\ddagger$ have opposite signs, the enthalpy change will dominate the kinetic effect, while the change in $\Delta S\ddagger$ will oppose it.

Equations (90) and (91) give a rough indication of the influence of pressure and temperature on ΔG. However, with the relatively large expansibilities found for enzyme systems, where the volume changes may be of the magnitude of 1 cm^3 mol^{-1} K^{-1}, it is clear that ΔV and ΔS can readily change sign within a temperature interval of 10K. This fact demands that great care be used in the estimation of the involved quantities, and that, if possible, Eq. (96) be used. The Maxwell relation

$$\left(\frac{\partial \Delta S}{\partial \Delta V}\right)_T = \left(\frac{\partial p}{\partial T}\right)_V = -\frac{(\partial \Delta V/\partial T)_p}{(\partial \Delta V/\partial p)_T} = -\frac{\partial \zeta}{\Delta \kappa} \tag{99}$$

suggests a relationship which has also been observed experimentally, namely, a linear correlation between ΔS and ΔV. This has been discussed (Hepler, 1965) for a very simple class of processes, the ionization of aqueous acids. The equation that describes the experimental relations is of the type

$$\Delta S = a + b\Delta V \tag{100}$$

The slope b may be interpreted as the value of $(\partial p/\partial T)_V$ of water at a pressure equal to the internal pressure in water. The intercept a seems to arise from the conventional choice of the pure liquid water as solvent standard state.

This relationship has also been found in rate processes, e.g., in organic reactions and also in enzyme-catalyzed reactions. For enzyme reactions the general rule seems to be that ΔS and ΔV have the same sign, although this may be an oversimplification. Intuitively, it is not very difficult to imagine that a process that involves an increase in volume also involves an increase in entropy. Data from statistical mechanical calculations also show that volume and entropy changes corresponding to changes in electrostatic properties of the molecules follow each other very closely (Morild and Larsen, 1978).

There are other interesting relationships, for example, the ones obtained from

$$dH = TdS + Vdp \tag{101}$$

which are

$$\left(\frac{\partial \Delta H}{\partial p}\right)_S = \Delta V \quad \text{and} \quad \left(\frac{\partial \Delta H}{\partial \Delta S}\right)_p = T \tag{102}$$

The latter describes in essence a phenomenon more commonly known as enthalpy–entropy compensation. One way of stating this is that the free energy changes in similar processes are kept within a narrow interval because ΔH and ΔS vary in proportion, and thereby compensate each other, through the equation

$$\Delta G = \Delta H - T\Delta S \tag{103}$$

There has been much debate about this particular linear relationship between ΔH and ΔS and its statistical significance (Krug et al., 1976). Many experimental researchers believe it is a real effect, that the relation is described by

$$\Delta H = a + b\Delta S \tag{104}$$

and that the so-called compensation temperature

$$b = \left(\frac{\partial \Delta H}{\partial \Delta S}\right)_p = T_c \tag{105}$$

may have a physical interpretation similar to a transition temperature. Note, however, that since thermodynamic potentials are interrelated by the Maxwell relationships, a correlation between any two potentials can be transformed to give the corresponding relationships between any other two. Thus, if ΔV is proportional to ΔS, and ΔS is proportional to ΔH, ΔV should also be proportional to ΔH. The combined use of pressure and temperature allows some of these correlations to be checked.

Indeed, some relationships between $\Delta V\ddagger$, $\Delta H\ddagger$, and $\Delta S\ddagger$ have been found for enzyme reactions. In the works of Somero and collaborators, these relationships have been interpreted in terms of salt effects on hydration energetics in the activation process.

A study of the temperature and pressure effects on the rate constant k must take Eq. (6) as a point of departure. A plot of $\ln k$ versus T^{-1} yields an experimental activation energy (which is not to be interpreted as the internal activation energy) through the relation

$$\ln k = \ln(k_B/h) + \ln T + \ln K\ddagger \tag{106}$$

$$\left(\frac{\partial \ln k}{\partial T}\right)_p = \frac{1}{T} + \left(\frac{\partial \ln K\ddagger}{\partial T}\right)_p = \frac{E_x}{RT^2}$$

$$E_x = RT + \Delta E\ddagger \tag{107}$$

Here, $\Delta E\ddagger$ is the increase of internal energy for the activation process. The connection with pressure comes with the introduction of the activation enthalpy

$$E_x = RT + \Delta H\ddagger - p\Delta V\ddagger \tag{108}$$

and the free energy can be written

$$\Delta G\ddagger = E_x - RT + p\Delta V\ddagger - T\Delta S\ddagger \tag{109}$$

The more trivial effect of temperature is the immediate disturbance of rates and equilibria due to adiabatic heating and cooling during compression and decompression. The maximum temperature increase that can be obtained in water by reversible compression is

$$\left(\frac{\partial T}{\partial p}\right)_S = \frac{\alpha TV}{C_p} = 1.86 \times 10^{-3} \text{ K bar}^{-1} \tag{110}$$

As most compressions are irreversible, the temperature increase can be expected to be somewhat larger. But in well-thermostated solid metal cells it is believed that the temperature changes will decay rapidly. With reasonably gentle onset of pressure the temperature effects should be controllable.

A temperature increase of 2K is not unrealistic for a pressure increase of 1000 bar. For a reaction rate with a $\Delta H\ddagger = 50$ kJ, this means an increase in the rate constant k of the order of 6%.

B. Ionic Strength and Salt Effects

1. Equilibrium

Biological macromolecules in aqueous solution are frequently sensitive to changes in the ionic strength, as addition of neutral salts influences both equilibrium and kinetic phenomena. Generally it is a matter of stabilization of a solute or part of a solute molecule through interactions between the solute and the surrounding solvent. Most often the various salt effects are mediated through influences on the thermodynamic activity coefficients γ_i. Some of the well-known equilibrium phenomena are changes in the solubility of solutes due to the salt addition, "salting in" and "salting out." If the activity coefficient of a solute in pure water is γ_i^0, while being γ_i in a salt solution, there is a nonideal free energy change as the solute is transferred from water to the salt solution

$$\Delta G = RT \ln(\gamma_i/\gamma_i^0) \tag{111}$$

As seen from Eq. (130) an activity coefficient may deviate significantly from unity at higher salt concentrations. The activity coefficient can therefore also be used as a measure of the deviation of the salt solution from a thermodynamically ideal solution. If the chemical potential of a solute in a (pressure-dependent) standard state of infinite dilution is μ^0, we find the standard partial molar volume from

$$\left(\frac{\partial \mu^0}{\partial p}\right)_T = V^0 \tag{112}$$

The chemical potential of the solute in a nonideal solution is given by

$$\mu = \mu^0 + RT \ln \gamma c \tag{113}$$

where c is the concentration of the solute. The partial molar volume of the solute at concentration c is given by

$$V = \left(\frac{\partial \mu}{\partial p}\right)_T = \left(\frac{\partial \mu^0}{\partial p}\right)_T + RT \left(\frac{\partial \ln \gamma}{\partial p}\right)_T \tag{114}$$

Then, we find the pressure dependence of the activity coefficient from

$$\left(\frac{\partial \ln \gamma}{\partial p}\right)_T = \frac{V - V^0}{RT} \tag{115}$$

$V - V^0$ is here the volume change of the solute as it is transferred from pure solvent to a nonideal solution, all at a pressure p.

One may also choose a pressure-independent standard state, e.g., at 1 bar. Then,

$$\left(\frac{\partial \mu^0}{\partial p}\right)_T \doteq 0 \tag{116}$$

and

$$\left(\frac{\partial \ln \gamma}{\partial p}\right)_T = \frac{V}{RT} \tag{117}$$

The interpretation of the pressure dependence of γ is now different, because the deviation from ideality at a pressure p includes not only contributions from the difference in concentration from c to pure solvent, but also the difference from the pressure p to 1 bar.

In moderately dilute solution, with a salt concentration c_s and a nonelectrolyte of concentration s_i (the solubility), the following expressions have been experimentally justified:

$$\ln \gamma_i = k_i s_i + k_s c_s \tag{118}$$

and

$$\ln \gamma_i^0 = k_i s_i^0 \tag{119}$$

where s_i^0 is the solubility of the nonelectrolyte in pure water. Since the chemical potential of a species is the same in all solutions in equilibrium with the pure substance, the activity a_i of the species must be the same:

$$a_i = \gamma_i s_i = \gamma_i^0 s_i^0 \tag{120}$$

From these equations we obtain the empirical Setschenow equation

$$\ln(s_i^0/s_i) = k_s c_s \tag{121}$$

which holds for small s_i and s_i^0.

The k_i is a parameter of the solute–water interaction, and the k_s is a parameter of the ion–solute interaction and is called the salting coefficient. However, molecular interactions between ion and solvent and between nonelectrolyte and solvent may also be important for its value. A positive value of k_s corresponds to salting out ($s_i^0 > s_i$) and a negative value of k_s corresponds to salting in ($s_i^0 < s_i$). Salting phenomena generally follow the lyotropic series of ions, and the effect of a given salt or of several salts is most often simply the sum of the effects of the constituent ions. This is the basis for some important experimental results obtained by Somero and co-workers (Low and Somero, 1975b,c; Somero et al., 1977; Greaney and Somero, 1980) with regard to the kinetic pressure effect on enzymes. Addition of salt has shown a profound influence on activation volumes, and may even result in a change of sign.

It is believed that these effects derive from salt influences on some process (associated with hydration) involved in establishing activation free energy values for enzymes, although not an active site phenomenon.

To understand the relation between salting in and salting out phenomena and pressure–volume quantities one may go back to the early theories of salt effects (McDevit and Long, 1952; Long and McDevit, 1952; Masterton and Lee, 1970). The first theories related the salt effect mainly to the influence of the nonelectrolyte on the dielectric constant of the solvent. For polar molecules this was confirmed by the observation that salting in increased with the dipole moment of the nonelectrolyte. The effect of various salts also followed an order similar to that found for nonpolar solutes. Nevertheless, these theories could not explain the variation of k_s with the nature of the salt, nor the salting in phenomenon shown by large ion electrolytes. The importance of the volume was pointed out by McDevit and Long (1952), who developed an approximate expression for k_s. The role of a nonpolar solute was assumed to consist mainly in occupying space, thereby modifying the ion–solvent interaction characteristic of the electrolyte solution in question. The salting in effect on nonelectrolytes was shown to increase with ionic size, and it was also observed that the salting out contribution was roughly proportional to the volume of the nonelectrolyte. By considering that nonpolar neutral molecules merely occupy volume and hence modify the ion–water interaction in a simple manner, McDevit and Long obtained the following limiting law:

$$k_s = \frac{V_i^0(V_s - V_s^0)}{2.3\beta RT} \tag{122}$$

Here, V_i^0 refers to the molar volume of the nonelectrolyte, V_s and V_s^0 are the intrinsic[3] and apparent molar volume of the salt, respectively, and β is the isothermic compressibility coefficient of the solution. Although this equation is strictly valid in the limit of $V_i \rightarrow 0$ and $c_s \rightarrow 0$, it works quite satisfactorily for small nonpolar solutes. The equation shows that the effect is greatest for nonelectrolytes of large molar volume V_i, and for salts that cause the largest electrostriction, $V_s - V_s^0$. A difficulty with this expression is that it is not always easy to evaluate the intrinsic volume of a salt (the mere volume, without

---

[3] Intrinsic volume is the hypothetical molar volume of the liquid salt at the temperature in question.

medium effects). A physical interpretation is that the electrostrictive compression of the solution makes it more difficult to insert the volume of a nonelectrolyte, and salting out occurs. A salt anion like perchlorate yields a "negative electrostriction," and salting in occurs.

Hey et al. (1976) also argue that is is the effect of the ions on the empty volume of the solvent which underlies the ion effects on macromolecules in aqueous solution.

This is related to the concept of internal pressure, which increases when salt is added to an aqueous solution. The increase in internal pressure resulting from the ion–solvent interaction then squeezes out the nonelectrolyte molecules from the solution.

When a solvent undergoes a small isothermal volume expansion, it does work against the cohesive forces which causes a change in the internal energy E of the solvent. The internal pressure is given by the relation (Dack, 1976)

$$p_i = \left(\frac{\partial E}{\partial V}\right)_T \tag{123}$$

and the cohesive forces create a pressure within a solvent of about 10^3–10^4 bar. The value of p_i for water is 1650 bar at 25°C, while a 3 M aqueous solution of NaBr exhibits a p_i value of 3000 bar. To get an idea of what it means energetically to increase the ionic strength, one has to perform studies of thermodynamic transfer quantities of different molecules from water to salt solutions. Many studies of ΔG_t, ΔH_t, and ΔS_t have been carried out, but unfortunately only a few are concerned with volumes of transfer, ΔV_t (Desrosiers and Desnoyers, 1976). From the available data it can be concluded that typically hydrophobic ions such as Bu_4N^+ have negative standard volumes of transfer. From water to NaCl solutions the difference in volume is at most -3 cm^3 mol^{-1} per unit concentration of salt (molar). Probably, this is a general tendency for most molecules when they are transferred from water to electrostricted solutions. It is also found that the volumes of nonelectrolytes in polar solvents increase with decreasing compressibility of the solvents. Since the compressibility of salt solutions tends to decrease with increasing salt concentration, the volumes of nonelectrolytes should be expected to increase slightly because of this, although the increase would in general be dominated by the decrease due to increased electrostriction.

Some polar nonelectrolytes show nonlinear Setschenow plots, and this behavior has been attributed to direct chemical interactions (Gordon, 1975). If a nonelectrolyte forms a complex ion with one of the ions of the salt, the solubility of a poorly soluble solute may increase linearly with salt concentration

$$s_i = s_i^0 + k_s c_s \tag{124}$$

instead of logarithmically, as in Eq. (121). This has been observed for model systems such as peptides and amides, but also for proteins, and the phenomenon still follows the lyotropic anion series. Anion binding probably involves several interaction mechanisms associated with general ion solvation. Close analogs occur in micellar electrolyte solutions and in ion-exchange resins. These equilibria are known to exert profound effects on conformational and aggregation equilibria in solutions of biological macromolecules.

If any equilibrium constant shows a behavior that can be described by the equation

$$\ln(K_i^0/K_i) = k_s c_s \tag{125}$$

a differentiation with respect to pressure (neglecting concentration changes) will give

$$\left(\frac{\partial \ln K_i^0}{\partial p}\right)_T - \left(\frac{\partial \ln K_i}{\partial p}\right)_T = k_s \left(\frac{\partial \ln k_s}{\partial p}\right)_T \tag{126}$$

and the volume change associated with the equilibrium in pure water, ΔV_i^0, will change into ΔV_i in a salt concentration c_s, according to

$$\Delta V_i = \Delta V_i^0 - k_s \Delta V_s \tag{127}$$

where ΔV_s is the volume change associated with the effect of salt on the equilibrium participant species. It can be interpreted as the volume change of transfer of the products from salt solution to water minus the volume change of transfer of the reactants from salt solution to water. If ΔV_s is negative, it can be interpreted as if the transfer volume of reactants is greater than that of products, and the salting coefficient k_s will increase with pressure, indicating an increasing salt effect.

2. Kinetics

Salt effects in kinetics are usually classified as primary or secondary, but there is much more to the subject than these special effects. The theoretical treatment of the primary salt effect leans heavily upon the transition state theory and the Debye–Hückel limiting law for activity coefficients. For a thermodynamic equilibrium constant one should strictly use activities a instead of concentrations (indicated by brackets).

$$K\ddagger = \frac{a_x\ddagger}{a_A a_B} = \frac{\gamma_x\ddagger}{\gamma_A \gamma_B} \frac{[X]\ddagger}{[A][B]} \tag{128}$$

Then, using the assumption that the reaction rate is proportional to the concentration of the activated complex together with the empirical expression for the rate, the observed rate constant can be expressed in terms of the theoretical rate constant k_3^0

$$k_3 = r(k_B T/h)K\ddagger(\gamma_A \gamma_B/\gamma_x\ddagger) = k_3^0(\gamma_A \gamma_B/\gamma_x\ddagger) \tag{129}$$

Finally, using the theoretical expression for the activity coefficient

$$\log \gamma\pm = -\frac{|z_1 z_2|A_\gamma I^{1/2}}{1 + aB_\gamma I^{1/2}} \tag{130}$$

where I is the ionic strength, z_1 and z_2 are the ionic charges, and a, A_γ, and B_γ are constants, one obtains

$$\log \frac{k_3}{k_3^0} = \frac{2z_1 z_2 A_\gamma I^{1/2}}{1 + aB_\gamma I^{1/2}} \tag{131}$$

which is the general expression for the effect of the ionic strength on reaction rates between charged species. Physically, added salt creates appropriate ion atmospheres about the activated complex, which stabilize ionic transition states with greater charge density or dipolar transition states with greater charge separation than the reactant molecules.

The secondary salt effect originates in the effect of salts upon equilibria that may be of importance to the rate, e.g., changing the activity of catalytically functioning species.

There are, however, many specific and anomalous kinetic salt effects, especially at higher salt concentrations, the origin of which lies in the effects on nonelectrolyte reactant activity coefficients. This is often the most interesting effect for enzymes, because the charge on the substrate is frequently zero, making the product $z_1 z_2$ also zero. The exact charge on the enzyme molecules can be difficult to determine if one is working at a pH removed from the isoelectric point of the enzyme.

In highly aqueous media, salt effects on nonelectrolyte–non-electrolyte reactions should normally be of the form (Gordon, 1975),

$$\log(k_3/k_3^0) = (k_s^A + k_s^B - k_s^X)c_s \tag{132}$$

Here, k_s^A, k_s^B, and k_s^X are the salting coefficients of A, B, and the activated complex X$\ddagger$. If all three species are nonpolar, the salt-induced medium effect will predominate in governing the ratio k_3/k_3^0. In water, this may specifically concern hydration phenomena of enzyme molecules, and in particular certain water-exposed groups in the molecular surface.

In the absence of charges and dipoles, short-range effects such as hydrogen bonding and hydrophobic hydration become important. The complex solute–solvent interactions in such cases, however, are not well understood. In addition there is the possibility of ion or ion-pair catalysis. Especially in cases with solvents of low dielectric constant, it has been observed that k_3 rather than log k_3 becomes linear in c_s

$$k_3 = k_3^0(1 + bc_s) \tag{133}$$

The analogy with the linear s_i–c_s dependence in equilibria, Eq. (124), on the one hand, and with ion catalysis of solvolysis at higher dielectric constants on the other, suggests that catalysis by direct complexation with added ion-pairs may occur. The resulting rate laws from experiment with organic reactions have been described by linear functions similar to Eq. (133).

The action of added ion-pairs may also be visualized as the establishment of an ion-pair atmosphere about a dipolar transition state. Simple thermodynamic treatment predicts linearity of log k_3 in c_s, but it has been shown that contributions of ion-pair–ion-pair repulsion, higher aggregation, and the effect of salt on the dielectric constant introduce curvature in the log k_3–c_s plot that is in the direction of a k_3–c_s dependence.

This is also the observation in experiments with salt addition to enzyme systems. The data points do not always form strictly linear curves, but it seems fair enough in most cases to describe the averaged functional form by

$$k_3 = k_3^0 + k_s c_s \tag{134}$$

It is to be expected that curves of this kind show irregularities when the complexity of the system and the drastic effect certain ions exert on activity coefficients are considered.

An intriguing observation is that a plot of $\Delta V\ddagger$, the activation volume as a function of salt concentration, almost invariably seems to be a mirror image of the rate plot. A typical diagram is shown in Fig. 9. However, it can be shown that the functional form of the rate constant necessarily must be reflected in the form of the activation volume in this case. According to what has been said hitherto, an equation

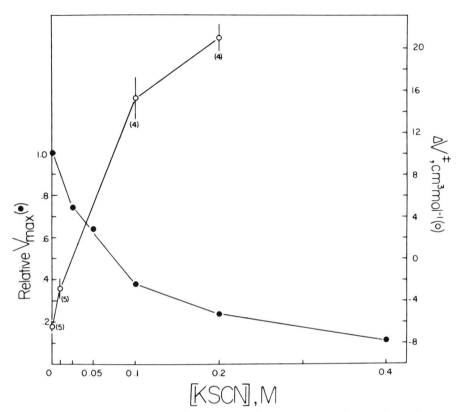

FIG. 9. Simultaneous plots of reaction rate (relative V_{max}) and activation volume ($\Delta V\ddagger$) for the lactate dehydrogenase system as functions of salt (KSCN) concentration. This mirror image may be explained by the similarity between Eq. (134) and Eq. (136). (By courtesy of Prof. G. N. Somero, unpublished.)

showing both the salt and the pressure dependence of the rate constant could be

$$k_3 = (k_3^{00} + k_s c_s) \exp(-p\Delta V_0\ddagger/RT) \tag{135}$$

Here, k_3^{00} is the value of k_3 at 1 bar and in pure water, and $\Delta V_0\ddagger$ is the activation volume found in pure water. The activation volume $\Delta V_s\ddagger$ at some salt concentration c_s is then found from

$$-\frac{\Delta V_s\ddagger}{RT} = \left(\frac{\partial \ln k_3}{\partial p}\right)_T = \left(\frac{\partial \ln(k_3^{00} + k_s c_s)}{\partial p}\right)_T - \frac{\Delta V_0\ddagger}{RT}$$

$$= \frac{k_3^{00}}{k_3^{00} + k_s c_s}\left(\frac{\partial \ln k_3^{00}}{\partial p}\right)_T$$

$$+ \frac{k_s c_s}{k_3{}^{00} + k_s c_s} \left(\frac{\partial \ln k_s c_s}{\partial p} \right)_T - \frac{\Delta V_0\ddagger}{RT}$$

$$= - \frac{k_s c_s}{k_3{}^{00} + k_s c_s} \frac{\Delta V_k}{RT} - \frac{\Delta V_0\ddagger}{RT}$$

$$\Delta V_s\ddagger = \Delta V_0\ddagger + \frac{\Delta V_k}{k_3{}^{00} + k_s c_s} k_s c_s \qquad (136)$$

Here, we have put

$$\left(\frac{\partial \ln k_s c_s}{\partial p} \right)_T = - \frac{\Delta V_k}{RT} \qquad (137)$$

and used the fact that

$$\left(\frac{\partial \ln k_3{}^{00}}{\partial p} \right)_T = 0 \qquad (138)$$

Equation (136) thus shows the same relationship between $\Delta V_s\ddagger$ and c_s as between k_3 and c_s, differing only with respect to the factor in front of the $k_s c_s$ term. For salting in salts, with $k_s < 0$, k_3 shows a decrease with c_s. For $\Delta V_s\ddagger$ to show the observed correlated increase, ΔV_k must be negative. ΔV_k is an effective or apparent volume change related to the pressure dependence of the salting coefficient k_s. Its physical interpretation is analogous to the corresponding volume change introduced above for equilibria, related to the interactions between enzyme and substrate and the salt solution. If the volume change is negative, it indicates that the effect of the salt upon the rate is enhanced with pressure. If it can be generally said that salting in is followed by increased hydration, and that hydration leads to a volume decrease, then a possible explanation of the salt effect on $\Delta V_s\ddagger$ can be put forward: The salting in hydration of near-surface groups of the enzyme that are normally well hydrated only in the transition state will remove a negative contribution to $\Delta V_s\ddagger$. Similarly, the salting ions will also eliminate a negative contribution to $\Delta G\ddagger$ due to salt effects on $\Delta H\ddagger$ and $\Delta S\ddagger$ (Greaney and Somero, 1980).

Specific ion requirements for many enzymes are very well documented, and it could be expected that such enzymes showed special relations between $\Delta V\ddagger$ and salt concentration. This also seemed to be the case in the experiments of Low and Somero (1975b). From 2.2 M ammonium sulfate, pyruvate kinase with specific cation (K^+, $NH_4{}^+$) binding sites gave much stronger titration effects for dialyzed samples than for freshly diluted samples. This indicated that retention of ions at the binding sites partly inhibited the strong $\Delta V\ddagger$ dependence on salt concentration.

C. Concentrations

1. Enzyme Concentration $[E]_0$

In pressure experiments one has to be very careful about the enzyme concentration. Usually, this concentration must be held within a narrow interval. If $[E]_0$ is too high, the catalytic process may proceed too rapidly to give linear rate curves. The substrate is quickly destroyed and the rate falls off at an early stage in the experiment, perhaps too early to get reliable measurements. It takes some time to put the pressure equipment together, and the measurement should last 5–10 times this dead-time. If the enzyme concentration is too low, the slope of the rate curve may be so small that it falls to practically zero while pressure is applied.

Another reason to be careful with enzyme concentration is the possibility of concentration-dependent dissociation of the enzyme into subunits. The fraction of active enzyme may depend on the concentration of enzyme present. Very low concentration may favor dissociation, while high concentrations may favor association. If the enzyme behaves in this way, the dissociation constant and its pressure dependence should be investigated first.

2. Substrate Concentration $[S]$

With the enzyme concentration fixed, it may be difficult to vary the substrate concentration to any large extent. Maybe only some limited range of substrate concentrations corresponds to linear rate curves. Sometimes the best combination is high substrate concentration and low enzyme concentration. But high substrate concentrations may have side-effects, e.g., inhibition. If this is so, the pressure effect will probably be very concentration-dependent.

D. Gas–Liquid Equilibria at High Pressure

A few gases may be involved in some enzyme reactions, e.g., CO_2 and O_2 as used by carbonic anhydrase and produced by catalase, respectively. If the presence of such dissolved gases affects rates and equilibria at ordinary pressure, their importance will increase at higher pressure. Henry's law says that the partial pressure of a gas above a solution is proportional to its mole fraction in the solution. At high pressure it is more correct to speak of the fugacity f of a gas, instead of partial pressure, in the same sense that one uses activity instead of concentration in solution calculations. In dilute solutions, the fugacity of the dissolved gas is given by

$$\lim_{x \to 0} (f/x) = H_y \qquad (139)$$

where x is the mole fraction and H_y is Henry's constant. According to Henry's law, the solubility of a gas increases with the partial pressure of the same gas. What happens when the gas partial pressure (fugacity) is constant and the total hydrostatic pressure increases is less well known.

Let us write the chemical potential of the gas,

$$\mu = \mu^0 + RT \ln f \tag{140}$$

At equilibrium the chemical potential should be equal in the gas and liquid phases. At uniform temperature and pressure, this leads to the same fugacities in the two phases. In the liquid, the fugacity may be related to the fugacity of a standard state, f^0

$$f = \gamma x f^0 \tag{141}$$

which defines H_y as

$$H_y = \gamma f^0 \tag{142}$$

If the standard state for f^0 is pressure-independent, we have from Eq. (117) that

$$\left(\frac{\partial \ln H_y}{\partial p} \right)_T = \left(\frac{\partial \ln \gamma}{\partial p} \right)_T = \frac{\overline{V}}{RT} \tag{143}$$

By integration of this equation, the relationship between the concentration of the dissolved gas and pressure in dilute solution can be found

$$\ln(f/x) = \ln H_y{}^0 + (1/RT) \int_1^P V \, dp \tag{144}$$

$H_y{}^0$ is Henry's constant in the standard state ($= f^0$). For constant concentration, the following expression results for the fugacity at a pressure p:

$$f_p = f^0 \exp(pV/RT) \tag{145}$$

if the partial molar volume can be considered constant over the pressure range. In some special cases this equation has been found to give results that deviate from the experimental results when an independently determined V is used. However, for most of the slightly soluble gases in water at ordinary temperature and not too high (<1500 bar) pressure the equation gives a good description of the experimental results.

The equation shows that the fugacity (the "activity") of the dissolved gas molecules increases exponentially with pressure. With

TABLE I
Solubility and Partial Molar Volume of Some Gases in Water
at 25°C and Atmospheric Pressure

	H_2	N_2	O_2	CO	CO_2	CH_4	He
Solubility ($x \times 10^4$)	0.15	0.12	0.23	0.18	7	0.24	0.13
Molar volume ($cm^3\ mol^{-1}$)	26	38	32	36	33	37	15

the molar volume of oxygen inserted, both calculations and experiments show that the fugacity for a fixed concentration of oxygen increases about four times when the pressure is increased from 1 to 1000 bar. Table I shows some values of solubility and partial molar volumes for some of the most common gases at atmospheric pressure and 25°C.

E. Buffer Ionization at High Pressure

In high-pressure biochemistry the buffer system should be chosen with care. As should be clear from Section IV,A, ionizations are followed by negative volume changes and therefore increase with pressure. The variation of pH with pressure depends on the volume of ionization ΔV_z, which can differ as much as ~ 30 cm^3 mol^{-1} from one buffer to another.

The ionization process

$$HA \rightarrow H^+ + A^-$$
(146)

is governed by the ionization constant

$$K_z = [H^+][A^-]/[HA]$$
(147)

which gives

$$pH = -\log H^+ = -\log K_z + \log([A^-]/[HA])$$
(148)

Differentiation with respect to pressure yields the ionization volume ΔV_z,

$$\left(\frac{\partial (pH)}{\partial p}\right)_T \simeq -\left(\frac{\partial \log K_z}{\partial p}\right)_T = \frac{\Delta V_z}{2.3RT}$$
(149)

Many buffers, such as phosphate, carbonate, and borate, have ionization volumes ranging from -24 to -32 cm^3 mol^{-1}. These buffers should be used only for reactions that are independent of pH to at least 0.5 unit below the pH at 1 bar. Very pH-sensitive reactions should be carried out in Tris buffer, which has an ionization volume of 1 cm^3 mol^{-1}.

If it is necessary to use one specific buffer, it is of course easy to calculate the pH at a given pressure when ΔV_z is known, and from a calibration curve of reaction rate versus pH correct for the rate change due to the pH perturbation. It seems comforting that there is a buffer such as Tris which keeps a pressure-independent pH. But one must be aware that pOH is not equally independent. Water also ionizes at high pressure, governed by

$$K_w = [H^+][OH^-] \tag{150}$$

with a volume change

$$\frac{\Delta V_w}{2.3RT} = -\left(\frac{\partial \log K_w}{\partial p}\right)_T = \left(\frac{\partial(pH)}{\partial p}\right)_T + \left(\frac{\partial(pOH)}{\partial p}\right)_T \tag{151}$$

If pH now is pressure-independent, this means that

$$\left(\frac{\partial(pOH)}{\partial p}\right)_T = \frac{\Delta V_w}{2.3RT} \tag{152}$$

the hydroxide ion concentration varies with pressure according to the ionization volume of water, -22.1 cm^3 mol^{-1}.

Although this may not be very important for enzyme reactions, it is critical for, for example, base-catalyzed reactions.

F. Isoenzymes

In an organism or a commercial enzyme preparation, there may be several isoenzymes or isozymes, i.e., multiple forms of a given enzyme. They can be detected and separated by gel electrophoresis because of differences in isoelectric pH values, and may differ slightly in amino acid composition. Although they all catalyze the same reaction, they may differ significantly in their K_m and W values, as well as in the specific reaction mechanism. It is clear that if such isozymes contribute more than a few percent to the total amount of enzyme used, this may have important consequences for reactions at high pressure. Probably, one or more of these isozymes may have very different volume changes in intermediate steps of the mechanism. This may then lead to erroneous interpretations of the volume changes of the main enzyme. If the purpose of the investigation is to find the overall response of an enzyme system to pressure *in vivo*, then isoenzymes do no harm. But when the detailed mechanism of a particular enzyme is to be clarified, thorough purification of the enzyme preparation is necessary.

VI. Association–Dissociation Equilibria of Enzymes

A. Protein Volumes

The volume of a protein can be considered to be made up of two or more different contributions—a constitutive or a compositional volume, a conformational volume, and a solvation volume. If the term constitutive is taken to mean the volume determined by the bond lengths and the van der Waals radii of the atoms in the protein, one can describe an envelope enclosing a region approximately equal to the protein volume (Rasper and Kauzmann, 1962). When a protein is in its native conformation, the secondary and tertiary structure will, in general, hinder a perfect packing of the atoms in the molecule. As a result, voids will occur in some parts of the molecule and compressed regions may occur in other parts. The net volume contribution from these voids and compressed regions of the molecule is termed the conformational volume. In addition to these, a solvation volume must be included. This is of course due to the electrostrictive effect of charged groups on the solvent and the volume change associated with the change in solvent structure around nonpolar groups. Furthermore, the solvent molecules may occupy some of the voids in the conformation volume, making these two terms, to some extent, inseparable.

From the volumes of the amino acid residues composing the protein one can calculate what may be termed the compositional volume (Zamyatnin, 1972). To find the volumes of the amino acid residues one can use the partial molar volume of the amino acids and subtract the electrostriction volume due to a pair of charges (NH_4^+ and COO^-). Then one obtains volumes under the condition that the amino acid residues in protein interact with solvent molecules in the same manner as they would interact with amino acids. The volumes will include the effect of interaction of charged and nonpolar groups of the amino acid side chains with solvent while the mutual interactions between the residues in the protein are not taken into account. In this way, the compositional volume should be equal to the sum of the constitutive and the solvation volume. Probably, the former is the more realistic. Concerning the solvation volume including electrostrictive effects, it has been demonstrated that proteins are amphoteric ions. At the isoelectric point large numbers of charged groups are present in a state not very different from that in simple carboxylic acids and amines. Ionization of the carboxyl groups of proteins is accompanied by a volume change similar to that of, for example, acetic acid. The ionization of the amino groups of proteins, however, is somewhat smaller than that observed in simple amines. The volume change associated with the ionization of a carboxylic acid is usually about -11 cm^3 mol^{-1}. If positively charged groups exist close to the group, the volume change is less, only -6 or -7 cm^3 mol^{-1}. If a negatively charged ion is located next to the group, the volume change may be about -20 cm^3 mol^{-1}.

The volume change associated with the protonation of an amino group, where a hydroxyl ion is also formed, is expected to be about -24 cm^3 mol^{-1}. In proteins this reaction involves volume changes that are only about two-thirds of this value, i.e., about -17 cm^3 mol^{-1}. Evidently the amino and imidazole groups in proteins are in a somewhat different environment from that of the same groups in small molecules. Of course, many special cases exist, where amino acids in special positions in the proteins give rise to abnormal volume changes (Rasper and Kauzmann, 1962; Kauzmann et al., 1962). Within a small pH range around neutrality, these reactions are reversible. But titrated to very high or very low pH values the proteins may be irreversibly denatured.

At ordinary pressure all the ionizable groups have their specific pK values and are present in ionized states according to these values. When pressure increases, we can expect all pK values to change, whereby the overall ionized state of the protein is changed. The whole hydration sheet may also be changed around the protein and conformational rearrangements may occur. This fact indicates that the volume of a protein may be very pressure-dependent.

A recent report by Gekko and Noguchi (1979) confirmed that of 14 globular proteins studied, all showed positive compressibilities. The results revealed that a large negative compression of the void compensates a positive compression due to the hydration of the proteins, resulting in a small positive value for β_s.

B. Reversible Changes

Pressure may cause several changes in enzymes, as well as some changes which are not directly associated with the catalytic process. These changes may include conformational changes and subunit dissociation–association processes. Pressures above 4000 bar may induce conformational changes to such an extent that the enzyme in effect becomes irreversibly denatured. These are dealt with in the next section. In this section we will deal with lower pressures and reversible processes, namely, interactions between subunits in quaternary structures. For most multimeric enzymes, the maintenance of the quaternary structure is necessary for proper functioning of the enzymes. Therefore, even if the dissociation process is not directly related to the catalytic process, it certainly has an indirect influence on this process.

In a pressure study involving a multimeric enzyme, it will in general not be possible to decide how much of the effect is due to direct influence of pressure on the catalytic process and how much of it is due to indirect influence through subunit dissociation and accompanying deactivation. Generally, a self-association reaction may be expressed in either of two equivalent forms:

$$A_1 + A_1 \rightleftarrows A_2$$
$$A_1 + A_2 \rightleftarrows A_3$$

$$. \quad . \quad .$$
$$. \quad . \quad .$$
$$. \quad . \quad .$$

$$A_1 + A_{i-1} \rightleftarrows A_i \qquad (153)$$

or

$$2A_1 \leftrightarrows A_2$$
$$3A_1 \leftrightarrows A_3$$

$$. \quad .$$
$$. \quad .$$
$$. \quad .$$

$$iA_1 \leftrightarrows A_i \qquad (154)$$

Any number of species can here be in equilibrium with each other. A_1 represents the monomer, taken to mean the lowest molecular weight species which participates in the equilibria. A_2 is the dimer, A_3 the trimer, A_4 the tetramer, etc.

The equilibria between the species can be represented by the chemical potentials either as

$$\mu_1 + \mu_{i-1} = \mu_i \qquad (155)$$

or

$$i\mu_1 = \mu_i$$

These are related to the molar concentration $[A_i]$ by the expression

$$\mu_i = \mu_i{}^0 + RT \ln \gamma_i[A_i] \qquad (156)$$

such that the free energy change for the ith reaction may be written

$$-\Delta G_i{}^{0\prime} = \mu_i{}^0 - \mu_i{}^0 - \mu_{i-1}{}^0 = RT \ln \frac{[A_1][A_{i-1}]}{[A_i]} \frac{\gamma_1\gamma_{i-1}}{\gamma_i} \qquad (157)$$

where $\Delta G_i^{0\prime}$ now is the standard free energy increment for dissociation. The (dissociation) equilibrium constant is defined as

$$K_{i(i-1)} = \frac{[A_1][A_{i-1}]}{[A_i]} \frac{\gamma_1 \gamma_{i-1}}{\gamma_i} \tag{158}$$

and thus

$$\Delta G_i^{0\prime} = -RT \ln K_{i(i-1)} \tag{159}$$

Corresponding expressions can be obtained when the alternative description (154) is used:

$$\Delta G_i^0 = -RT \ln K_i = -RT \ln \frac{[A_1]^i}{[A_i]} \frac{\gamma_1^i}{\gamma_i} \tag{160}$$

Comparison between these two expressions indicates that

$$K_i = K_{i(i-1)} K_{(i-1)(i-2)} \cdots K_{32} K_{21} \tag{161}$$

and if

$$\left(\frac{\partial \ln K_{i(i-1)}}{\partial p} \right)_T = -\frac{\Delta V}{RT} \tag{162}$$

then

$$\left(\frac{\partial \ln K_i}{\partial p} \right)_T = -\frac{(i-1)\Delta V}{RT} \tag{163}$$

From a study involving several multimeric and monomeric enzymes, Penniston (1971) concluded that subunit dissociation must be considered as the major determinant of the effect of pressure on enzymic systems. This view may not be generally accepted.

Let us consider a dimeric enzyme A_2 which may dissociate into monomers A_1 in the usual way,

$$A_2 \rightleftarrows 2A_1 \tag{164}$$

with the equilibrium constant

$$K_{21} = [A_1]^2 / [A_2] \tag{165}$$

and a dissociation volume

$$\Delta V_{21} = 2V(A_1) - V(A_2) \tag{166}$$

The activation volume associated with the pure catalytic process is $\Delta V\ddagger$, and the rate is simply given by

$$v = k[A_2] \tag{167}$$

that is, only the dimer is catalytically active. If, for simplicity, the activity coefficients are neglected, this equation may now be written

$$v = (k/K_{21})[A_1]^2 \qquad (168)$$

The values of the rate and equilibrium constants at atmospheric pressure are denoted with a superscript 0, and the expression for the pressure-dependent rate will be

$$v_p = \frac{k^0 \exp(-p\Delta V\ddagger/RT)}{K_{21}^0 \exp(-p\Delta V_{21}/RT)} [A_1]^2$$

or

$$v_p = (k^0/K_{21}^0)[A_1]^2 \exp[-p(\Delta V\ddagger - \Delta V_{21})/RT] \qquad (169)$$

From this equation it is obvious that the rate v_p at a pressure p is dependent on both the volumes $\Delta V\ddagger$ and ΔV_{21}. If $\Delta V\ddagger > 0$ and $\Delta V_{21} < 0$, they will co-act, and the rate will be more strongly reduced with pressure than if only the volume change $\Delta V\ddagger$ appeared.

If the rate is dependent on both an active dimer and a less active monomer, the situation is a little more complicated,

$$v = k_1[A_1] + k_2[A_2] \qquad (170)$$

and

$$v_p = (k_2^0/K_{21}^0)[A_1]^2 \exp[-p(\Delta V_2\ddagger - \Delta V_{21})/RT] \\ + k_1^0 \exp(-p\Delta V_1\ddagger/RT)[A_1] \qquad (171)$$

If both monomer and dimer experience the same activation volume, $\Delta V_1\ddagger = \Delta V_2\ddagger = \Delta V\ddagger$, this equation reduces to

$$v_p = [(k_2/K_{21}^0)[A_1] \exp(p\Delta V_{21}/RT) + k_1^0] \exp(-p\Delta V\ddagger/RT)[A_1] \qquad (172)$$

From spectrophotometric or centrifugal measurements it will often be possible to deduce K_{21}^0 and ΔV_{21}. Assume that a fraction α of the total concentration of A_2 is dissociated into A_1,

$$A_2 \leftrightarrows 2A_1 \qquad (173)$$
$$1 - \alpha \qquad 2\alpha$$

If A_1 has a molar extinction coefficient ϵ_1 and A_2 has a molar extinction coefficient ϵ_2, the instantaneous extinction coefficient ϵ is given by

$$\epsilon = [(1 - \alpha)\epsilon_2 + 2\alpha\epsilon_1]/(1 + \alpha) \qquad (174)$$

From measurement of ϵ, we may obtain α as

$$\alpha = (\epsilon - \epsilon_2)/(2\epsilon_1 - \epsilon_2 - \epsilon) \qquad (175)$$

and the equilibrium constant is

$$K_{21} = (2\alpha)^2/(1 - \alpha) \tag{176}$$

Measurements of this constant at various pressures yield ΔV_{21} by means of Eq. (12).

It is well known that appreciable pressures may be generated in a water column in an ultracentrifuge. If the surface of the column is located at a distance r_0 from the rotation axis, where there is a pressure p_0, the pressure p_r at a distance r in the column is (Kim *et al.*, 1977)

$$p_r = p_0 + \tfrac{1}{2}\rho\omega^2(r^2 - r_0^2) \tag{177}$$

when the density of the column is ρ and the angular rotation frequency is ω. From an analysis of the results from the measurements at different frequencies, it is possible to determine ratios between dissociation constants, and thereby the volume change

$$\Delta V_{21} = \frac{2RT}{\rho\omega^2(r^2 - r_0^2)} \ln \frac{K_{21}(r)}{K_{21}(r_0)} \tag{178}$$

From such measurements one then obtains ΔV_{21} and can use this in Eq. (172) to find $\Delta V\ddagger$. Then one is able to decide the different contributions to the pressure effect on the rate from the two cooperating volumes.

Penniston (1971) seems to have found that as a rule the activity of monomeric enzymes is stimulated by pressure while the activity of multimeric enzymes is inhibited. This is certainly true in many cases, but it is doubtful that it is generally valid. Table II includes a compilation of results from experimental studies of enzyme reactions under pressure. It does not imply any critical examination of the experimental works and it is probably not exhaustive. The significance of the activation volumes listed varies very much and has not always been stated. Most often, the significance is in the interval ± 2 to ± 5 cm^3 mol^{-1}. The values of the activation volumes may be very pressure-dependent, but are usually taken at atmospheric pressure, and often they have been calculated by this author from article figures. For these reasons, one should always consult the original article for complete information. Nevertheless, interesting features can be determined from this table. One of the main purposes of the presentation has been to investigate the generality of Penniston's statement that subunit dissociation is favored by pressure and is the major determinant of the pressure effect on enzymes. The number of subunits of a particular enzyme may vary from one species to another and since the numbers given here are taken from the Handbook, they may

TABLE II

The Activation Volumes of Some Enzymes Studied under Pressure

Enzyme	Species[a]	pH	T (°C)	Sub-units[b]	$\Delta V\ddagger$[c] (cm³ mol⁻¹)	References
Acetylcholinesterase	Coryphanea hippurus	7.5	25	4	28	Hochachka et al. (1975a)
	Antimora rostrata	7.5	25	4	21	Hochachka et al. (1975a)
Alcohol dehydrogenase	Baker's yeast	7.5	24	4	1	Morild (1977a)
	Horse liver	7.5	25	2	-28	Morild (1977b)
Alkaline phosphatase	Chicken intestine	—	26	2	14	Penniston (1971)
	Calf intestine	10.4	40	2	24	Mohankumar and Berger (1972)
Argininosuccinate lyase	Bovine liver	7.5	27	2	25	Penniston (1971)
Apartase	Escherichia coli	7.0	37–56	4	5–24	Haight and Morita (1962)
ATPase, mitochondrial	Beef heart	8.75	25	4	30	Penniston (1971)
ATPase ETP$_H$	Beef heart	—	25	4	22	Penniston (1971)
ATPase, mitochondrial	Rat liver	—	27	4	23	Penniston (1971)
ATP-³²P$_i$ exchange	Beef heart	—	27		40	Penniston (1971)
	Rat liver	—	27		42	Penniston (1971)
ATP-phosphoribosyltransferase	E. coli	7.0	25	2	50	E. Morild and H. Kryvi (Unpublished)
Ca²⁺-ATPase	Coryphaneoides	7.6	25		-6	Dreizen and Kim (1971)
Catalase	E. coli	7.0	25	4	5	Morild and Ølmheim (1980)
Chymotrypsin	—	7.7	14.8		-14	Werbin and McLaren (1951a)
Chymotrypsinogen	—	7.6	25		21	Curl and Jansen (1950)
α-Chymotrypsin	—	7.7	20		-6	Neuman and Lockyer (1974), Neuman et al. (1975)
Citrate synthase	Antimora rostrata	7.5	6	2	175	Hochachka et al. (1975b)
	Squid muscle	7.8	25	2	8	Hochachka et al. (1975c)

(Continued)

TABLE II (*Continued*)

Enzyme	Species[a]	pH	T (°C)	Sub-units[b]	ΔV‡[c] (cm³ mol⁻¹)	References
Dextranase	—	5.2	25		-4	Ludwig and Greulich (1978)
Dextransucrase	*Leuconostoc mesenteroides*	5.2	25		-14	Ludwig and Greulich (1978)
Formic dehydrogenase	*E. coli*	5.6–9.0	30		0	Borrowman (1950)
Formic dehydrogenase	—	7.0	27.5		4	Morita (1957)
Fructose diphosphatase	Coryphaneoides	7.7	2	4	0	Hochachka *et al.* (1971a)
	Coryphaneoides	8.75	2	4	-40	Hochachka *et al.* (1971b)
	Salmo gairdneri	8.75	3	4	-40	Hochachka *et al.* (1971b)
Fumarase	Baker's yeast	6.5	23	4	28	Andersen and Broe (1972)
Glyceraldehyde-phosphate dehydrogenase		7.5	25	2	60	Schmid *et al.* (1975)
Glycogen phosphorylase a	Rabbit muscle	6.7	25	4	16	E. Morild and H. Kryvi (Unpublished)
Glycogen phosphorylase b	Rabbit muscle	6.7	25	4	-12	E. Morild and H. Kryvi (Unpublished)
Inorganic pyrophosphatase	*Bacillus stearothermophilus*	7.9	90			Morita and Mathemeier (1964)
Invertase	—	7.1	40	4	-69	Eyring *et al.* (1946)
Invertase	—	7.0	30	4	-8	Eyring *et al.* (1946)
Invertase	—	4.7	25	4	-4	Ludwig and Greulich (1978)
Isocitrate dehydrogenase	*Antimora rostrata*	8.0	26	2	10	Moon and Storey (1975)
	Pig heart	8.0	15	2	26	Low and Somero (1975c)
Lactate dehydrogenase	Rabbit muscle	7.0	25	4	0	Schmid *et al.* (1975)
	Antimora rostrata	7.5	25	4	0	Baldwin *et al.* (1975)
	Trematomus borchgrevinki	7.5	5	4	0	Low and Somero (1975a)
	Halibut	7.5	10	4	5–35	Somero *et al.* (1977)
	Hippoglossus hippoglossus	7.5	5	4	0	Low and Somero (1975a)
	Coryphaneoides	7.5	5	4	-4	Low and Somero (1975a)

Enzyme	Source					Reference
	Rabbit muscle	7.5	15	4	−4	Low and Somero (1975a)
	Rattail	7.5	3	4	0	Moon et al. (1971a)
	Sebastolobus altivelis	7.5	5	4	8	Siebenaller and Somero (1978)
	Sebastolobus alascanus	7.5	5	4	13	Siebenaller and Somero (1978)
	Cypselurus heterurus	8.6	15–25	4	+	Gillen (1971)
	Lagodon rhomboides	8.6	15–25	4	+	Gillen (1971)
	Lepomis macrochirus	8.6	15–25	4	+	Gillen (1971)
	Myctophid	8.6	15–25	4	+	Gillen (1971)
	Photonectes margarita	8.6	15–25	4	+	Gillen (1971)
	Pseudoscopelus sp.	8.6	15–25	4	+	Gillen (1971)
	Scopelogadus mizolepis	8.6	15–25	4	+	Gillen (1971)
Luciferase	—	—	22	1	22	Laidler (1951)
Lysozyme	—	6.2–8.7	35		−10–−24	Holyoke and Johnson (1951)
Malate dehydrogenase	Marine bacterium	8.5	25	4	12	Berger (1974)
	Pig heart	7.8	15	2	−14	Low and Somero (1975c)
	B. stearothermophilus	8.5	28–101	4	—	Morita and Haight (1962)
	E. coli	7.0	27.5	4	2	Morita (1957)
Malic enzyme	Squid muscle	7.0	25	4	28	Storey et al. (1975a,b)
Mitochondrial dehydrogenase	Allomyces macrogynus	7.3–8.5	26.5	2		Hill and Morita (1964)
Myokinase	Rabbit muscle	8.0	23		−7	Penniston (1971)
NADase	Neurospora crassa	—	26		2	Penniston (1971)
Na-K-ATPase	Antimora rostrata	7.1	6	1	−25	Moon (1975)
	Oncorhyncus kisutch	7.1	6	1	47	Moon (1975)
5′ Nucleotidase	Snake venom	8.75	25		−18	Penniston (1971)
Pancreatic amylase	—	—	22	2	−28	Laidler (1951)
Pancreatic lipase	—	—	22		13	Laidler (1951)
Pepsin	—	4.8	35.5		5	Matthews et al. (1940)
Pepsin	—	5.9	25		9	Curl and Jansen (1950)
Pepsin	—	—	22		22	Laidler (1951)

(Continued)

TABLE II (Continued)

Enzyme	Species[a]	pH	T (°C)	Sub-units[b]	$\Delta V t^c$ (cm³ mol⁻¹)	References
Peroxidase	Horseradish	—	23	3	-2	Penniston (1971)
Phosphofructokinase	Oligoplites mundus	8.0	30	4	-15	Moon et al. (1971a)
Phosphofructokinase	Rattail	8.0	3	4	-11	Moon et al. (1971a)
	Rattail	8.0	28	4	-46	Moon et al. (1971a)
	Rattail	8.0	28	4	-46	Moon et al. (1971a)
	Oligoplites mundus	8.0	3	4	0	Moon et al. (1971a)
Pyruvic carboxylase	Rat liver	8.0	27	4	47	Penniston (1971)
Pyruvate kinase	Trematomus borchgrevinki	7.5	5	4	11	Low and Somero (1975a)
	Scorpaena gutatta	7.5	5	4	15	Low and Somero (1975a)
	Coryphaneoides	7.5	5	4	24	Low and Somero (1975a)
	Sebastolobus alt.	7.5	5	4	24	Low and Somero (1975a)
	Cancer magister	7.5	5	4	13*	Low and Somero (1975b)
	Trematomus borchgrevinki	7.5	5	4	-3*	Low and Somero (1975b)
	Skipjack tuna heart	7.5	5	4	15*	Low and Somero (1975b)
	Mugil cephalus	7.5	5	4	15*	Low and Somero (1975b)
	Bufo marinus	7.5	5	4	20*	Low and Somero (1975b)
	Rabbit	7.5	5	4	25*	Low and Somero (1975b)
	Chicken	7.5	5	4	26*	Low and Somero (1975b)
	Coryphaneoides	7.5	3	4	53	Mustafa et al. (1971)
	Coryphaneoides	7.5	3	4	44	Moon et al. (1971c)
	Squid muscle	7.2	25	4	32	Storey and Hochachka (1975b)
	Squid muscle	7.2	10	4	53	Storey and Hochachka (1975b)

Enzyme	Species					Reference
Ribonuclease	*Ectreposebastes imus*	7.5	3	4	29	Moon et al. (1971c)
	Oligoplites mundus	7.5	3	4	47	Moon et al. (1971c)
	Salmo gairdneri	7.5	3	4	47	Moon et al. (1971c)
Ribonuclease	Bovine pancreas	5.0	3.0	2	−20	Williams and Shen (1972)
Salivary amylase	—	—	22	2	−22	Laidler (1951)
Succinate dehydrogenase	*E. coli*	7.0	27.5	2	27	Morita (1957), Morita and ZoBell (1956)
Sucrase	—	—	22		−8	Laidler (1951)
Superoxide dismutase	—	7.0	25	2	30	Morild and Ølmheim (1981)
Trypsin	—	7.3	25–35		−5−−10	Werbin and McLaren (1951b)
Trypsin	—	8.1	25		−36	Fraser and Johnson (1951)
Xanthine oxidase	Buttermilk	7.0	25		−40	Morild and Ølmheim (1981)

[a] The species given are those given in the reference, and may sometimes be mentioned by their English name and sometimes by their Latin name.

[b] The numbers of subunits of the enzymes used are rarely given in the references. Those listed are taken from the "Handbook of Biochemistry and Molecular Biology, Proteins" (G. D. Fasman, ed.), Vol. II, p. 325. CRC, Cleveland, 1975, and may not be valid for the species investigated.

[c] The numbers listed are very suspect, because they may be valid for a narrow pressure range only. Usually, this range is close to one atmosphere. The volume change $\Delta V\ddagger$ may be very pressure-dependent, and may even change sign. The volumes denoted by an asterisk are so-called "structural" volumes.

TABLE III

Protein Subunit Dissociation–Association Volumes and Protein Denaturation Volumes[a]

Protein	Species	Type	pH	T (°C)	ΔV_b†	References
Poly-L-valylribonuclease	—	Aggregation	6.8	39	259	Kettman et al. (1966)
Actin	—	Association	8.1	25	391	Ikkai and Ooi (1966)
Collagen	—	Aggregation	7.5	0	0.8×10^{-3} cm³ g⁻¹	Cassel and Christensen (1967)
Myosin A and B	Rabbit muscle	Precipitation	7.0	5	34–50	Rainford et al. (1965)
β-Casein	Milk	Association	—	—	—	Payens and Heremans (1969)
Lactate dehydrogenase	Pig heart	Dissociation	—	—	—	Jaenicke and Koberstein (1971)
Ribosomes	Rat liver	Association	—	—	—	Hauge (1971)
Ribosomes	Escherichia coli	Association	—	—	—	Van Diggelen et al. (1971)
Ribosomes	Escherichia coli	Association	—	—	—	Schulz et al. (1976a)
Ribosomes	Escherichia coli	Aggregation	7.3	22	240	Schulz et al. (1976b)
Glutamate dehydrogenase	Bovine liver	Association	7.8	25	21	Heremans (1974)
Lactate dehydrogenase	Rabbit muscle	Dissociation	7.0	5	-62	Schmid et al. (1979)
Lactate dehydrogenase	Rabbit muscle	Aggregation	7.0	5	-97	Schmid et al. (1979)
Globulin	Human serum	Denaturation	6.0	65	—	Johnson and Campbell (1946)
Ovalbumin	Egg white	Denaturation	4.8	32	—	Suzuki et al. (1963)
Albumin	Horse serum	Denaturation	4.8	31	—	Suzuki et al. (1963)
Ribonuclease A	—	Denaturation	2.8	42.5	-36	Gill and Glogovsky (1965)
Ribonuclease A	—	Denaturation	2.0	27.5	-43	Brandts et al. (1970)
Ribonuclease A	—	Denaturation	4.0	47.4	-7.5	Brandts et al. (1970)
Chymotrypsinogen	—	Denaturation	2.07	20.5	-110	Hawley (1971)
Chymotrypsinogen	—	Denaturation	2.0	20.5	-94	Hawley and Mitchell (1975)
Chymotrypsinogen	Bovine	Denaturation	7.6	23	-31.2	Li et al. (1976)
Lysozyme	Egg white	Denaturation	7.6	23	-19.7	Li et al. (1976)
Metmyoglobin	—	Denaturation	5.0	20	-100	Zipp and Kauzmann (1973)

[a] The volumes (cm³ mol⁻¹) are sometimes activation volumes and sometimes equilibrium volumes.

not always be valid for the investigated species. For some enzymes, the number does not seem to be known at all. However, there are many counterexamples to Penniston's statement. At least five dimeric and ten tetrameric enzymes have their reaction rates increased by pressure and some monomeric enzymes have their rates decreased. Either the subunits do not always dissociate at high pressure, or this dissociation is not the major determinant of the pressure effect. Possibly, both statements may be true. The pressure effect will also always depend on the catalytic process and the volumes of intermediate complexes. From Table II it is clear that a specific catalytic process may show very different pressure behavior when catalyzed by enzymes from different species, although their subunit number may be the same. The enzymes from deep sea species do not seem to show special preferences as to the nature of their quaternary structure. This has been discussed by Hochachka and Somero (1973).

It is true that most quaternary structures are disrupted by high pressures, but there seems to be no *a priori* reason why this always should be so, and there is hitherto only a vague theoretical basis for this. Penniston argues on the basis of measurements on small model systems, which have shown negative volume changes when isolated hydrophobic and ionic groups are exposed to water. Hvidt (1975), however, has shown that more appropriate model systems suggest small and positive volume changes when buried aliphatic groups are brought into contact with water. The volume changes in some association–dissociation reactions of protein subunits are given in Table III.

C. Irreversible Changes

A reversible equilibrium can usually be detected in either of two ways: (a) In the transition region, the attainment of a time-independent value of the physical observable implies the establishment of a reversible equilibrium. (b) The reversal of solution conditions will lead to a reversal in the value of the physical observable if the transition under study is under thermodynamic control. The observable may be the optical density of the solution, and the solution conditions may be determined by temperature and pressure. An irreversible change is encountered if the physical observable cannot attain a time-independent value in the transition region, nor be reversed by reversal of the solution conditions (Brandts *et al.*, 1970).

Such changes are often detected when enzymes are exposed to pressures above 4000 bar. Very high pressures (> 6000 bar) almost invariably denature the enzymes, while moderate pressures (≈ 1000 bar) may even stabilize the native forms. There exist examples of reversible transitions between native (N) and denatured (D) forms,

$$N \leftrightarrows D \qquad (179)$$

but most such processes, driven to the right with increasing pressure, are irreversible. There has, however, been some discussion as to whether this transition proceeds in one or several steps (Li *et al.*, 1976).

Before detailed enzyme reaction studies are carried out under pressure, a check on irreversible denaturation should be made. If the enzymatic activity is observed at saturating conditions, hysteresis effects should appear in raising and lowering of pressure if denaturation occurs.

The cooperation between all factors determining the tertiary structure of a protein is not well understood. If it were, it would be possible to predict the native structure of a protein molecule from a knowledge of the primary structure, the sequence of the amino acids in the backbone polypeptide chain. It is still generally held that one of the most important determinants of the native protein conformation is the tendency of the backbone to fold in such a way that the nonpolar side chains avoid contact with the surrounding aqueous phase. This has been confirmed by X-ray analysis of protein crystals. Many authors have therefore believed that the denaturation of a protein by pressure was more or less identical to the transfer of the hydrophobic core to the aqueous surroundings, and that this process should be followed by a negative volume change as in model processes. However, many of the denaturation investigations have given results inconsistent with the expected thermodynamic consequences of exposing hydrophobic groups (Bello, 1978). According to the model data, large negative ΔVs, of the order of hundreds of cubic centimeters per mole should be expected. The data of the known denaturation investigations were reviewed some time before, and are now collected in Table III. The volume changes are all rather small and negative. A review of protein denaturation was given by Kauzmann (1974), and later two short discussions were presented by Heremans (1978, 1979).

Results from thermal denaturation and heat capacity studies have shown that the proteins are not necessarily completely unfolded in this process. The volume observations also suggest that the denatured state is not one in which all hydrophobic groups are exposed to water. But the results can also be understood from the effect of close polar and electrostatic groups interacting with the water structure surrounding the hydrophobic groups. The volume change is heavily dependent on the density of the hydrocarbon core and on the concentration and interaction of side chains along the backbone of the protein. The presence of polar and charged groups has been shown to result in smaller magnitude of ΔVs.

D. Time-Dependent Changes

It may happen that the enzyme is denatured during a kinetic pressure experiment. If the denaturation rate is much faster than the reaction rate under study, it is clear that difficulties arise. A denaturation rate much slower than the reaction rate may be tolerated and easily corrected for by means of calibration measurements. When the denaturation rate and the reaction rate are comparable, some exact calculations should be carried through, showing the extent of the influence of the denaturation on the measured reaction rate. This can be done as follows:

Assuming that the protein denaturation and inactivation is first-order, the reduction of the concentration of active enzyme is given by

$$d[E]/dt = -k_D[E] \tag{180}$$

where k_D is some denaturation rate constant to be separately determined. The concentration of active enzyme at a time t, taking the initial concentration to be $[E]_0$, is

$$[E]_t = [E]_0 \exp(-k_D t) \tag{181}$$

It is assumed that the reaction rate v_t is proportional to the concentration $[E]_t$, so that

$$v_t = k[E]_t = k[E]_0 \exp(-k_D t) \tag{182}$$

Now, both rate constants k and k_D must be pressure-dependent and associated with certain activation volumes $\Delta V\ddagger$ and $\Delta V_D\ddagger$, respectively. As usual, these enter in the following way,

$$v_t(p) = k^0 \exp(-p\Delta V\ddagger/RT)[E]_0 \exp[-k_D^0 \exp(-p\Delta V_D\ddagger/RT)t] \tag{183}$$

and we see that the denaturation activation volume factor appears doubly exponentially. From this simple example it is clear that the denaturation activation volume factor has a tremendous influence on the reaction rate. It is also clear that the two activation volumes in general can be determined in two different experiments. First, $\Delta V_D\ddagger$ can be determined in an experiment using only enzyme, and then $\Delta V\ddagger$ can be determined in an experiment using both enzyme and substrate. $\Delta V_D\ddagger$ can be determined, for example, from light scattering at high pressure.

In the simple case of only one active enzyme species, Eq. (182) can be integrated to give the product concentration $[c]$ at the time t at a pressure p (Greulich and Ludwig, 1977)

$$[c]_t = \int_0^t k[E]_0 \exp(-k_D t) \, dt \qquad (184)$$

where k and k_D now have the values at pressure p. The result is

$$[c]_t = (k[E]_0/k_D)[1 - \exp(-k_D t)] \qquad (185)$$

Here, $k[E]_0 k_D^{-1}$ is the final product concentration at $t = \infty$, $[c]_\infty$, and we can write

$$\ln([c]_\infty - [c]_t) = \ln(k[E]_0 k_D^{-1}) - k_D t \qquad (186)$$

In this way k_D may be found from a plot of $\ln([c]_\infty - [c]_t)$ against t. If this is done at several pressures, the pressure dependence of k_D may be evaluated, in this case in one type of experiment.

It is also possible that the denaturation is not complete, i.e., that the denatured state is a less active form of the enzyme. If we denote the active and less active species of the enzyme by E_a and E_1, we would have

$$d[E_a]/dt = -k_D[E_a] \qquad (187)$$

and

$$d[E_1]/dt = k_D n[E_a] \qquad (188)$$

assuming that the process taking place is

$$E_a \rightarrow nE_1 \qquad (189)$$

where n is some small integer. As before

$$[E_a]_t = [E_a]_0 \exp(-k_D t) \qquad (190)$$

and now, also

$$[E_1]_t = n[E_a]_0[1 - \exp(-k_D t)] \qquad (191)$$

In this case, the rate is given by

$$\begin{aligned}
v_t &= k_a[E_a]_t + k_1[E_1]_t \\
v_t &= k_a[E_a]_0 \exp(-k_D t) + nk_1[E_a]_0[1 - \exp(-k_D t)] \\
&= [E_a]_0\{k_a \exp(-k_D t) + nk_1[1 - \exp(-k_D t)]\} \\
&= [E_a]_0[nk_1 + (k_a - nk_1) \exp(-k_D t)] \qquad (192)
\end{aligned}$$

With associated activation volumes $\Delta V_a\ddagger$ and $\Delta V_1\ddagger$, this equation becomes

$$v_t(p) = [E_a]_0\{nk_1^0 \exp(-p\Delta V_1\ddagger/RT) + [k_a^0 \exp(-p\Delta V_a\ddagger/RT)$$
$$- nk_1^0 \exp(-p\Delta V_1\ddagger/RT)] \exp[-k_D^0 \exp(-p\Delta V_D\ddagger/RT)t]\} \quad (193)$$

In this case it will be extremely difficult to calculate $\Delta V_a\ddagger$ and $\Delta V_1\ddagger$. If $\Delta V_a\ddagger = \Delta V_1\ddagger = \Delta V\ddagger$, the equation reduces to

$$v_t(p) = [E_a]_0[nk_1^0 \exp(-p\Delta V\ddagger/RT) + (k_a^0 - nk_1^0) \exp\{-(p\Delta V\ddagger/RT)$$
$$+ [k_D^0 \exp(-p\Delta V_D\ddagger/RT)]t\}]$$
$$= [E_a]_0[nk_1^0 + (k_a^0 - nk_1^0) \exp\{-[k_D^0 \exp(-p\Delta V_D\ddagger/RT)]t\}]$$
$$\times \exp(-p\Delta V\ddagger/RT) \quad (194)$$

with reasonable hope of calculating $\Delta V\ddagger$.

Another possibility is that the denaturation of different forms of the enzyme proceeds with different rate constants (Greulich and Ludwig, 1977). As a simple example, let us consider the two forms: free enzyme E and the substrate complex ES. Now, the denaturation rate of the total enzyme concentration $[E]_{tot} = [E] + [ES]$ is

$$\frac{d[E]_{tot}}{dt} = -k_{D1}[E] - k_{D2}[ES] = -k_D[E]_{tot} \quad (195)$$

The experimental rate constant is then

$$k_D = k_{D1} \frac{[E]}{[E]_{tot}} + k_{D2} \frac{[E]}{[E]_{tot}} \quad (196)$$

From the usual expression of the Michaelis–Menten kinetics

$$v = k[ES] = \frac{k[E]_{tot}}{1 + K_m/[S]} \quad (197)$$

one obtains

$$\frac{[E]_{tot}}{K_m + [S]} = \frac{[E]}{K_m} = \frac{[ES]}{[S]} \quad (198)$$

so that

$$k_D = k_{D1} \frac{K_m}{K_m + [S]} + k_{D2} \frac{[S]}{K_m + [S]}$$
$$= \frac{k_{D1}K_m + k_{D2}[S]}{K_m + [S]} \quad (199)$$

Suppose now that $k_{D2} = 0$, which means that the ES complex is not denatured by pressure. The inverse of k_D is the denaturation half-life τ

$$\tau = \frac{1}{k_D} = \frac{1}{k_{D1}} + \frac{1}{k_{D1}K_m}\,[S]$$

$$\tau = \tau_{D1} + \frac{\tau_{D1}}{K_m}\,[S] \tag{200}$$

which can be plotted against [S] for different pressures. If this becomes a straight line, the assumption that the ES complex is protected may be right. However, this procedure requires that also K_m is evaluated at all pressures.

Another method to determine time-dependent properties is pressure jump relaxation. In a simple equilibrium between two states A and X,

$$A \underset{k_{-1}}{\overset{k_1}{\rightleftarrows}} X \tag{201}$$

It can easily be shown that the relaxation time is defined by

$$\tau = 1/(k_1 + k_{-1}) \tag{202}$$

By disturbing the equilibrium at a pressure p with a sudden release of pressure information can be obtained about association–dissociation processes going on at high rates. This subject is not pursued further here. For more information, see Kegeles and Ke (1975), Davis and Gutfreund (1976), Tai *et al.* (1977), Clegg *et al.* (1975), Kegeles (1978), and Halvorson (1979).

VII. Concluding Remarks

It has been stressed that three main processes in an enzyme system may be affected by pressure, namely, binding, catalysis, and possibly denaturation. None of these processes seems to be characterized by especially large or small volume changes, and accordingly none can be said to be the major determinant of the pressure effect in general. Probably, the relative contribution from these processes to the overall pressure effect will always vary from enzyme to enzyme. However, by using the outlined theoretical framework, it should be possible to determine the main contributions in most cases. Such a separation of the involved volume changes would be most interesting for a future analysis of pressure effects.

Of the activation volumes given in Table II, 70 values are either zero or positive and 31 values are negative. This indicates that two-thirds of all enzymes are inhibited by pressure. However, great care must be taken when general conclusions are drawn.

Many of the investigations are carried out at low temperatures and/or at pH values deviating somewhat from neutrality. It has earlier been emphasized that the influence from factors such as temperature, pressure, pH, concentrations, ionic strength, and salt effects must be considered when pressure effects are discussed. This is because conditions that are optimal to one enzyme may be intolerable to another. After all, we are dealing with enzymes from microorganisms, cold-blooded, and warm-blooded animals. Such complications make comparison of pressure effects on enzymes nearly meaningless.

Perhaps it would be most informative to study the enzymes at their respective physiological conditions, at least for the purpose of predicting the effect of pressure on a living organism. Nevertheless, deliberate changes of conditions may be valuable when pressure is used to probe structure and mechanism. As we learn to understand the role of protein structure in the determination of pressure effects, we may be able to predict the effects on one enzyme from knowledge of the effect on another. Then, we may be able to gain information about pressure effects on human enzymes.

ACKNOWLEDGMENTS

I thank professors A. Ben-Naim, T. Brun, W. Kauzmann, W. J. le Noble, R. C. Neumann, Jr., G. N. Somero, and A. A. Yayanos for stimulating discussions and their hospitality during my stay in the United States.

REFERENCES

Andersen, B., and Broe, E. (1972). *Acta Chem. Scand.* **26**, 3691.
Asano, T., and Le Noble, W. J. (1978). *Chem. Rev.* **78**, 407.
Baldwin, J., Storey, K. B., and Hochachka, P. W. (1975). *Comp. Biochem. Physiol. B* **52**, 19.
Bello, J. (1978). *J. Phys. Chem.* **82**, 1607.
Ben-Naim, A. (1980). "Hydrophobic Interactions." Plenum, New York.
Berger, L. R. (1974. *Proc. Int. Conf. High Pressure, 4th, 1974*, Kyoto, p. 639.
Borrowman, S. R. (1950). Ph.D. Thesis, University of Utah, Salt Lake City.
Brandts, J. F., Oliveira, R. J., and Westort, C. (1970). *Biochemistry* **9**, 1038.
Buchanan, J., and Hamann, S. D. (1953). *Trans. Faraday Soc.* **49**, 1425.
Cassel, J. M., and Christensen, R. G. (1967). *Biopolymers* **5**, 431.
Clegg, R. M., Elson, E. L., and Maxfield, B. W. (1975). *Biopolymers* **14**, 883.
Curl, A. L., and Jansen, E. F. (1950). *J. Biol. Chem.* **185**, 713.
Dack, M. R. J. (1976). *Aust. J. Chem.* **29**, 779.
Davis, J. S., and Gutfreund, H. (1976). *FEBS Lett.* **72**, 199.
Desrosiers, N., and Desnoyers, J. E. (1976). *Can. J. Chem.* **54**, 3800.
Distéche, A. (1972). *Symp. Soc. Exp. Biol.* **26**, 27.

Dreizen, P., and Kim, H. D. (1971). *Am. Zool.* **11**, 513.

Drude, P., and Nernst, W. (1894). *Z. Phys. Chem.* **15**, 79.

Eckert, C. A. (1972). *Annu. Rev. Phys. Chem.* **23**, 239.

Eyring, H., Johnson, F. H., and Gensler, R. L. (1946). *J. Phys. Chem.* **50**, 453.

Fraser, D., and Johnson, F. H. (1951). *J. Biol. Chem.* **190**, 417.

Gekko, K., and Noguchi, H. (1979). *J. Phys. Chem.* **83**, 2706.

Gill, S. J., and Glogovsky, R. L. (1965). *J. Phys. Chem.* **69**, 1515.

Gillen, R. G. (1971). *Mar. Biol.* **8**, 7.

Glasstone, S., Laidler, K. J., and Eyring, H. (1941). "The Theory of Rate Processes." McGraw-Hill, New York.

Gordon, J. E. (1975). "The Organic Chemistry of Electrolyte Solutions." Wiley, New York.

Greaney, G., and Somero, G. N. (1980). *Biochemistry* **24**, 5322.

Greulich, K. O., and Ludwig, H. (1977). *Biophys. Chem.* **6**, 87.

Grob, C. A. (1976). *J. Am. Chem. Soc.* **98**, 920.

Haight, R. D., and Morita, R. Y. (1962). *J. Bacteriol.* **83**, 112.

Halvorson, H. R. (1979). *Biochemistry* **18**, 2480.

Hamann, S. D. (1974). *In* "Modern Aspects of Electrochemistry" (B. E. Conway and J. O'M. Bockris, eds.), No. 9, p. 47. Plenum, New York.

Hardmann, M. J., Coates, J. H., and Gutfreund, H. (1978). *Biochem. J.* **171**, 215.

Hauge, J. G. (1971). *FEBS Lett.* **17**, 168.

Hawley, S. A. (1971). *Biochemistry* **10**, 2436.

Hawley, S. A., and Mitchell, R. M. (1975). *Biochemistry* **14**, 3257.

Hepler, L. G. (1965). *J. Phys. Chem.* **69**, 965.

Heremans, K. (1974). *Proc. Int. Conf. High Pressure, 4th, 1974* p. 627.

Heremans, K. (1978). *In* "High Pressure Chemistry" (H. Kelm, ed.), p. 467. Reidel Publ., Dordrecht, Netherlands.

Heremans, K. (1979). *In* "High Pressure Science and Technology" (K. D. Timmerhaus and M. S. Barber, eds.), Vol. 1, p. 699. Plenum, New York.

Hey, M. J., Clough, J. M., and Taylor, D. J. (1976). *Nature (London)* **262**, 807.

Hill, E. P., and Morita, R. Y. (1964). *Limnol. Oceanogr.* **9**, 243.

Hochachka, P. W. (1971). *Am. Zool.* **11**, 425.

Hochachka, P. W., ed. (1976). "Biochemistry at Depth." Pergamon, Oxford.

Hochachka, P. W., and Somero, G. N. (1973). "Strategies of Biochemical Adaptation." Saunders, Philadelphia, Pennsylvania.

Hochachka, P. W., Schneider, D. E., and Kuznetsov, A. (1970). *Mar. Biol.* **7**, 285.

Hochachka, P. W., Berisch, H. W., and Marcus, F. (1971a). *Am. Zool.* **11**, 437.

Hochachka, P. W., Schneider, D. E., and Moon, T. W. (1971b). *Am. Zool.* **11**, 479.

Hochachka, P. W., Storey, K. B., and Baldwin, J. (1975a). *Comp. Biochem. Physiol. B* **52**, 13.

Hochachka, P. W., Storey, K. G., and Baldwin, J. (1975b). *Comp. Biochem. Physiol. B* **52**, 43.

Hochachka, P. W., Storey, K. B., and Baldwin, J. (1975c). *Comp. Biochem. Physiol. B* **52**, 193.

Holyoke, E. D., and Johnson, F. H. (1951). *Arch. Biochem. Biophys.* **31**, 41.

Hvidt, A. (1975). *J. Theor. Biol.* **50**, 245.

Ikkai, T., and Ooi, T. (1966). *Science* **152**, 1756.

Jenner, G. (1975). *Angew. Chem.* **87**, 186.

Johnson, F. H., and Campbell, D. H. (1946). *J. Biol. Chem.* **163**, 689.

Johnson, F. H., and Eyring, H. (1970). *In* "High Pressure Effects on Cellular Processes" (A. M. Zimmerman, ed.), p. 2. Academic Press, New York.

Johnson, F. H., Eyring, H., and Stover, B. J. (1974). "The Theory of Rate Processes in Biology and Medicine." Wiley, New York.

Kauzmann, W. (1974). *Proc. Int. Conf. High Pressure, 4th, 1974* p. 619.

Kauzmann, W., Bodanzsky, A., and Rasper, J. (1962). *J. Am. Chem. Soc.* **84**, 1777.

Kegeles, G. (1978). *In* "Methods in Enzymology" (C. H. W. Hirs and S. N. Timasheff, eds.), Vol. 48, p. 308. Academic Press, New York.

Kegeles, G., and Ke, C. (1975). *Anal. Biochem.* **68**, 138.

Kettman, M. S., Nishikawa, A. H., Morita, R. Y., and Becker, R. R. (1966). *Biochem. Biophys. Res. Commun.* **22**, 262.

Kim, H., Deonier, R. C., and Williams, J. W. (1977). *Chem. Rev.* **7**, 659.

Kohnstam, G. (1970). *Prog. React. Kinet.* **5**, 335.

Krug, R. R., Hunter, W. G., and Grieger, R. A. (1976). *J. Phys. Chem.* **80**, 2335.

Kuntz, I. D., and Kauzamnn, W. (1974). *Adv. Protein Chem.* **28**, 239.

Laidler, K. J. (1951). *Arch. Biochem.* **30**, 226.

Laidler, K. J., and Bunting, P. S. (1973). "The Chemical Kinetics of Enzyme Action." Oxford Univ. Press (Clarendon), London and New York.

le Noble, W. J. (1967). *Prog. Phys. Org. Chem.* **5**, 207.

le Noble, W. J., Guggisberg, H., Asano, T., Cho, L., and Grob, C. A. (1976). *J. Am. Chem. Soc.* **98**, 920.

Li, T., Hook, J. W., Drickamer, H. G., and Weber, G. (1976). *Biochemistry* **15**, 5571.

Long, F. A., and McDevit, W. F. (1952). *Chem. Rev.* **51**, 119.

Low, P. S., and Somero, G. N. (1975a). *Comp. Biochem. Physiol. B* **52**, 67.

Low, P. S., and Somero, G. N. (1975b). *Proc. Natl. Acad. Sci. U.S.A.* **72**, 3014.

Low, P. S., and Somero, G. N. (1975c). *Proc. Natl. Acad. Sci. U.S.A.* **72**, 3305.

Ludwig, H., and Greulich, K. O. (1978). *Biophys. Chem.* **8**, 163.

McDevit, W. F., and Long, F. A. (1952). *J. Am. Chem. Soc.* **74**, 1773.

Macdonald, A. G. (1975). "Physiological Aspects of Deep Sea Biology." Cambridge Univ. Press, London and New York.

Masterton, W. L., and Lee, T. P. (1970). *J. Phys. Chem.* **74**, 1776.

Matthews, J. E., Dow, R. B., and Anderson, A. K. (1940). *J. Biol. Chem.* **135**, 597.

Millero, F. J. (1971). *Chem. Rev.* **71**, 147.

Millero, F. J. (1972). *In* "Water and Aqueous Solutions" (R. A. Horne, ed.), p. 519. Wiley, New York.

Mohankumar, K. C., and Berger, L. R. (1972). *Anal. Biochem.* **49**, 336.

Moon, T. W. (1975). *Comp. Biochem. Physiol. B* **52**, 59.

Moon T. W., and Storey, K. B. (1975). *Comp. Biochem. Physiol. B* **52**, 51.

Moon, T. W., Mustafa, T., and Hochachka, P. W. (1971a). *Am. Zool.* **11**, 467.

Moon, T. W., Mustafa, T., and Hochachka, P. W. (1971b). *Am. Zool.* **11**, 473.

Moon, T. W., Mustafa, T., and Hochachka, P. W. (1971c). *Am. Zool.* **11**, 502.

Morild, E. (1977a). *Biophys. Chem.* **6**, 351.

Morild, E. (1977b). *J. Phys. Chem.* **81**, 1162.

Morild, E. (1981). *Acta Chem. Scand., Ser. A.* In press.

Morild, E., and Aksnes, G. (1981). *Acta Chem. Scand., Ser. A.* In press.

Morild, E., and Larsen, B. (1978). *J. Chem. Soc., Faraday Trans. 2* **74**, 1778.

Morild, E., and Ølmheim, J. E. (1981). To be published.

Morild, E., and Tvedt, I. (1978). *Acta Chem. Scand., Ser. B* **32**, 593.

Morita, R. Y. (1957). *J. Bacteriol.* **74**, 251.

Morita, R. Y., and Haight, R. D. (1962). *J. Bacteriol.* **83**, 1341.

Morita, R. Y., and Mathemeier, P. F. (1964). *J. Bacteriol.* **88**, 1667.

Morita, R. Y., and ZoBell, C. E. (1956). *J. Bacteriol.* **71**, 668.

Mustafa, T., Moon, T. W., and Hachachka, P. W. (1971). *Am. Zool.* **11**, 451.

Neuman, R. C., and Lockyer, G. (1974). *Proc. Int. Conf. High Pressure, 4th, 1974* p. 635.

Neuman, R. C., and Pankratz, R. P. (1973). *J. Am. Chem. Soc.* **95**, 8372.

Neuman, R. C., Owen, D., and Lockyer, G. D. (1975). *J. Am. Chem. Soc.* **98**, 2982.

Oakenfull, D., and Fenwick, D. E. (1977). *Aust. J. Chem.* **30**, 741.

Payens, T. A. J., and Heremans, K. (1969). *Biopolymers* **8**, 335.

Penniston, J. T. (1971). *Arch. Biochem. Biophys.* **142**, 322.

Rainford, P., Noguchi, H., and Morales, M. (1965). *Biochemistry* **4**, 1958.

Rasper, J., and Kauzmann, W. (1962). *J. Am. Chem. Soc.* **84**, 1771.

Schmid, G., Lüdemann, H. D., and Jaenicke, R. (1975). *Biophys. Chem.* **3**, 90.

Schmid, G., Lüdemann, H. D., and Jaenicke, R. (1979). *Eur. J. Biochem.* **97**, 407.

Schulz, E., Jaenicke, R., and Knoche, W. (1976a). *Biophys. Chem.* **5**, 253.

Schulz, E., Lüdemann, H. D., and Jaenicke, R. (1976b). *FEBS Lett.* **64**, 40.

Siebenaller, J., and Somero, G. N. (1978) *Science* **201**, 255.

Somero, G. N., Neubauer, M., and Low, P. S. (1977). *Arch. Biochem. Biophys.* **181**, 438.

Storey, K. B., and Hochachka, P. W. (1975a). *Comp. Biochem. Physiol. B* **52**, 169.

Storey, K. B., and Hochachka, P. W. (1975b). *Comp. Biochem. Physiol. B* **52**, 187.

Storey, K. B., Baldwin, J., and Hochachka, P. W. (1975a). *Comp. Biochem. Physiol. B* **52**, 165.

Storey, K. B., Mustafa, T., and Hochachka, P. W. (1975b). *Comp. Biochem. Physiol. B* **52**, 183.

Stranks, D. R. (1974). *Pure Appl. Chem.* **38**, 303.

Suzuki, K., and Taniguchi, Y. (1972). *Symp. Soc. Exp. Biol.* **26**, 103.

Suzuki, K., Miyosawa, Y., and Suzuki, C. (1963). *Arch. Biochem. Biophys.* **101**, 225.

Tai, M. S., Kegeles, G., and Huang, C. (1977). *Arch. Biochem. Biophys.* **180**, 537.

Terasawa, S., Itsuki, H., and Arakawa, S. (1975). *J. Phys. Chem.* **79**, 2345.

Van Diggelen, O. P., Ostrom, H., and Bosch, L. (1971). *FEBS Lett.* **19**, 115.

Weale, K. E. (1967). "Chemical Reactions at High Pressures." Spon, London.

Werbin, H., and McLaren, A. D. (1951a). *Arch. Biochem. Biophys.* **31**, 285.

Werbin, H., and McLaren, A. D. (1951b). *Arch. Biochem. Biophys.* **32**, 325.

Williams, R. K., and Shen, C. (1972). *Arch. Biochem. Biophys.* **152**, 606.

Zamyatnin, A. A. (1972). *Prog. Biophys. Mol. Biol.* **24**, 109.

Zipp, A., and Kauzmann, W. (1973). *Biochemistry* **12**, 4217.

THE ANATOMY AND TAXONOMY OF PROTEIN STRUCTURE

By JANE S. RICHARDSON[1]

Department of Anatomy, Duke University, Durham, North Carolina[2]

[1] The copyright for the schematic backbone drawings in this article (Figs. 1, 14, 52, 53, 62, 71–86, 87b, 88, 89c, 90b and d, 92b and c, 94b, 96b, 102b, 103b, 104b, and 105–108) is held by Jane S. Richardson. Upon application to her, she will make these figures available, free of charge, for nonprofit scientific or educational use.

[2] Mailing address: 213 Medical Sciences IA, Duke University, Durham, North Carolina 27710.

167

ADVANCES IN
PROTEIN CHEMISTRY, Vol. 34

PROTEIN ANATOMY

I. Background

A. Introduction

X-Ray crystallography is a technically sophisticated but conceptually simple-minded method with the great advantage that, to a first approximation, its results are independent of whatever preconcep-

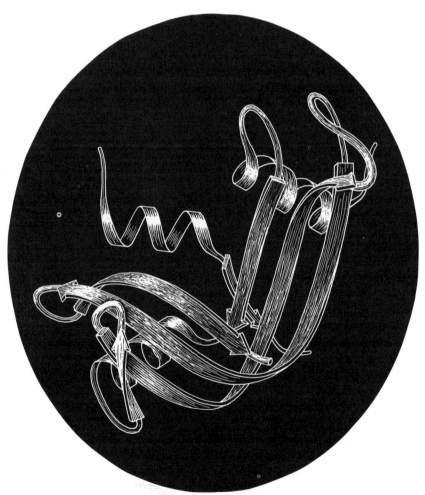

FIG. 1. Schematic drawing of the polypeptide backbone of ribonuclease S (bovine pancreatic ribonuclease A cleaved by subtilisin between residues 20 and 21). Spiral ribbons represent α-helices and arrows represent strands of β sheet. The S peptide (residues 1–20) runs down across the back of the structure.

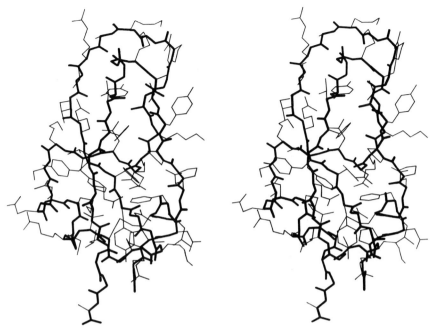

FIG. 2. Stereo drawing of all nonhydrogen atoms of basic pancreatic trypsin inhibitor. The main chain is shown with heavy lines and side chains with thin lines.

tions we bring to the task. This was very fortunate in the case of proteins, because it is unlikely that we could ever have successfully made the jump to such elegant and complex structures as those shown in Figs. 1 and 2 if we had been obliged to rely on more logical and indirect methods. For small inorganic and organic molecules indirect inference had succeeded magnificently, so that X-ray crystallography provided no startling revelations but only a prettier and more accurate picture of what was already known. However, even after knowing what the answer should look like for proteins, 20 years of effort has failed to derive three-dimensional protein structures from spectroscopic and chemical data or from theoretical calculations.

Before the first X-ray results, protein structure was visualized in terms of analogies based on chemistry and mathematics. The models proposed were relatively simple and extremely regular, such as geometrical lattice cages (Wrinch, 1937), repeating zigzags (Astbury and Bell, 1941), and uniform arrays of parallel rods (Perutz, 1949). In light of these very reasonable expectations, the low-resolution X-ray structure of myoglobin (Kendrew et al., 1958) came as a considerable shock. Kendrew, in describing the low-resolution model (see Fig. 3), says "Perhaps the most remarkable features of the molecule are its complexity and its lack of symmetry. The arrangement seems to be

FIG. 3. Electron density contours of sperm whale myoglobin at 6 Å resolution.

almost totally lacking in the kind of regularities which one instinc-
tively anticipates." Perutz was even more outspoken about his initial
disappointment: "Could the search for ultimate truth really have revealed so hideous and visceral-looking an object?" (Perutz, 1964).

In the last 20 years we have learned to appreciate the aesthetic
merits of protein structure, but it remains true that the most apt meta-
phors are biological ones. Low-resolution helical structures are indeed "visceral," and high-resolution electron-density maps (for instance, see Fig. 13) are like intricate, branched coral, intertwined but
never touching. β sheets do not show a stiff repetitious regularity but
flow in graceful, twisting curves, and even the α-helix is regular more
in the manner of a flower stem, whose branching nodes show the
influences of environment, developmental history, and the evolution
of each separate part to match its own idiosyncratic function.

The vast accumulation of information about protein structures provides a fresh opportunity to do descriptive natural history, as though
we had been presented with the tropical jungles of a totally new
planet. It is in the spirit of this new natural history that we will attempt to investigate the anatomy and taxonomy of protein structures.

B. Amino Acids and Backbone Conformation

A protein, of course, is a polypeptide chain made up of amino acid
residues linked together in a definite sequence. Amino acids are
"handed" (except for glycine, in which the normally asymmetric α-
carbon has two hydrogens), and naturally occurring proteins contain
only L-amino acids. That handedness has far-reaching effects on protein structure, as we shall see, and it is very useful to be able to distin-

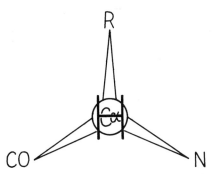

FIG. 4. The "corncrib": A mnemonic for the handedness of atomic positions around the asymmetric α-carbon in naturally occurring L-amino acids. Looking down on the α-carbon from the direction of the hydrogen atom, the other branches should be CO—R—N, reading clockwise (i.e., carbonyl, side-chain R, then main-chain N).

guish the correct form easily. A simple mnemonic for that purpose is the "corncrib," illustrated in Fig. 4. Looking from the hydrogen direction, the other substituents around the α-carbon should read CO—R—N in a clockwise direction (R is the side chain). Threonine and isoleucine have handed β-carbons. A mnemonic for both of them is that if you are standing on the backbone with the hydrogen direction of the β-carbon behind you, then your left arm is the heavier of the two branches (the longer chain in Ile and the oxygen in Thr).

The sequence of side chains determines all that is unique about a particular protein, including its biological function and its specific three-dimensional structure. Each of the side groups has a certain "personality" which it contributes to this task. Histidine is the only side chain that titrates near physiological pH, making it especially useful for enzymatic reactions. Lys and Arg are normally positively charged and Asp and Glu are negatively charged; those charges are very seldom buried in protein interiors except when they are serving some special purpose, as in the activity and activation of chymotrypsin (Blow et al., 1969; Wright, 1973). Asparagine and glutamine have interesting hydrogen-bonding properties, since they resemble the backbone peptides. The hydrophobic residues provide a very strong driving force for folding, through the indirect effect of their ceasing to disrupt the water structure once they are buried (Kauzmann, 1959); they also, however, affect the structure in a highly specific manner because their extremely varied sizes and shapes must all be fitted together in very efficient packing (Lee and Richards, 1971). Proline has stronger stereochemical constraints than any other residue, with only one instead of two variable backbone angles, and it lacks the normal backbone NH for hydrogen bonding. It is both disruptive to regular secondary structure and also good at forming turns in the poly-peptide chain, so that in spite of its hydrophobicity it is usually found at the edge of the protein. Glycine has three different unique capabilities: as the smallest side group (only a hydrogen), it is often required where main chains must approach each other very closely; Gly can assume conformations normally forbidden by close contacts of the β-carbon; and it is more flexible than other residues, making it valuable for pieces of backbone that need to move or hinge.

The basic geometry of amino acid residues is quite well determined from small-molecule crystal structures (see Momany *et al.*, 1975). In terms of the accuracy of protein structure determinations, all of the bond lengths are invariant. Bond angles are also essentially invariant, except perhaps for τ, the backbone N—Cα—C angle (see Fig. 5). The α-carbon is tetrahedral, which would give 110°, but there are indications from accurately refined protein structures (e.g., Deisenhofer and Steigemann, 1975; Watenpaugh *et al.*, 1979) that τ can some-

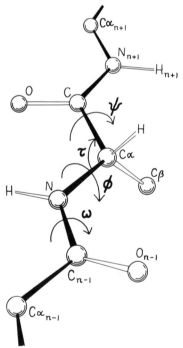

FIG. 5. A key to standard nomenclature for the atoms and the more important bond angles and dihedral angles along the polypeptide backbone. Atoms of the central residue are without subscripts.

times stretch to larger values in order to accommodate other strains in
the structure. The dihedral angle ω at the peptide is very close to
180° (producing a trans, planar peptide with the neighboring α-
carbons and the N, H, C, and O between them all lying in one plane),
but there is evidence that ω can also vary slightly in real structures.
Cis peptides, with ω = 0°, can occur perhaps 25% of the time in prolines but essentially never for any other residue. The proline ring is
not quite flat, and occasionally protein structures are now being refined accurately enough to determine the direction of ring pucker
(e.g., Huber et al., 1974). In the following discussions we will for the
most part ignore possible effects such as proline ring pucker and variation in τ and ω.

The remaining dihedral angles are the source of essentially all the
interesting variability in protein conformation. As shown in Fig. 5,
the backbone dihedral angles are φ and ψ in sequence order on either
side of the α-carbon, so that φ is the dihedral angle around the N—Cα
bond and ψ around the Cα—C bond. The side chain dihedral angles
are χ_1, χ_2, etc. The four atoms needed to define each dihedral angle
are taken either along the main backbone or out the side chain, in sequence order: N, Cα, C, N define ψ and N, Cα, Cβ, Cγ define χ_1. The
sign, or handedness, of any dihedral angle is defined as shown in Fig.
6: looking directly down the central bond (from either direction) and
using the front bond as a stationary reference to define 0°, then the dihedral angle is positive if the rear bond is clockwise from 0° and negative if it is counterclockwise. The choice of reference atom
(IUPAC-IUB, 1970) for side chain branches is made according to consistent chemical conventions, but it produces confusing results for the

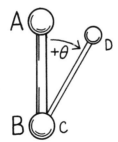

FIG. 6. Standard convention for defining dihedral angles, using four atoms in sequence order either along the main chain or along the major branch of the side chain.
Looking along the bond between the central two atoms (in either direction), use the end
atom in front as the 0° angle reference. Then the dihedral angle (marked θ) is measured
by the relative position of the end atom in back (positive if clockwise, negative if counterclockwise) with respect to the reference atom position.

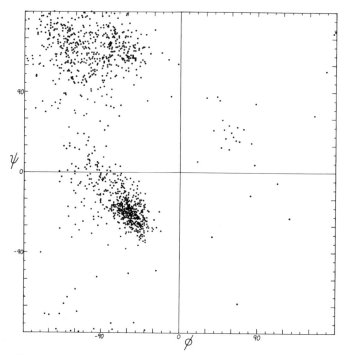

FIG. 7.   Plot of main chain dihedral angles ϕ and ψ (see Fig. 5 for definition) experimentally determined for approximately 1000 nonglycine residues in eight proteins whose structures have been refined at high resolution (chosen to be representative of all categories of tertiary structure).

branched β-carbon residues since χ_1 of 180° for Val puts its two Cγ atoms in the same position that the branches of Ile or Thr would occupy for $\chi_1 = -60°$.

The parameters ϕ and ψ are the most important ones.   An extremely useful device for studying protein conformation is the Ramachandran plot (Ramachandran *et al.*, 1963) which plots ϕ and ψ.   Figure 7 plots ϕ vs ψ for each nonglycine residue in eight of the most accurately determined protein structures (also picked to be representative of the various structure categories); Fig. 8 plots the glycine ϕ vs ψ from 20 proteins.   The glycine plot is approximately symmetrical around the center, because glycine can adopt both right-handed and left-handed versions of any allowed conformation; however, there are some deviations from that symmetry, such as the different shapes and positions of the left- and right-handed α clusters.

[Cautionary note: the conventions for naming and displaying ϕ and ψ have been changed twice.   The original version in Ramachandran *et*

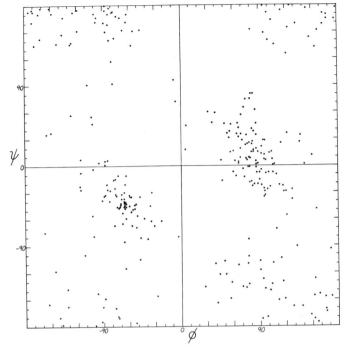

FIG. 8. Plot of main chain dihedral angles ϕ and ψ experimentally determined for the glycines in 20 high-resolution protein structures.

al. (1963) defined ψ (called ϕ') in the same way as it is now used but defined ϕ as $\phi + 180°$, so that the Ramachandran plot (with 0°,0° at the bottom left) had the α-helix in the upper left quadrant. Between 1966 and 1970, Ramachandran plots looked the same way they do now, but 0°,0° was at the bottom left and the numerical values of ϕ and ψ both differed by 180° from the current convention (e.g., Watson, 1969; Dickerson and Geis, 1969). Now 0°,0° is in the center of the ϕ,ψ plot, so that taking the mirror image of a conformation corresponds to inverting the numerical ϕ,ψ values through zero. For the current set of conventions, refer to the IUPAC-IUB Commission on Biochemical Nomenclature (1970).]

Theoretical calculations can provide a rather good understanding of these observed ϕ,ψ distributions. The first approach is to calculate what conformations are allowed without bump of hardsphere atoms of van der Waals radius. Figure 9 is a "derivation diagram" of the allowed regions, showing which pair of atoms is responsible for each forbidden zone (from Mandel *et al.*, 1977). Four large regions sym-

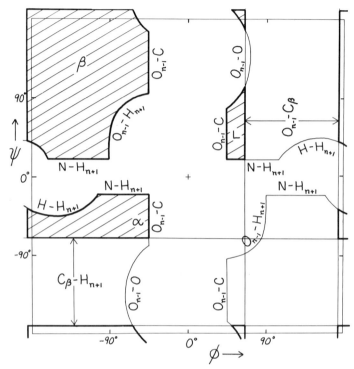

FIG. 9. "Derivation diagram" showing which atomic collisions (using a hard-sphere approximation) produce the restrictions on main chain dihedral angles ϕ and ψ. The crosshatched regions are allowed for all residues, and each boundary of a prohibited region is labeled with the atoms which collide in that conformation. Atom names are the same as in Fig. 5. Adapted from Mandel *et al.* (1977), with permission.

metrical around 0°,0° are allowed for glycine. The presence of a β-carbon produces a bump with the carbonyl oxygen of residue $n - 1$ that is a function only of ϕ and not ψ and a bump with the NH of residue $n + 1$ that depends only on ψ and not ϕ. When the resulting vertical and horizontal disallowed strips are removed from the Ramachandran plot, one is left with fairly large regions around the β and the right-handed α conformations and a small region of left-handed α (Fig. 9). This outline fits the distribution observed in proteins (Fig. 7) fairly well, except for the rather frequent occurrence of residues in the bridge between the α and β regions. That bridge region becomes allowed if the C—Cα—N bond angle τ at the α-carbon is increased (e.g., Ramachandran and Sasisekharan, 1968), or if the grazing bump between N_i and H_{i+1} is otherwise softened. Detailed conformational energy calculations for alanine dipeptides (e.g., Maigret *et al.*, 1971;

Zimmerman and Scheraga, 1977a) can reproduce the observed distribution in most respects, in spite of omission of all long-range and medium-range interactions.

Another useful type of representation for protein structures is the diagonal plot. It is a matrix with the amino acid sequence number along both axes, in which either distance between the respective α-carbons or contact between the respective residues is plotted for each possible pair of residues (see Fig. 10). The diagonal plot is probably the most successful method yet devised of quantitatively mapping the

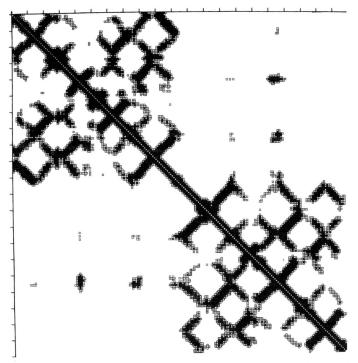

FIG. 10. Diagonal plot of close Cα–Cα distances for an immunoglobulin light chain. Sequence number increases across the top and down the side of the square matrix, and a matrix cell is darkened if the two α-carbons whose intersection it represents are sufficiently close together in the three-dimensional structure. The matrix is exactly symmetrical across the diagonal. The upper left quadrant shows contacts internal to the variable domain, the lower right quadrant those internal to the constant domain, and the off-diagonal quadrants show the rather sparse contacts between the two domains. Bands perpendicular to the diagonal are produced by chain segments running antiparallel to each other (in this case, β strands). Diagonal plot courtesy of Michael Liebman.

chain folding in three dimensions onto the plane (stereo drawings are neither rigorously two-dimensional nor explicitly quantitative). The large-scale structural features (except for handedness, and perhaps twist) have their counterpart in the diagonal plot: a helix gives a pronounced thickening along the diagonal, for instance, and a pair of antiparallel β strands produce a narrow stripe perpendicular to the diagonal. The appearance of each of the major structure types discussed in Sections III,B–E is fairly clear on diagonal plots, although less distinctive than in three dimensions. For example, the division into two well-separated domains with similar internal structures is extremely obvious in Fig. 10; the first layer of squares out from the diagonal indicates antiparallel organization (fairly narrow bands for β structure, as in this case, and wider bands if the elements were α-helices), the strong bands in the second layer are produced by the Greek key topology (see Section III,B), and the third layer is produced by closure of the barrel. Diagonal plot representations provide crucial simplifications of a number of computational problems (e.g., Kuntz *et al.*, 1976; Tanaka and Scheraga, 1977; Remington and Matthews, 1978), and they seem to be an especially useful tool for those people who are more at home with an algebraic than with a geometrical representation.

C. Levels of Error

The following analysis and discussion of protein structure is based almost exclusively on the results of three-dimensional X-ray crystallography of globular proteins. In addition, one structure is included that was determined by electron diffraction (purple membrane protein), and occasional reference is made to particularly relevant results from other experimental techniques or from theoretical calculations. Even with this deliberately restricted viewpoint the total amount of information involved is immense. Millions of independent parameters have been determined by protein crystallography, and the relationships among almost any subset of them are of potential interest. A major aim of the present study is to provide a guide map for use in exploring this forest of information.

One issue which needs to be discussed before starting the analysis is the problem of evaluating levels of probable error. X-Ray crystallography has a relatively high degree of inherent reliability, because it basically amounts merely to obtaining a picture of the protein. Serious mistakes or experimental difficulties usually produce recognizably unintelligible garbage rather than misleading artifacts. However, there are many minor inaccuracies or problems of interpretation that can affect reliability of the final coordinates. Also, there is now an enormous difference in accuracy between the best and the worst-determined structures: increasing numbers of large proteins are being solved for which the ordered diffraction pattern may not extend beyond 3.5 Å, while on the other hand it is now not uncommon for a protein structure to receive exhaustive least-squares refinement out to 1.5 Å resolution. The problem of valid error estimation has not yet been solved even for a given refinement technique, mainly because it is difficult to estimate the likelihood of occasional large mistakes in assigning starting coordinates which might not be correctible by refinement. There are now a few cases in which the same structure was independently refined by different methods from independently determined starting coordinates (e.g., Huber *et al.*, 1974; and Chambers and Stroud, 1979, for trypsin), or where two subunits related by noncrystallographic symmetry were refined independently (e.g., Mandel *et al.*, 1977, for cytochrome c), so that we may soon develop some empirically based error-estimation procedures. So far the main conclusions from such comparisons are that temperature factors are good indicators of relative error level within a structure and that the standard deviation between independent, well-refined structures is very small (perhaps 0.1 or 0.2 Å) for at least 90 or 95% of the atoms, but there are occasional quite large disagreements (as much as several angstroms) that fall well outside the tail of the normal distribution for the smaller errors. For well-refined structures, then, the temperature factor (called "B") is inversely proportional to the relative accuracy of a given atom, or group, position. In the extreme case, an atom that refined to the maximum allowed temperature factor or that was in zero electron density has an essentially undetermined position, and quite probably is actually disordered in the protein. In addition to the relative local error level, one must bear in mind that there is always a small but finite probability that the position is grossly wrong, even for an apparently well-determined group. This probability is almost vanishingly small for a structure refined at, say, 1.5 Å to a residual of 15%, but if the residual were 25 or 30% or the data only went out to 3 Å resolution, then the likelihood of occasional large errors is quite substantial.

There are also some general rules of thumb that can be used to guess at error levels in unrefined and lower resolution structures. A first fundamental problem is to judge when there might be mistakes in the chain tracing that involve incorrect connectivity of the backbone. In a survey of 47 independent chain tracings of novel proteins which have been either confirmed or disconfirmed by further evidence, all of the tracings at 2.5 Å resolution or better were correct, whether the sequence was known or not. Below 3.5 Å resolution the sequence is irrelevant; with luck, an occasional structure can be traced reliably if it is simple and helical (e.g., Hendrickson et al., 1975). For the resolution range between 2.5 and 3.5 Å, knowledge of the sequence makes considerable difference: only 20% of the structures with known sequences had to be rearranged, while two-thirds of those without sequences had at least one connectivity change. Placement of all the major structural features is correct even when connectivity is not. Assignment of secondary structure elements is apt to be conservative in initial structure reports, so that the helices and β strands initially cited are almost invariably confirmed but additional elements may be recognized later.

In structures for which complete coordinates have been determined but not refined, error levels can be estimated according to position in the protein and what parameter is in question. Quite uniformly, main chain atoms are located more exactly than side chains and interior side chains are better determined than exposed ones. In general, positional parameters are more reliably known than dihedral angles. Ring plane orientation is much easier to determine for Trp, Tyr, and Phe than for His, because the electron density for a five-membered ring is nearly round at lower than about 2 Å resolution. Some parameters are especially prone to an occasional large error. If the carbonyl oxygen showed up clearly in the electron density, then ϕ and ψ are determined accurately, but if the carbonyl oxygen was not visible, then the orientation of the peptide is quite uncertain: in many cases it can flip by 180° without affecting positions of the surrounding α-carbons and side chains to any noticeable degree. Peptide rotation that is approximately independent of the surrounding chain can be seen between type I and type II tight turns (see Fig. 30). Peptide rotation involves a coupled change of ψ_n and ϕ_{n+1} by equal and opposite amounts. There may occasionally be true disorder of a peptide orientation in the protein, as has been suggested by dynamic calculations for several external peptides in pancreatic trypsin inhibitor (McCammon et al., 1977). ϕ and ψ are generally less accurately known for glycine than for other residues, because the β-carbon is not present in the map to help determine conformation. Another parameter subject to occasional large ambiguities is χ_1. It is not too unusual, for instance, for the side chain electron density of a valine to show definite elongation parallel to the backbone direction but with no clear indication to which side the β-carbon protrudes. Of the two possible χ_1 values one is staggered and one is eclipsed. If the crystallographer picks the staggered χ_1 value he greatly improves his chances of being correct, but he is undermining the validity of future attempts at empirical determination of χ_1 distributions. When the β-carbon is unbranched, the electron density sometimes extends out straight with no indication of the elbow bend at Cβ, in which case χ_1 is also difficult to determine.

In summary, there are three important generalizations about error estimation in protein crystallography. The first is that the level of information varies enormously as a function primarily of resolution, but also of sequence knowledge and extent of refinement. The second generalization is that no single item of information is completely immune from possible error. If the electron density map is available or indicators such as temperature factors are known from refinement, then it is possible to tell which parameters are most at risk. The third important generalization is that errors occur at a very low absolute rate: 95% of the reported information is completely accurate, and it represents a detailed and objective storehouse of knowledge with which all other studies of proteins must be reconciled.

II. Basic Elements of Protein Structure

A. Helices

The α-helix is *the* classic element of protein structure. A single α-helix can order as many as 35 residues whereas the longest β strands include only about 15 residues, and one helix can have more influence on the stability and organization of a protein than any other individual structure element. α-Helices have had an immense influence on our understanding of protein structure because their regularity makes them the only feature readily amenable to theoretical analysis.

The α-helix was first described by Pauling in 1951 (Pauling *et al.*, 1951) as a structure predicted to be stable and favorable on the basis of the accurate geometrical parameters he had recently derived for the peptide unit from small-molecule crystal structures. This provided the solution to the long-standing problem of explaining the strength and elasticity of the α-keratin structure and accounting for the appearance of its X-ray fiber diffraction pattern. Helices had frequently been proposed before as the α structure, but none of them could adequately match the diffraction pattern because they had been limited by the implicit assumption that a regular helix would necessarily have an integral number of amino acid residues per turn. In fact, as Pauling first realized, the α-helix has 3.6 residues per turn, with a hydrogen bond between the CO of residue n and the NH of residue $n + 4$ (see Fig. 11). The closed loop formed by one of these hydrogen

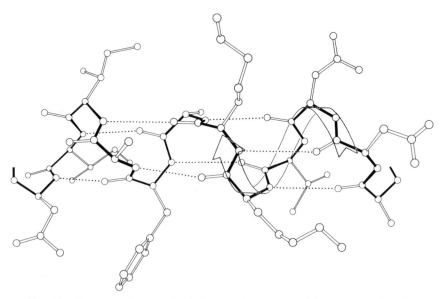

FIG. 11. Drawing of a typical α-helix, residues 40–51 of the carp muscle calcium-binding protein. The helical hydrogen bonds are shown as dotted lines and the main chain bonds are solid. The arrow represents the right-handed helical path of the backbone. The direction of view is from the solvent, so that the side groups on the front side of the helix are predominantly hydrophilic and those in the back are predominantly hydrophobic.

bonds and the intervening stretch of backbone contains 13 atoms (including the hydrogen), as illustrated in Fig. 12. In the usual nomenclature for describing the basic structure of polypeptide helices, the α-helix is known as the 3.6_{13}-helix, where 3.6 is the number of residues per turn and 13 is the number of atoms in the hydrogen-bonded loop. The rise per residue along the helix axis is 1.5 Å.

The α-helix received strong experimental support when Perutz (1951) found the predicted 1.5 Å X-ray reflection from hemoglobin crystals and from tilted fibers of keratins. The final conclusive demonstration of the α-helix in globular protein structure came from the high-resolution X-ray structure of myoglobin (Kendrew *et al.*, 1960). It was shown that the myoglobin helices matched Pauling's calculated structure quite closely, and also that they were all right-handed (for L-amino acids, the left-handed α-helix has a close approach between the carbonyl oxygen and the β-carbon). It is easy to determine that, for instance, Fig. 11 is right-handed: if the curled fingers of the right hand are turned in the direction of their tips (as if tightening a screw) and the whole hand is moved in the direction of the outstretched

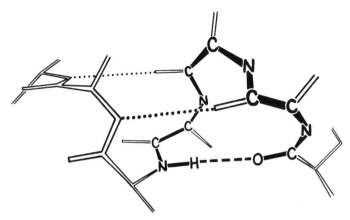

FIG. 12. Illustration of the 13-atom hydrogen-bonded loop which determines the subscript in the description of the α-helix as a 3.6_{13}-helix (the 3.6 refers to the number of residues per turn). The 13 atoms are those in the shortest covalently connected path which joins the ends of a single hydrogen bond (the hydrogen is one of the 13 atoms): . . . O—C— N—Cα—C—N—Cα—C— N—Cα—C—N—H . . . .

thumb, then a right-handed helical path is traced out. Handedness is an enormously influential parameter in protein structure; most features for which handedness can be defined prefer one sense to the other, and the α-helix is only the first of many examples we will encounter.

Figure 13 shows the electron density map at 2 Å resolution for one of the α-helices in staphylococcal nuclease. Bumps for the carbonyl oxygens are clearly visible; they point toward the C-terminal end of the helix, and are tipped very slightly outward away from the helix axis. At the top, in the last turn of the helix, there is a carbonyl tipped still further outward and hydrogen-bonded to a solvent molecule (marked with an asterisk). Side chain atoms or waters frequently bond to free backbone positions in the first or last turn of a helix, and hydrogen bonds with water are even more favorable for carbonyls than for NH groups (see Section II,H).

With 3.6 residues per turn, side chains protrude from the α-helix at about every 100° in azimuth. Since the commonest location for a helix is along the outside of the protein, there is a tendency for side chains to change from hydrophobic to hydrophilic with a periodicity of three to four residues (Schiffer and Edmundson, 1967). This trend can sometimes be seen in the sequence, but it is not strong enough for reliable prediction by itself. Different residues have weak but definite preferences either for or against being in α-helix: Ala, Glu, Leu,

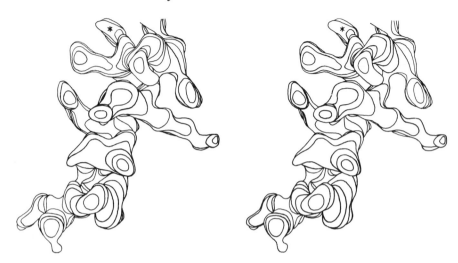

FIG. 13.   Stereo drawing of one contour level in the electron density map at 2 Å resolution for the residue 54–68 helix in staphylococcal nuclease.   Carbonyl groups point
up, in the C-terminal direction of the chain; the asterisk denotes a solvent peak bound
to a carbonyl oxygen in the last turn.   Side chains on the left (including a phenylalanine
and a methionine) are in the hydrophobic interior, while those on the right (including
an ordered lysine) are exposed to solvent.

and Met are good helix formers while Pro, Gly, Tyr, and Ser are very
poor (Levitt, 1977).   α-Helices were central to all the early attempts to
predict secondary structure from amino acid sequence (e.g., Davies,
1964; Guzzo, 1965; Prothero, 1966; Cook, 1967; Ptitsyn, 1969; Kotelchuk and Scheraga, 1969; Pain and Robson, 1970) and they are still the
feature that can be predicted with greatest accuracy (e.g., Schulz *et al*.,
1974b; Chou and Fasman, 1974; Lim, 1974; Matthews, 1975; Maxfield

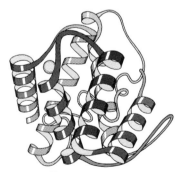

Thermolysin domain 2

FIG. 14.   Schematic drawing of the backbone of an all-helical tertiary structure: domain 2 of thermolysin.

FIG. 15. Stereo drawing of a bent helix (glyceraldehyde-phosphate dehydrogenase residues 146–161) with an internal proline. The proline ring produces steric hindrance to the straight α-helical conformation as well as having no NH group available for a hydrogen bond. A proline is the commonest way of producing a bend within a single helix, as well as occurring very frequently at the corners between helices.

and Scheraga, 1976; Nagano, 1977b; Wu *et al.*, 1978). As much as 80% of a structure can be helical, and only seven proteins are known that have no helix whatsoever. Figure 14 shows the second domain of thermolysin, a structure that is predominantly α-helical.

The backbone conformational angles for right-handed α-helix are approximately $\phi = -60°$, $\psi = -60°$, which is in a favorable and relatively steep energy minimum for local conformation, even ignoring the hydrogen bonds. α-Helices are certainly the most regular pieces of structure to be found in globular proteins, but even so they show significant imperfections. There can be slight bends in the axis of a helix, of any amount from almost undetectable up to about 20° (e.g., Anderson *et al.*, 1978), either with or without a break in the pattern of hydrogen bonding. One of the most obvious ways to produce such a bend is with a proline. Proline fits very well in the first turn of an α-helix but anywhere further on it not only is missing the hydrogen bond donor but also provides steric hindrance to the normal conformation. It is rare but certainly not unknown in such a position (see Fig. 15). An α-helix is almost invariably made up of a single, connected stretch of backbone (as opposed, for instance, to the backbone changeovers seen for double-helix in transfer RNAs: Holbrook *et al.*, 1978). Almost the only known exception to this rule is the interrupted helix from subtilisin that is shown in Fig. 16.

The generally regular, repeating conformation in the α-helix places all of the charge dipoles of the peptides pointing in the same direction along the helix axis (positive toward the N-terminal end). It has been shown (Hol *et al.*, 1978) that the overall effect is indeed a significant net dipole for the helix, in spite of shielding effects. The helix dipole may contribute to the binding of charged species to the protein: for example, negative nucleotide phosphates, which are typically found near the N-termini of helices.

The only other principal helical species besides the α-helix which occurs to any great extent in globular protein structure is the 3_{10}-helix

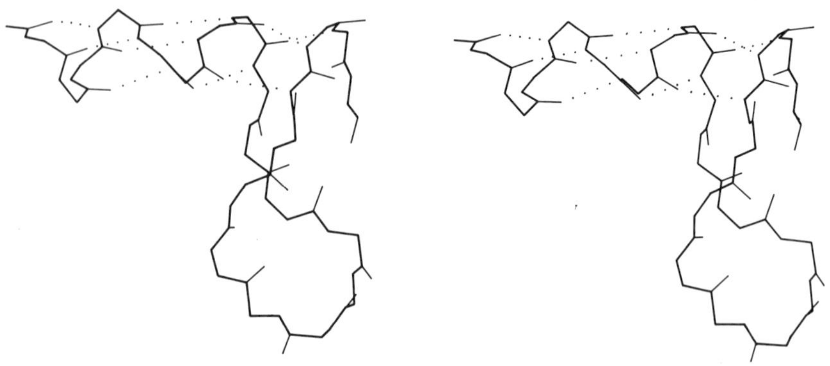

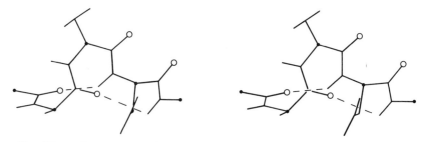

FIG. 16. An unusual interrupted helix from subtilisin (residues 62–86), in which the helical hydrogen bonds continue to a final turn that is formed by a separate piece of main chain. Such interrupted helices (broken on one side of the double helix) are apparently a fundamental feature of nucleic acid structure as illustrated by tRNA, but are exceedingly rare in protein structure.

(see Fig. 17), with a three-residue repeat and a hydrogen bond to residue $n + 3$ instead of $n + 4$. Its backbone conformational angles are approximately $\phi = -60°$, $\psi = -30°$, within the same energy minimum as the α-helix. However, for a long periodic structure the 3_{10}-helix is considerably less favorable than the α-helix in both local conformational energy and hydrogen bond configuration. In the refinement of rubredoxin at 1.2 Å resolution, Watenpaugh *et al.* (1979) found that bond angles along the main chain were significantly distorted in all four of the regions that have two successive 3_{10}-type hydrogen bonds. Long 3_{10} helices are very rare but short pieces of approximate 3_{10}-helix occur fairly frequently. Two consecutive residues in 3_{10} conformation form a good tight turn (see Section II,C), and three consecutive 3_{10} residues forming two interlocked tight turns is also fairly common. But another important location for short bits of

FIG. 17. A short segment of 3_{10} helix from carbonic anhydrase (residues 159–164). Main chain carbonyl oxygens are shown as open circles.

3_{10}-helix is at the C-terminal end of an α-helix. It is quite common for the last helical turn to tighten up, with hydrogen bonds back to residue $n - 3$ or else bifurcated hydrogen bonds to both $n - 3$ and $n - 4$ (e.g., Watson, 1969). Nemethy *et al.* (1967) showed that this arrangement is not necessarily quite like 3_{10}-helix; they described the α_{II}-helix for this sort of position, which retains the helical parameters of an α-helix but tilts the peptide so that the NH points more inward toward the helix axis and at the same time points more toward the $n - 3$ than the $n - 4$ carbonyl. The conformations in real proteins show somewhat of a mixture between the α_{II} tilt and the 3_{10} tightening. Figure 18 shows an example. 3_{10} or α_{II} conformation does not tend to occur nearly as often at the N-termini of α-helices. The reason is that the tighter loop with $n + 3$-type hydrogen bonds requires the group involved to move closer to the helix axis, either by tilting (α_{II}) or by tightening the helix (3_{10}). This motion is easy for the NH group but not for the CO: neighboring carbonyl oxygens would come too close together.

Another frequent feature of the C-termini of helices is a residue (usually glycine) in left-handed α conformation with its NH making a hydrogen bond to the CO of residue $n - 5$ (see Schellman, 1980); this often follows a residue with the 3_{10} or α_{II} bonding described above.

A few other helical conformations occur occasionally in globular protein structures. The polyproline helix, of the same sort as one strand out of a collagen structure, has been found in pancreatic trypsin inhibitor (Huber *et al.*, 1971) and in cytochrome c_{551} (Almassy and Dickerson, 1978). An extended "ε helix" has been described as occurring in chymotrypsin (Srinivasan *et al.*, 1976). In view of the usual variability and irregularity seen in local protein conformation it is unclear that either of these last two helix types is reliably distinguishable from simply an isolated extended strand; however, the presence of prolines can justify the designation of polyproline helix.

The ways in which α-helices pack against one another were initially described by Crick (1953) as "knobs into holes" side chain packing which could work at either a shallow left-handed crossing angle or a

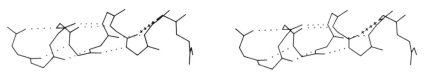

FIG. 18. An example of the α_{II} conformation at the end of the A helix in myoglobin (residues 8–17). The normal α-helical hydrogen bonds are shown dotted, while the tighter α_{II} bond is shown by crosses.

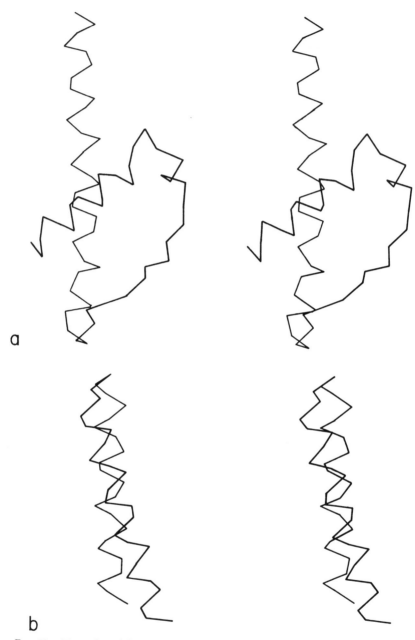

FIG. 19. Examples of the two commonest types of helix–helix contact: (a) Class II (from hexokinase) with an interhelix angle of about −60°; (b) Class III (from myohemerythrin) with an interhelix angle of about +20°.

steeper right-handed one. Helix–helix interactions have recently been analyzed in more detail by several different groups, using quite varied approaches and points of view. Chothia *et al*. (1977) considered the helix contact angles at which ridges formed by rows either of $n,n + 3$ or of $n,n + 4$ side chains can pack against each other. They predict three classes (I, II, and III) of contact at angles of -82, -60, and $+19°$, respectively (the angle is handed but does not consider direction of the helices). For 25 cases they find a distribution consistent with these classes, although there is better discrimination between classes II and III than between I and II. Richmond and Richards (1978) determine contact residues by calculating solvent-accessible area lost on bringing helix pairs together, and model the interactions using helices of close-packed spheres. They find contact classes that match the packing of Chothia's classes II and III, but for approximately perpendicular helices (class I) they find a favorable contact only if the two central residues are glycine or alanine and pack directly on top of each other. In globins the helix axes are about 2 Å closer together for steeply angled contacts than for nearly parallel ones, which have a long contact surface between relatively large residues. Figure 19 shows stereo drawings of class II and class III helix contacts. Efimov (1977, 1979) also considers side chain packing as the determinant for helix contacts, but from a rather different theoretical perspective. He first considers what side chain conformations will allow close packing of neighboring hydrophobic side chains on a single helix, then considers how to close-pack side chains of hydrophobic patches on the buried side of two parallel or antiparallel helices, then finally considers the angles for packing together two layers of helices by matching two of the relatively flat hydrophobic surfaces produced in the second step.

Each of these approaches has its advantages; the contact nets drawn by Chothia *et al*. are the only version that explicitly shows the actual (rather than idealized) residue contacts, but they have made correlations only with the one variable of contact angle. Efimov has obtained a very interesting regularity that successfully predicts side chain conformation at the right and left edges of hydrophobic strips, but has not considered either the interactions directly in between helix pairs in his first step or the possibility that close (as opposed to distant hydrophobic) contacts could occur at steep angles. Richmond and Richards have the advantage of identifying residue contacts in a way that is not influenced by theoretical preconceptions, and they have considered side chain identity (although not conformation) in detail. Because of the great local variability of side chain size and packing and because relatively few examples have yet been analyzed, it is obviously possible to describe a given contact as fitting quite different idealized models. The current large data set of proteins shows a strong tendency for class III (shallow) interactions to be antiparallel and for parallel helix interactions to be class II. It seems likely that the antiparallel up and down helix bundle structures (see Section III,B) would be composed of paradigm class III interactions, and the doubly wound α/β structures (see Section III,C) would contain paradigm class II interactions, but none of the 15 proteins analyzed by the above three methods happen to fall into either of those categories. If multiple examples of paradigm classes II and III contacts can be analyzed and compared, it may then be possible to define a meaningfully distinct class of perpendicular contacts.

B. β Structure

The other major structural element found in globular proteins is the β sheet. Historically, it was first observed as the β, or extended, form of keratin fibers. An approximate understanding of the molecular structure involved was achieved much earlier for the β than for the α structure, because repeat distances along the fiber showed that the backbone must be almost fully extended, which did not leave very much choice of conformation even when the details of backbone geometry were not well known. Astbury described the β structure in 1933 as straight, extended chains with alternating side chain direction and hydrogen bonds between adjacent antiparallel chains. Pauling and Corey (1951) described the correct hydrogen-bonding patterns for both antiparallel and parallel β sheet, and also realized that the sheets were "pleated," with α-carbons successively a little above and below the plane of the sheet. Some features of β structure, such as its characteristic twist, were not recognized until after several β sheets had been seen in three-dimensional protein structures.

β sheet is made up of almost fully extended strands, with ϕ,ψ angles which fall within the wide, shallow energy minimum in the upper left quadrant of the Ramachandran plot (see Figs. 7 and 9). β strands can interact in either parallel or antiparallel orientation, and each of the two forms has a distinctive pattern of hydrogen bonding. Figures 20 and 21 illustrate examples of antiparallel and parallel β sheets from real protein structures. The antiparallel sheet has hydrogen bonds perpendicular to the strands, and narrowly spaced bond pairs alternate with widely spaced pairs. Looking from the N- to C-terminal direction along the strand, when the side chain points up the narrow pair of H bonds will point to the right. Parallel sheet has evenly spaced hydrogen bonds which angle across between the strands.

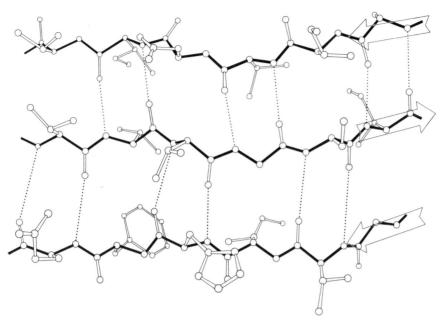

FIG. 20. An example of antiparallel β sheet, from Cu,Zn superoxide dismutase (residues 93–98, 28–33, and 16–21). Arrows show the direction of the chain on each strand. Main chain bonds are shown solid and hydrogen bonds are dotted. In the pattern characteristic of antiparallel β sheet, pairs of closely spaced hydrogen bonds alternate with widely spaced ones. The direction of view is from the solvent, so that side chains pointing up are predominantly hydrophilic and those pointing down are predominantly hydrophobic.

Within a β sheet, as within an α-helix, all possible backbone hydrogen bonds are formed. In both parallel and antiparallel β sheet, the side groups along each strand alternate above and below the sheet, while side groups opposite one another on neighboring strands extend to the same side of the sheet and are quite close together. These close side chain pairs on neighboring strands show preferences for having hydrophobic groups together, unlike charges together, and branched β-carbons next to unbranched β-carbons (in antiparallel sheet), but none of these preferences are stronger than 2 to 1. Lifson and Sander (1980a,b) have shown that specific residue pairs on neighboring strands recognize each other, over and above simple grouping by polarity, but again they comment on the fact that the correlations are not as strong as one would have expected. As an example of the kind of factors involved, let us examine the interactions of a pair of side chains with branched β-carbons on neighboring strands of β sheet. Valine and isoleucine have a rather strong conformational preference (better

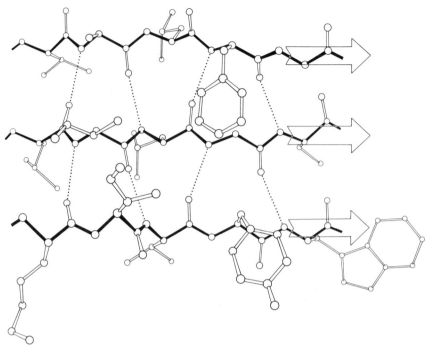

FIG. 21. An example of parallel β sheet, from flavodoxin (residues 82–86, 49–53, and 2–6). In the pattern characteristic of parallel β sheet, the hydrogen bonds are evenly spaced but slanted in alternate directions. Since both sides of the sheet are covered by other main chain (as is almost always true for parallel sheet), side groups pointing in both directions are predominantly hydrophobic except at the ends of the strands.

than two-thirds of the cases) for the χ_1 orientation staggered relative to the main chain (Janin *et al.*, 1978). Since the relation between adjacent parallel strands is a translation, neighboring Val or Ile residues in the preferred conformation "cup" against each other back-to-front in a very favorable packing. Since the relationship between adjacent antiparallel strands is twofold, in that case a pair of side chains with the preferred χ_1 angle will either pack back-to-back leaving unfilled space or else front-to-front, which produces a collision unless the main chain conformation is adjusted. The effects of these restrictions can indeed be seen in the patterns of residue-pair occurrence, but only weakly. Looking at the actual pairs of, for instance, Val-Val or Val-Ile in antiparallel sheet, one finds either that one of the side chains has adopted an unfavorable χ_1 angle so that the two can pack well (as in the upper left corner of Fig. 20) or else the main chain has twisted to put the β-carbons at an optimum distance (e.g., when a Leu-Val pair in chymotrypsin becomes a Val-Val pair in elastase, the β-carbons move 0.65 Å further apart). This in turn, of course, shows one reason why the χ_1 preference is not stronger or the ϕ,ψ angles more regular. In general, the impression one takes away from this kind of examination is that the protein is balancing so many factors at the same time that there are always ways to compensate for any individual problem. Thus studies of individual parameters uncover only weak regularities in spite of the strength of the overall packing constraints. Looking at long strings of adjacent side chains across the centers of large sheets, such as shown in the stereo figures of Lifson and Sander (1980b), one sees a stronger expression of the packing difference between antiparallel and parallel sheets: Ile-Leu-Val-Leu and Val-Ala-Thr-Gly-Ile in elastase and Ala-Ile-Ala-Val, Ala-Ile-Leu-Ile-Ala, and Ser-Thr-His-Val-Ser in concanavalin A, versus Val-Val-Ile-Val-Val-Val and Ile-Val-Ile in glyceraldehyde-phosphate dehydrogenase domain 1 and Val-Val-Ile, Val-Val-Val, and Ile-Ile-Val in triosephosphate isomerase.

β strands can combine into either a pure parallel sheet, a pure antiparallel sheet, or a mixed sheet with some strand pairs parallel and some antiparallel. If the assortment were random there would be very few pure sheets, but in fact there is a strong bias against mixed sheets (Richardson, 1977), perhaps because the two types of hydrogen bonding need slightly different peptide orientations. Only about 20% of the strands inside β sheets have parallel bonding on one side and antiparallel on the other.

Parallel β sheet is in general a good deal more regular than antiparallel. If ϕ,ψ angles are plotted for both types of sheet, as for instance in Nagano (1977a), the parallel residues cluster rather tightly while the antiparallel ones spread over the entire quadrant. Parallel β structure almost never occurs in sheets of less than five total strands, whereas antiparallel β structure often occurs as a twisted ribbon of just two strands. Figure 22 shows such a two-stranded antiparallel β ribbon. Parallel β sheets and the parallel portions of mixed sheets are always thoroughly buried, with other main chain (often α-helices) protecting them on both sides. Antiparallel sheets, on the other hand, typically have one side exposed to solvent and the other side buried, so that they often show an alternation of side chain hydrophobicity in the amino acid sequence. β sheets in general show a tendency toward greater hydrophobicity for the central than for the edge strands of the sheet (Sternberg and Thornton, 1977c). These three requirements of parallel β sheets (regularity, size, and protection) all suggest that parallel β structure is less stable than antiparallel (Richardson,

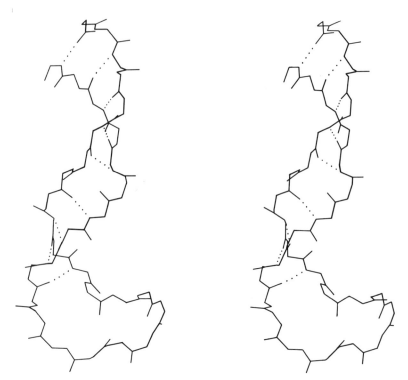

FIG. 22.   An example of a long two-stranded ribbon of antiparallel β structure, from lactate dehydrogenase (residues 263–294).   Side chains are not shown; hydrogen bonds are dotted.   As is typical of isolated two-stranded ribbons, the chains show a very strong twist (180° in about five residues).

1977), since it apparently needs the cooperativity of an extensive hydrogen-bond network (see Sheridan *et al.*, 1979) and also seems to need those hydrogen bonds shielded from water.   (It is actually possible to shield the backbone with large hydrophobic side chains, but those are not the residues that would occur on an exposed surface.) Mixed β sheets tend to have the general appearance characteristic of their predominant H-bonding type.   Sheets that are approximately half and half, such as carboxypeptidase or carbonic anhydrase, tend to look like parallel sheets because they require substantial protection on both sides.   Figure 23 is a schematic drawing of a typical parallel-type β sheet structure in a protein.

One of the most conspicuous features of β sheet as it occurs in the known protein structures is its twist (Chothia, 1973).   This twist always has the same handedness, although it has unfortunately been

FIG. 23.   Schematic drawing of the backbone of flavodoxin, a protein in which a parallel β sheet is the dominant structural feature.   The sheet (represented by arrows) is shown from one edge, so that the characteristic twist can be seen clearly.

described by two conflicting conventions in the literature.   If defined in terms of the angle at which neighboring β strands cross each other, then the twist is left-handed (e.g., Quiocho *et al.*, 1977; Shaw and Muirhead, 1977); if defined in terms of the twist of the hydrogen-bonding direction or of the peptide planes as viewed along a strand, then the twist is right-handed (e.g., Schulz *et al.*, 1974a; Chothia *et al.*, 1977).   We will use the right-handed definition in this article, because it is meaningful even for an isolated strand.   Figure 23 shows the side view of a β sheet in which the twist is obvious.

There is of course no a priori reason to expect the flat $n = 2$ conformation to be especially favored for handed amino acids; however, the exact mechanism by which L-amino acids favor right-handed strand twist is not entirely obvious and has been explained in several different ways.   Detailed calculations of local conformational energy (e.g., Zimmerman and Scheraga, 1977a) always place the minimum well off to the right of the $n = 2$ line of a flat strand (see Dickerson and Geis, 1969) although the minimum is a very broad, shallow one. Chothia (1973) points out that probabilistic effects will produce a right-handed average twist, since many more of the accessible conformations within the general β area on the ϕ,ψ plot lie to the right of the $n = 2$ line.   Raghavendra and Sasisekharan (1979) have found that inclusion of H bond and nonbonded interactions between a pair of antiparallel β strands produces a considerably deeper calculated energy minimum in the right-handed region.   There is some evidence from small-molecule peptide crystal structures (Ramachandran, 1974) of a systematic tetrahedral distortion at the peptide nitrogen, and Weatherford and Salemme (1979) have shown that the combination of that distortion with optimal β sheet hydrogen bond geometry would favor a right-handed strand twist. In the known structures, β strand twist varies from close to 0° per residue to about 30° per residue, with the highest values for two-stranded ribbons (see Fig. 22) and generally lower values the more strands are present and the longer they are. This indicates some degree of conflict between the requirements for optimal hydrogen bonding and for lowest local conformational energy.

Once it has been decided what β strands belong in a given sheet (a process involving occasional subjective decisions for marginal cases), then it is possible to give a simple and unambiguous description of the topological connectivity of those strands in the sheet (Richardson, 1976, 1977). Each connection between two β strands must fall into one of two basic categories: hairpin connections in which the backbone chain reenters the same end of the β sheet it left, and "crossover" connections in which the chain loops around to reenter the sheet on the opposite end (see Fig. 24). Each connection is named according to how many strands it moves over in the sheet and in which direction, with an "x" added for crossover connections. Thus, a "+1" is a hairpin and a "+1x" a crossover connection between nearest-neighbor strands; a "+2" is a hairpin and a "+2x" is a crossover connection that skips past one intervening strand in the sheet, and so on. The conformation of the connecting loop is irrelevant to this topological designation. Nearest-neighbor connections of ±1 and ±1x are by far the most common, occurring about three times as frequently as all other connection types put together (Richardson, 1977; Sternberg and Thornton, 1976).

The topology of an n-stranded β sheet can be specified by a list of its

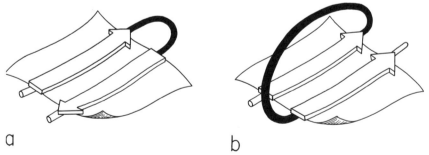

a   b

FIG. 24.   The two major sorts of connection between β strands: (a) a "hairpin," or same-end, connection (this example is type +1, to a nearest-neighbor strand); (b) a "crossover," or opposite-end, connection (this one is type +1x).

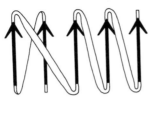

$$-|x, +2x, +|x, +|x$$

FIG. 25.   A topological schematic diagram of the connectivity in the parallel β sheet of flavodoxin.   Arrows represent the β strands; thin-line connections lie below the plane of the sheet and fat connections above it.   No attempt is made to indicate the length or conformation of the connecting chains (most of them are helical) or the twist of the β sheet.   The topology can also be specified by a sequential list of the connection types: in this case, $-1x, +2x, +1x, +1x$.

$n - 1$ connections, starting from the N-terminus.   For example, flavo-doxin (Fig. 23) can be described as either $+1x$, $-2x$, $-1x$, $-1x$, or $-1x$, $+2x$, $+1x$, $+1x$ (absolute value of the signs is not meaningful, since the sheet could be turned upside down).   We will use connection types to describe and classify β sheets, and will also use a simplified kind of topology diagram (see Fig. 25) which views the sheet from above.   There is another type of topology diagram also common in the literature which views the sheet end-on (see Levitt and Chothia, 1976); the topology is less explicit but more features of the three-dimensional structure are retained.   That is a significant advantage in the cases in which it works best, but since adherence to the convention forces substantial distortions in some proteins, we will use separate diagrams for the three-dimensional structure and for the topology in the overall survey (see Sections III,A–E).

Crossover connections have a handedness (see Fig. 26), since they

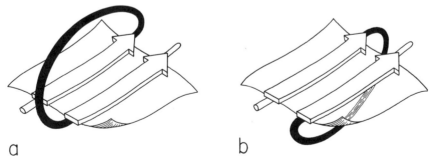

a   b

FIG. 26.   (a) A right-handed $+1x$ crossover connection; (b) a left-handed $+1x$ crossover connection.

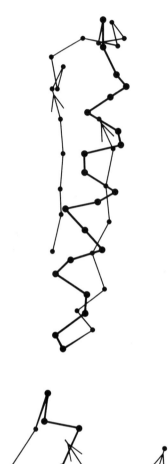

FIG. 27a

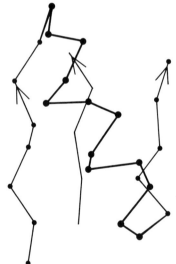
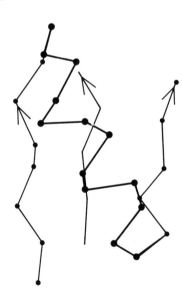

FIG. 27b

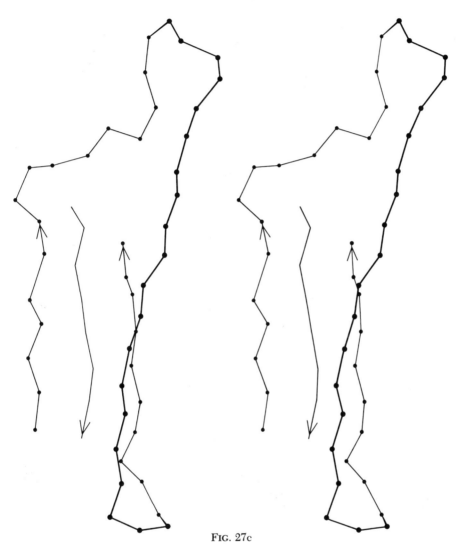

FIG. 27c

FIG. 27. Examples of particular crossover connections: (a) a right-handed + 1x, residues 200–242 from carboxypeptidase A; (b) a right-handed + 2x, residues 109–133 from papain; (c) a right-handed + 2x, residues 169–214 from concanavalin A.

form a loose helical turn from one strand, up (or down) and around, and back into the next strand. Essentially every one of the crossover connections in the known protein structures regardless of the length or conformation of the connecting loop, is right-handed (Richardson, 1976; Sternberg and Thornton, 1977a). There is one really well-authenticated left-handed crossover in subtilisin and one in glucosephosphate isomerase in a region where the chain connectivity is not completely certain (Shaw and Muirhead, 1977), while there are many more than a hundred right-handed crossovers. Over half of the crossover connections have at least one helix in the connecting strand, and in many of those cases the helix packs against one or both of the β strands it connects (see Fig. 27a). Sternberg and Thornton (1976) have explained the handedness by the fact that β sheet twist makes the right-handed connection shorter and more compact (as can be seen in Fig. 26). Nagano (1977a) has explained the handedness by the preferred packing angles of a helix against a β strand, which again would allow more compact and shorter corners (between the α and β elements) in the right-handed form. Both of these explanations are sure to be important contributing causes of crossover handedness, but they are limited to the relatively short, straightforward examples with tight corners. The large number of crossover connections which are too long, start off in the wrong direction, or do not pack against the β sheet (see Fig. 27a and c for examples) show almost as strong a handedness constraint as the more classic cases. In an attempt to account for these long examples, Richardson (1976) proposed a hypothetical folding scheme for crossover connections by which the twist of a long extended strand or of a helix flanked by extended chains is transferred to the crossover loop as the backbone curls up (see Fig. 28). However it is achieved, the right-handedness of crossover connections is the dominant factor controlling the appearance of both singly wound and doubly wound parallel α/β structures (see Section III,C). Crossover connections are also fairly common in antiparallel β sheet.

Parallel β structure usually forms large, moderately twisted sheets such as in Fig. 23, although occasionally it rolls up into a cylinder with helices around the outside (e.g., triosephosphate isomerase). Large antiparallel sheets, on the other hand, usually roll up either partially (as in the first domain of thermolysin or in ribonuclease) or completely around to join edges into a cylinder or "barrel." Occurrence, topology, and classification of β barrels will be discussed in Section III,D, but here we will consider the interaction between the β sheets on opposite sides of the barrel, especially in terms of the angle at which opposite strands cross.

β barrels may be made up of as few as 5 or as many as 13 strands. Their interiors are packed with hydrophobic side chains, which are found to have the same average side chain volume as for a normal amino acid composition. There are no large barrels filled with tryptophans or small ones filled with alanines, presumably because mutation to change the size of even as many as two or three residues at once would still produce a bad fit. The cross sections of all the barrels

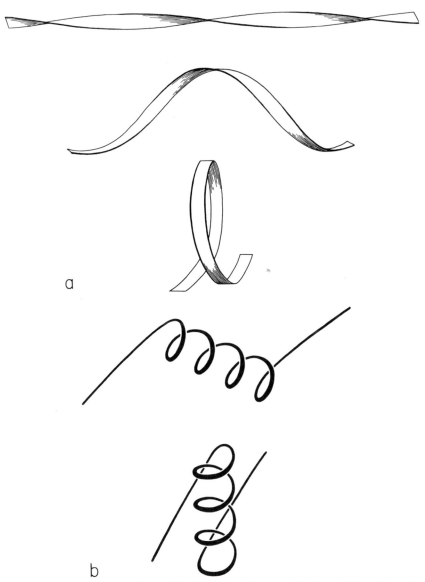

FIG. 28. Illustration of possible folding schemes which would produce the handedness of crossover connections as a consequence of (a) the handedness of twist of an initial β ribbon, or (b) the handedness of an initial α-helix.

look remarkably alike, regardless of strand number, with a slight flattening in one direction. Figure 29 shows examples of cross sections from real β barrels with different numbers of strands. The nearly constant appearance is obtained by varying the degree of strand twist

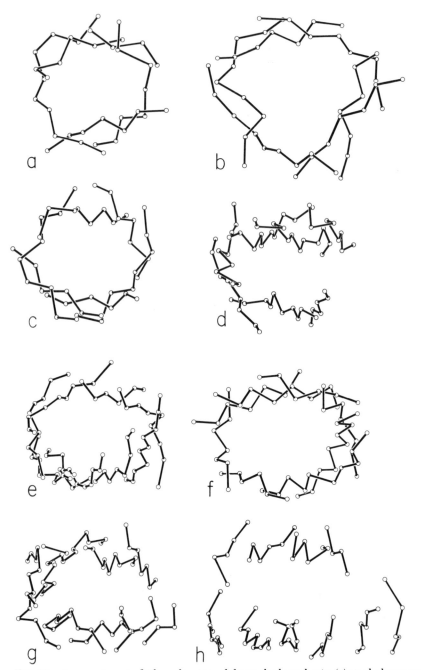

FIG. 29. An assortment of β barrels, viewed down the barrel axis: (a) staphylococcal nuclease, 5-stranded; (b) soybean trypsin inhibitor, 6-stranded; (c) chymotrypsin, 6-stranded; (d) immunoglobulin (McPC603 C_H1) constant domain, 7-stranded; (e) Cu,Zn superoxide dismutase, 8-stranded; (f) triosephosphate isomerase, 8-stranded; (g) im-

around the barrel. Like the bias-woven finger-bandages which tighten around a finger when stretched, a barrel with a given number of strands has a smaller diameter the less twist it has. Twist can be measured by the angle at which strands on opposite sides of the barrel cross one another; that angle averages 95° for 5- and 6-stranded antiparallel barrels, 40° for 7- and 8-stranded ones, and 30° for 9- through 13-stranded ones. Barrel diameter can also be maintained with fewer strands by separating one or more strand pairs further apart than normal hydrogen-bonding distance; this is a very pronounced effect in plastocyanin, for instance, which has a very low twist angle for an 8-stranded barrel. Eight-stranded parallel barrels are more twisted (averaging 75°) than 8-stranded antiparallel ones because all of their strands are hydrogen-bonded and more regular. Beyond eight or nine strands the twist cannot decrease any further and the barrel cross section simply flattens more, keeping the same short axis (11–12 Å).

C. Tight Turns

Tight turns (also known as reverse turns, β turns, β bends, hairpin bends, 3_{10} bends, kinks, widgets, etc.) are the first and most prevalent type of nonrepetitive structure that has been recognized. While helices and β structure have the property that approximately the same ϕ,ψ angles are repeated for successive residues, pieces of nonrepetitive structure have a particular succession of different ϕ,ψ values for each residue, so that the concept of residue position within the structure is more influential than in a repeating structure. Of course, no startlingly new local conformations are available: most residues are either approximately α type or β type, with occasional left-handed α-type residues which are usually but not always glycines. However, by combining those three basic conformations in various orders, allowing for the considerable variation available within each of the conformational minima, and utilizing various patterns of hydrogen-bonding and side chain position, an enormous number of quite different structures are possible even within a stretch as short as three or four residues.

Tight turns were first recognized from a theoretical conformational analysis by Venkatachalam (1968). He considered what conformations were available to a system of three linked peptide units (or four successive residues) that could be stabilized by a backbone hydrogen bond between the CO of residue n and the NH of residue $n + 3$. He

---

munoglobulin (McPC603 V_H) variable domain, 9-stranded; (h) tomato bushy stunt virus protein domain 3, 10-stranded. Twist decreases significantly as strand number increases, but cross section stays nearly constant.

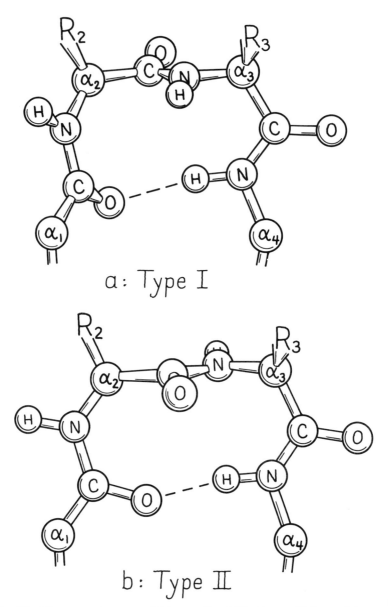

FIG. 30. The two major types of tight turn (I and II). In type II (bottom), R_3 is generally glycine.

found three general types, one of which (type III) actually has repeating ϕ, ψ values of $-60°$, $-30°$ and is identical with the 3_{10}-helix. The other two types are nonhelical and fold the chain back on itself around a rather square corner so that the first and fourth α-carbons are only about 5 Å apart, as seen in Fig. 30. The backbone at either end of type I or II turns is in approximately the right position to continue in an antiparallel β ribbon. Type I turns have approximately $\phi_2 = -60°$, $\psi_2 = -30°$, $\phi_3 = -90°$, $\psi_3 = 0°$, and type II approximately $\phi_2 = -60°$, $\psi_2 = 120°$, $\phi_3 = +90°$, $\psi_3 = 0°$; these two types are related to one another by a 180° flip of the central peptide unit. Types I and III are identical for residue 2 and differ by only 30° in ϕ_3 and ψ_3 (compare Fig. 31a and c).

Types I′ and II′ (see Figs. 31 and 32) are the mirror images (for backbone conformation) of types I and II, with the inverse ϕ, ψ values of those given above. For types II, II′, and I′ the dihedral angles are such that for one or both of the central positions glycine is strongly preferred. In the rather common type II turn, for instance, the carbonyl oxygen of the middle peptide is too close to the β-carbon of the side chain in position 3 (see Fig. 30b), but the bump is relieved if residue 3 is glycine. For type II′ the bump is between $C\beta$ of residue 2 and the NH of the middle peptide. A survey by Chou and Fasman (1977) that identified and characterized 459 tight turns in actual protein structures found that 61% of the type II turns had a glycine in position 3. Type II′ turns strongly prefer glycine in position 2; types I′ and III′ prefer glycine in position 2, but in the actual cases observed seem to adjust conformation slightly rather than have glycine in position 3.

In addition to the above three turn types and their mirror images, Lewis et al. (1973) defined five additional types which both they and Chou and Fasman (1977) find can account for all observed cases (outside of helix) where the α-carbons of residues n and $n + 3$ are less than 7 Å apart. Type V is a rather unusual departure from type II which has $\phi_2 = -80°$, $\psi_2 = +80°$, $\phi_3 = +80°$, $\psi_3 = -80°$, and type V′ is its mirror image. Type VI has a cis-proline in position 3; the cis-proline turn was very elegantly demonstrated by Huber and Steigemann (1974) in the refinement of the Bence-Jones protein REI (see Fig. 33b). Type VII has either ϕ_3 near 180° and $\psi_2 < 60°$ or else $\phi_3 < 60°$ and ψ_2 near 180°. Type IV is essentially a miscellaneous category, which includes any example with two of the dihedral angles more than 40° away from ideal values for any of the other types.

In order to evaluate the occurrence and distinctness of the major turn types as found empirically in protein structures, Figs. 35 through

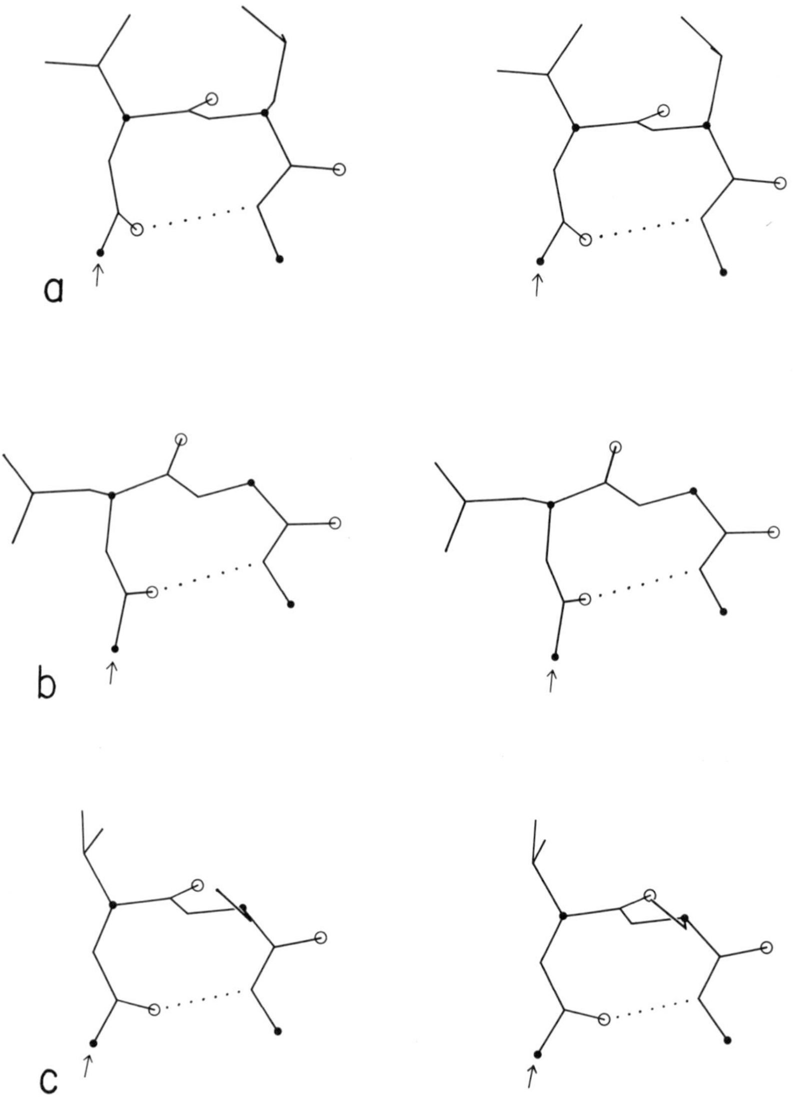

FIG. 31. Stereo drawings of particular examples of type I (a), I' (b), and III (c) turns from the known protein structures. (a) Thermolysin 12–15 (Gly-Val-Leu-Gly); (b) papain 183–186 (Glu-Asn-Gly-Tyr); (c) flavodoxin 34–37 (Asn-Val-Ser-Asp).

37 plot ϕ,ψ values found for turns in Chou and Fasman (1977). Figure 35 shows that types I and III form a single tight cluster even for position 3; their ideal ϕ,ψ values are so close that they could be distinguished only in the most highly refined protein structures. We would suggest eliminating type III as a distinct category. The ideal values

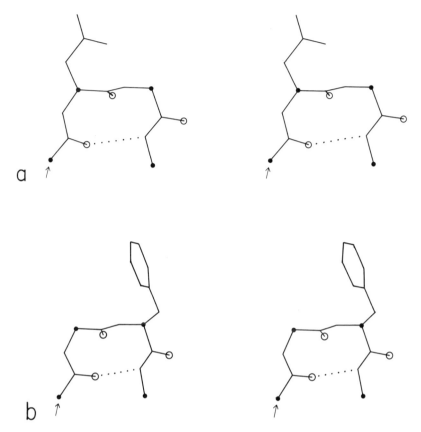

FIG. 32.   Stereo drawings of particular examples of types II (a) and II' (b) turns from the known protein structures.   (a) Concanavalin A 43–46 (Gln-Asp-Gly-Lys); (b) carboxypeptidase A 277–280 (Tyr-Gly-Phe-Leu).

for an inclusive type I category could either be left as they are and would include essentially all the type III examples, or else ψ_3 could be changed to about $-10°$ to be closer to the center of the total cluster of values.   There is a rather large number of "nonideal" type I turns that occur at the top in Fig. 35b; it might perhaps be productive to group them as type Ib (since their ϕ_3,ψ_3 values are in the β region). Some of these turns have an overall "L" shape (similar to Fig. 34a) and some look like a type I turn with the third peptide flipped over.

Figure 36 shows good clusters for types II and II' but no evidence of definable type V or V' examples, which we would also suggest eliminating as separate categories.

Figure 37 plots ϕ and ψ for the type VI (*cis*-proline) turns.   Although it is a small sample, there is very strong evidence for two distinct con-

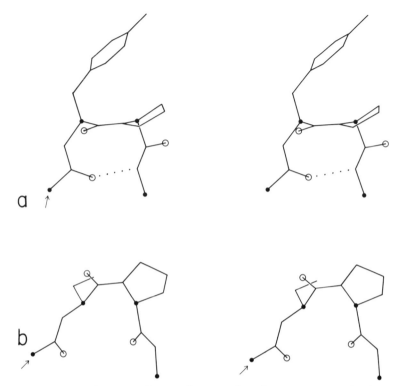

FIG. 33.   Stereo drawings of particular examples of types VIa (a) and VIb (b) *cis*-proline turns.   (a) Ribonuclease S 91–94 (Lys-Tyr-Pro-Asn); (b) Bence-Jones protein REI 6–9 (Gln-Ser-Pro-Ser).

formations which would be considerably easier to distinguish in an electron density map than to tell *cis*- from *trans*-proline in the first place.   One of the conformations (which could be called type VIa) has approximately α φ,ψ values for the proline, has a "concave" orientation of the middle peptide and the proline ring relative to the overall curve of the turn (see Fig. 33a), and typically is hydrogen-bonded. The other conformation (which includes the original examples found by Huber and could be called type VIb) has approximately β φ,ψ values for the proline, has a "convex" orientation of the middle peptide and the proline ring (see Fig. 33b) and is usually not hydrogen-bonded.

Since type VII turns are defined by only two angles and can have two different values for those, they vary greatly in appearance.   Of the nine observed examples at least half are questionable (for instance, staphylococcal nuclease 47–50 is at the end of a partially disordered loop, and for rubredoxin 46–49 the unconstrained refinement has

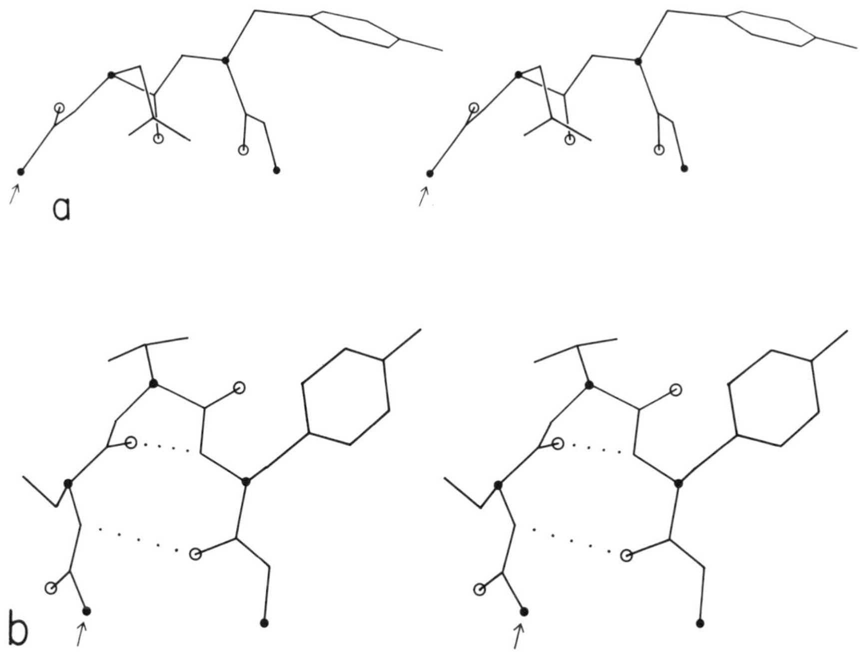

FIG. 34. Stereo drawings of particular examples of type VII (a) and of γ turns (b).
(a) Ribonuclease S 23–26 (Ser-Asn-Tyr-Cys); (b) thermolysin 25–27 (Ser-Thr-Tyr).

placed the atoms of the middle peptide out of line just enough so that the ordinary definition of ϕ and ψ is meaningless). Therefore type VII also seems unjustifiable as a distinct category. Some of the type VII turns (and also some type IVs) fall at the edge of the cluster of ϕ,ψ values seen for type Ib (see above), and perhaps could be included in that group.

In summary, then, tight turns can be rather well described by a set of categories consisting of types I, I′, II, II′, VIa, VIb, and miscellaneous (IV), with the possible addition of type Ib.

In order to demonstrate what the various types of turns actually look like, Figs. 31 through 34 show stereo views of turn examples from real structures that have ϕ,ψ angles very close to the defining values for each type. Type III is illustrated for completeness, but it cannot be distinguished from type I by inspection unless it is part of a continuing 3_{10}-helix. Types IV and V are not shown, because type IV is a miscellaneous category and there are no ideal cases of type V (see Fig. 36). The turns are all shown in approximately the same standard orientation: with the mean plane of the four α-carbons in the plane of the

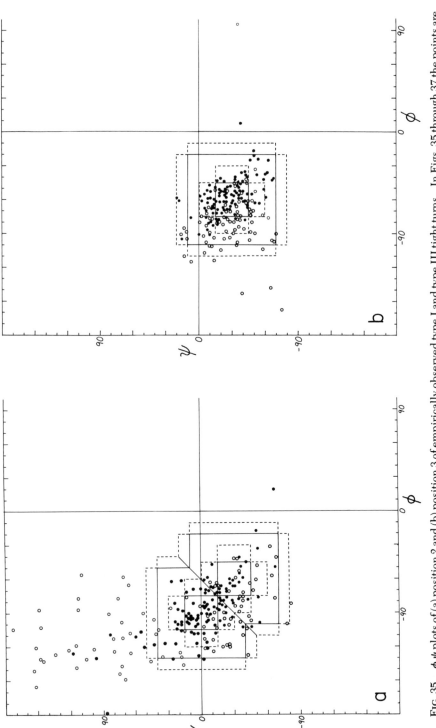

FIG. 35. ϕ, ψ plots of (a) position 2 and (b) position 3 of empirically observed type I and type III tight turns. In Figs. 35 through 37 the points are plotted from Chou and Fasman (1977); the outer dotted lines represent the limits of the turn type as defined in that reference and the inner dotted lines represent the limits used in Lewis *et al.* (1973); turns with hydrogen bonds are plotted as solid circles and those without as open circles.

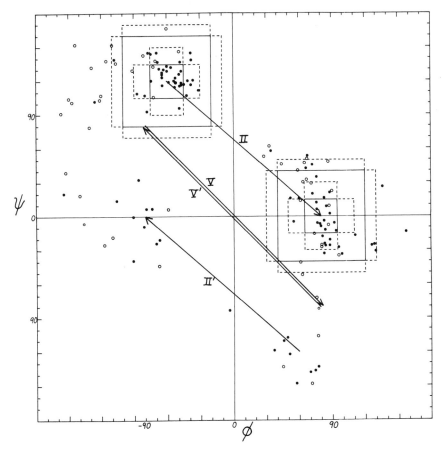

FIG. 36. ϕ,ψ plot for positions 2 and 3 of tight turns type II, II', V, and V'. Arrows go from ideal values of position 2 to position 3 for each turn type.

page and the chain entering at the lower left. In this orientation the β-carbons are always at or above the plane for types I, I', II, and II' turns (since only the backbone conformation can be mirrored). The virtual-bond dihedral angle defined by the four α-carbons is close to 0° for type II or II' and is somewhat positive (averaging about + 45°) for type I and somewhat negative for type I'. All four types have the third peptide essentially in the mean α-carbon plane. If the carbonyl oxygens are visible in an electron density map, then these four turn types can be fairly readily distinguished. In types I and II' the second carbonyl oxygen points approximately 90° down from the plane, while in types II and I' it points approximately 90° up. The first oxygen points nearly 90° down from the center of the plane in type I,

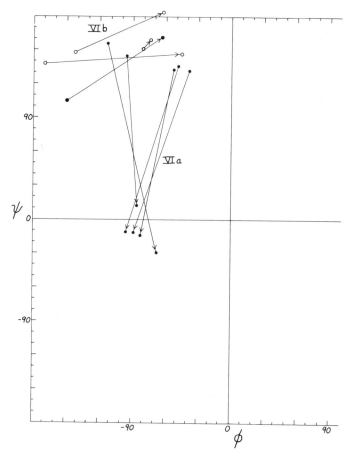

FIG. 37. ϕ,ψ plot for the *cis*-proline (type VI) turns from Chou and Fasman (1977), plus the two examples in the Bence-Jones protein REI. Arrows point from position 2 to position 3 (the proline) for each example. The two conformational groups are labeled as VIa and VIb.

nearly 90° up in type I′, slightly up in type II, and slightly down in type II′. The position of the second carbonyl oxygen, then, distinguishes between types I and II (or I′ and II′), while the position of the first carbonyl oxygen distinguishes types I vs II′ (or II vs I′). For either distinction intermediate cases should be rare, because they lie in a strongly prohibited region of the ϕ,ψ map.

The simple conception of a tight turn as approximately planar with a linear hydrogen bond is fairly accurate for type II. However, even an "ideal" type I turn is decidedly nonplanar, with the NH and CO of its hydrogen bond almost perpendicular to each other. That oxygen is in the plane of the last three α-carbons, but the first α-carbon and peptide are swung up out of the plane, producing the 45° virtual dihedral angle.

One additional sort of tight turn involving only three residues has been described theoretically (Nemethy and Printz, 1972) and also observed at least once in a protein structure (Matthews, 1972). This is the γ turn, which has a very tight hydrogen bond across a seven-atom ring between the CO of the first residue and the NH of the third (see Fig. 34b). It also can continue with a normal β sheet hydrogen bond between the NH of residue 1 and the CO of residue 3. Residues 1 and 3 are not far from the usual β conformation, while $\phi_2 = 70°$ and $\psi_2 = -60°$.

Although the presence of the hydrogen bond led to the initial characterization of tight turns by Venkatachalam (1968), hydrogen-bonding was dropped as a necessary condition as soon as any surveys were done on known protein structures (e.g., Crawford *et al.*, 1973), because numerous examples were found outside plausible hydrogen-bonding distance but with otherwise very turnlike conformation. About half of the turns listed in Chou and Fasman (1977) are hydrogen-bonded (by the criterion that O_1 to N_4 is less than 3.5 Å). Apparently the various turn conformations are sufficiently favorable so that they do not require stabilization by the hydrogen bond. This should not be surprising, since the turn types essentially consist of the basic α, β, and left-handed glycine conformations in various combinations. Also, since turns typically occur at the surface, peptides can hydrogen bond to solvent when not bonded to each other.

There are a number of characteristic residue preferences for tight turns. The most general is a strong tendency for turn residues to be hydrophilic (e.g., Kuntz, 1972; Rose, 1978), which might reflect inherent conformational preferences but is more probably a result of the almost universal location of turns at the protein surface where they interrupt or join together segments of secondary structure that are more internal. Glycines are quite common in tight turns, as can be inferred from their preference in types II, I', and II'; however, glycine is actually not quite as common as would be implied by energy calculations (see Chou and Fasman, 1977). Proline is also common in turns; besides the *cis*-proline turn, proline also fits well in position 2 of types I, II, and III turns and position 3 of type II'. About two-thirds of the Pro-Asn and Pro-Gly sequences in the known protein structures are found as the middle two residues of a tight turn (Zimmerman and Scheraga, 1977b).

Tight turns can combine with other types of structure in a number of ways. In addition to their classic role of joining β strands, they often occur at the ends of α-helices (see Section II,A). A type II turn forms a rather common combination next to a G1 β bulge (see Section II,D). Isogai *et al.* (1980) have surveyed the occurrence of successive tight turns, which either form approximately helical features or else form more complex chain reversals than single turns.

In addition to the approach described at length above, a number of authors have adopted a very useful but much looser approach to defining turns (see Kuntz, 1972; Levitt and Greer, 1977; Rose and Seltzer, 1977). Instead of a detailed conformational analysis in terms of conformational angles, hydrogen bonds, etc., these authors want a concept of turns that can be defined systematically and reproducibly from preliminary α-carbon coordinates and that (except for Levitt and Greer) is meant to include also the larger and more open direction-changes in the polypeptide chain. Kuntz looks at direction changes for $C\alpha_n - C\alpha_{n+1}$ vectors versus $C\alpha_{n+x} - C\alpha_{n+x+1}$ where x is 2 (for usual tight turns) or more; turns have direction changes greater than 90° and short $C\alpha_n$ to $C\alpha_{n+x}$ distances. Levitt and Greer assign turns to nonhelical, non-β segments for which the virtual dihedral angle defined by four successive α-carbons is between $-90°$ and $+90°$. Rose and Seltzer define turns as local minima in the radius of curvature calculated from points $C\alpha_{n-2}$, $C\alpha_n$, and $C\alpha_{n+2}$, with the modification that turns correspond only to places where the chain in both directions cannot be fitted inside a single rigid cylinder of 5.2 Å diameter. Of course, the conformational approach of Lewis *et al.* (1973) defining detailed turn types must start with a general definition also, which happens to be a $C\alpha_n$ to $C\alpha_{n+3}$ distance less than 7 Å. Conversely, any of the above general definitions could be used as a starting point from which to examine detailed conformations.

Both of these approaches have their strengths and their limitations. A great many people have deplored the fact that there is not very close agreement between the sets of turns (or of any other structure type) identified by any two different people (or computer programs) for a given protein. Each author suggests that the problem can be solved if everyone else accepts his definitions and criteria. Not only does this seem unlikely to happen, but it would probably be very undesirable. The major reason for the discrepancies is that each author has a different set of purposes for which he wants to use the structural characterization, and each is working from a data base of structures known to widely varying degrees of detail and reliability. Even the standard IUPAC-IUB conventions (1970) recognize two different definitions of α-helix, for instance: one based on ϕ,ψ values and the other on hydrogen-bonding.

The looser sort of turn definition derived from Cα positions might be the most useful one for studying protein folding and perhaps even for predicting turns. It can be applied to the largest available data base since it requires only α-carbon coordinates. However, in a very real sense it is not appropriate by itself for the accurately determined structures because it is incapable of taking into account the enormous amount of additional information they contain. For example, Levitt and Greer calculate possible hydrogen bonds from just Cα positions, which is a useful way of extending partial information; however, where atomic coordinates are available it turns out that a small but significant fraction of those bonds are definitely not present. Some form of detailed, conformational definition for turns is clearly needed for purposes such as energy calculations, examination of side chain influences, or correlation with spectroscopic observations such as CD or Raman. The detailed definitions will probably be increasingly fruitful as the number of highly refined protein structures increases; one severe problem at present is that they have usually been applied uncritically to all available sets of complete atomic coordinates, regardless of the fact that very few of the structures are known well enough for the distinction of 30° in dihedral angle between types I and III to have much meaning, and that in places where carbonyl oxygens were not visible in the electron density map even the distinction between types I and II is ambiguous. The best solution is probably to continue using whatever turn criteria are most appropriate for a given purpose but to state the criteria explicitly and to give careful consideration to selection of a data base.

By any sort of definition, turns are an important feature of protein structure. Kuntz (1972) found 45% of protein backbone in turns or loops; Chou and Fasman (1977) found 32% of protein chain in turns (counting four residues per turn); and Zimmerman and Scheraga (1977b) found 24% of the nonhelical residues in turns (counting only the central dipeptide). There are also some particular proteins whose structure appears heavily dependent on turns: Fig. 38 shows high-potential iron protein (Carter *et al.*, 1974), with the 17 turns in 85 residues indicated and their location at the surface evident.

Large portions of most protein structures can be described as stretches of secondary structure (helices or β strands) joined by turns, which provide direction change and offset between sequence-adjacent pieces of secondary structure. Tight turns work well as α–α and α–β joints, but their neatest application is at a hairpin connection

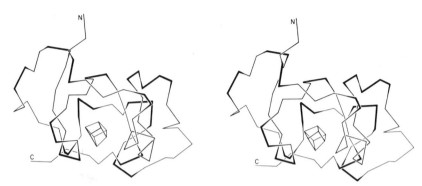

FIG. 38. Stereo drawing of the polypeptide backbone of high-potential iron protein. Tight turns are shown with their central peptide as a dark line. The box in the center represents the iron–sulfur cluster.

between adjacent antiparallel β strands, where the hydrogen bond of the turn is also one of the β sheet hydrogen bonds, as in Fig. 39.

It has often been suggested (e.g., Lewis *et al.*, 1971; Rose *et al.*, 1976) that turns can provide a decisive influence in directing the process of protein folding to the native conformation, since they seem ideally suited to help specify as well as encourage the decisive long-range interactions that form tertiary structure. As one simple example of the way in which turns can direct and specify other interactions, let us consider what happens when a tight turn immediately joins two adjacent antiparallel strands, as shown in Fig. 39. During the folding process it must be determined whether strand B will lie to the left or to the right of strand A. If the position of the tight turn is shifted by one residue along the sequence, then the turn must be made in the opposite direction to preserve hydrogen-bonding and amino acid handedness, as shown in the two parts of the figure. If the first strand runs from bottom to top and if the side chain in position 1 of the tight turn points toward you, then the second strand must lie to the right; if the side chain in position 1 of the turn points away, then the second strand must lie to the left. Since there are strong positional preferences for the residues in tight turns, it seems plausible that they can actually exert this sort of influence on the folding of neighboring strands.

D. Bulges

The β bulge (Richardson *et al.*, 1978) is a small piece of nonrepetitive structure that can occur by itself in the coil regions, but which most often occurs, and is most easily visualized, as an irregularity in an-

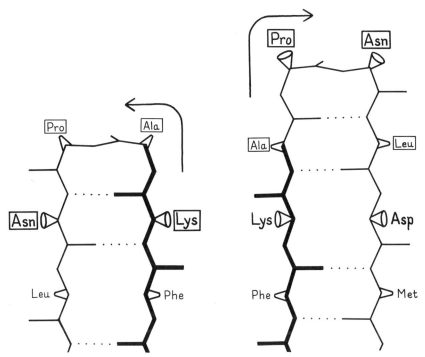

Fig. 39. An example of the effects of shifting the sequence location of a tight turn by one residue. The reference strand Phe-Lys-Ala is in the same position for both cases and is shown in heavy lines; the 4 turn residues are boxed. When the turn is at residues Lys-Ala-Pro-Asn the second β strand must lie to the left of the first, while if the turn is shifted to residues Ala-Pro-Asn-Leu the second β strand must lie to the right of the first. For the sequence illustrated here, the right-hand position would be preferred.

tiparallel β structure. A β bulge is defined as a region between two consecutive β-type hydrogen bonds that includes two residues on one strand opposite a single residue on the other strand. Figure 40 shows a β bulge from trypsin. The two residues on the bulged strand are called positions 1 and 2, and the one on the opposite strand position X. Sometimes the hydrogen bond to the CO of position X is forked, coming from the NH groups of both positions 1 and 2; the bulge in Fig. 40 shows this feature.

Only about 5% of the β bulges are between parallel strands, and most of the antiparallel ones are between a closely spaced (see Section II,B) rather than a widely spaced pair of hydrogen bonds. The additional backbone length of the extra residue on the longer side is accommodated partly by bulging that strand to the right and toward

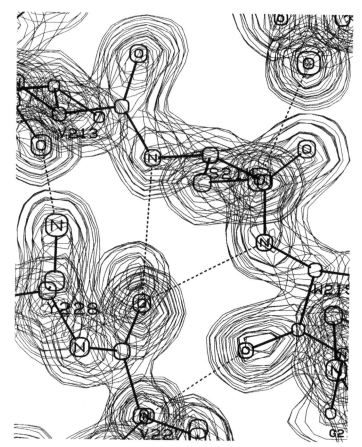

FIG. 40. A classic β bulge: the model and electron density from refined trypsin residues Ser-214, Trp-215, and Val-227. Courtesy of Chambers and Stroud.

you as seen in Fig. 40 and partly by putting a slight bend in the β sheet. Like tight turns, bulges affect the directionality of the polypeptide chain, but in a much less drastic manner. β bulges are not as common as tight turns, but over a hundred examples are known (see Richardson *et al.*, 1978, for a listing of 91).

β bulges can be classified into several different types, which are illustrated schematically in Fig. 41. By far the commonest is the "classic" β bulge, which occurs between a narrow pair of hydrogen bonds on antiparallel strands and has the side chains of positions 1, 2, and X all on the same side of the β sheet (see Fig. 41a). Residue 1 is in approximately α-helical conformation (averaging $\phi_1 = -100°$,

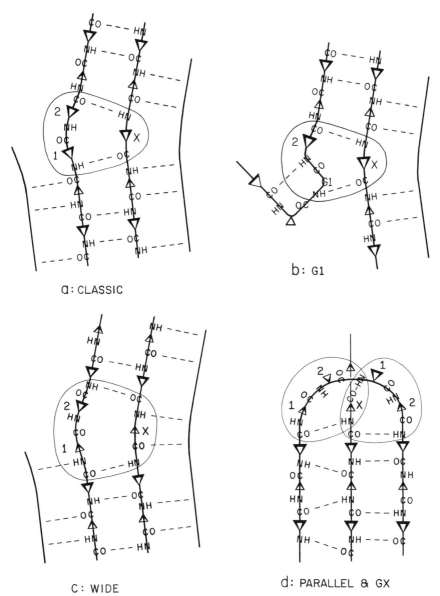

a: CLASSIC

b: G1

C: WIDE

d: PARALLEL & GX

FIG. 41. Diagrammatic illustrations of various types of β bulges (circled): (a) a classic β bulge; (b) a G1 β bulge, with the associated type II tight turn; (c) a wide β bulge; (d) a + 2x connection which forms a parallel β bulge on the left side and a Gx bulge on the right. Positions 1, 2, and X of the bulge are labeled. Small triangles represent side chains that are below the sheet, and larger triangles those that are above it.

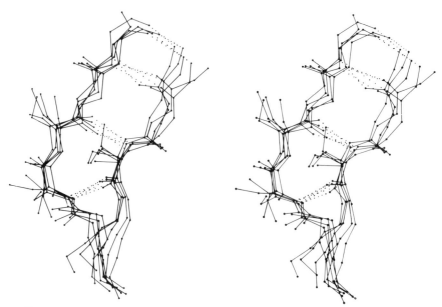

FIG. 42. Five superimposed examples of classic β bulges, in stereo: chymotrypsin Phe-41, Cys-42 opposite Leu-33; chymotrypsin Ala-86, Lys-87 opposite Lys-107; concanavalin A Leu-107, Ser-108 opposite Ala-196; carbonic anhydrase C Ile-90, Gln-91 opposite Val-120; and staphylococcal nuclease Ile-15, Lys-16 opposite Lys-24. Here and in Fig. 43 side chains are shown (out to Cγ) only for the three positions within the bulge. At the very bottom of this figure only the backbone is shown, where the two strands overlap in this projection.

$\psi_1 = -45°$) and residues 2 and X in approximately normal β conformation (averaging $\phi_2 = -140°$, $\psi_2 = 160°$, and $\phi_x = -100°$, $\psi_x = 130°$). Figure 42 is a stereo drawing of five examples of classic β bulges superimposed on one another. Note that the carbonyl oxygens on either side of residue 1 both point in about the same direction, as is typical of α-helical conformation, while the carbonyls surrounding residues 2 and X point opposite each other, as in β structure. Figure 42 also shows that a classic bulge locally accentuates the normal right-handed twist (see Section II,B) of the β strands.

The next most common type is the G1 bulge, illustrated schematically in Fig. 41b. It also lies between a narrow pair of hydrogen bonds, and position 1 is almost invariably a glycine because of its backbone conformation: $\phi_1 \simeq 85°$, $\psi_1 \simeq 0°$ and $\phi_2 \simeq -90°$, $\psi_2 \simeq 150°$. More than half of the G1 bulges are found within an interlocking structure in which the glycine in position 1 of the G1 bulge is also the required glycine in position 3 of a type II tight turn (see preceding section). The plane of the tight turn and its hydrogen bond is almost perpendicular to the plane of the G1 bulge. This combined structure has a consistent handedness which is dictated by the requirements of the three hydrogen bonds. A set of G1 bulges with their associated tight turns are shown superimposed in stereo in Fig. 43. For G1 bulges, the side chain of the glycine (if it had one) would be on the opposite side of the β sheet from those in positions 2 and X. In contrast to classic bulges, G1 bulges seldom have continued β structure on the bottom end (as seen in Fig. 41b or 43); when there is an associated tight turn the bulged strand enters at a very sharp angle, and in many of the remaining cases there is a short connection between the two strands of the bulge (so that position 1 equals either $x + 3$ or $x + 4$).

"Wide type" β bulges are those that occur between a widely spaced pair of hydrogen bonds on antiparallel strands, as shown schematically in Fig. 41c. They apparently are much less constrained than narrow bulges, since they occur with a great variety of backbone conformations; however, they do not occur as often. Even more unusual are the Gx bulges (so named because they often have a glycine in position X) with a hydrogen-bonding pattern similar to that shown in Fig. 41d, and the parallel bulges, which can take several forms, one of which is also illustrated in Fig. 41d.

Bulges have the general property of causing the normal β sheet al-

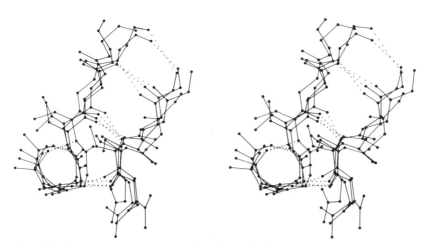

FIG. 43. Four superimposed examples of G1 β bulges with associated type II tight turns: trypsin Gly-133, Thr-134 opposite Ile-162; elastase Gly-204, Gly-205 opposite Thr-221; Bence-Jones REI V_L Gly-16, Asp-17 opposite Leu-78; and cytochrome c Gly-37, Arg-38 opposite Trp-59. The tight turn is the small hydrogen-bonded loop at the lower left; its plane is approximately perpendicular to the plane of the bulge. The α-carbon in the lower right corner of the loop is the required glycine in position 3 of the turn and position 1 of the bulge.

ternation of side chain direction to be out of register on the two ends of one of the bulge strands. The bulge can be thought of as turning over one-half of the out-of-register strand; the classic bulge accomplishes this by changing the ψ angle of the residue in position 1 by about 180°, which moves it from the β conformational region to the α region. Once half of the strand has flipped over, it must shift sideways by one residue along the other strand in order to hydrogen bond; that shift produces the bulge of two residues opposite one.

Bulges, as well as tight turns, are very often found at active sites, probably because they have a strictly local but specific and controlled effect on side chain direction. Another possible function for β bulges would be as a mechanism for accommodating a single-residue insertion or deletion mutation without totally disrupting the β sheet. There seem to be several such cases in the immunoglobulins, the clearest of which is a one-residue insertion in the C_H1 domain of Fab'NEW relative to the sequence of McPC603 C_H1: the NEW structure has a bulge in the middle of a long pair of β strands, while in McPC603 those strands form regular β structure all the way along.

The other general property of β bulges is that they alter the direction of the backbone strands forming them; classic bulges also accentuate the right-handed twist of the strands. This means that bulges are often useful for shaping large features of β sheet and/or extended hairpin loops. In antiparallel β barrels, for instance (see Section II,B), an extremely strong local twist is needed for closing barrels as small as five or six strands. Bulges in chymotrypsin, trypsin, elastase, staphylococcal nuclease, papain domain 2, and probably soybean trypsin inhibitor are strategically located at the sharpest corners in the

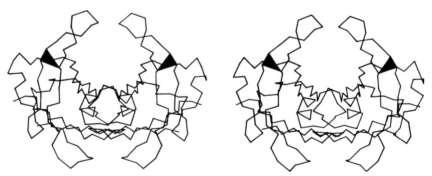

FIG. 44. Stereo view of the prealbumin dimer. The black triangles are β bulges which help to turn outward the β ribbons that form the loops proposed as a possible site for binding double-helical DNA.

β strands. Extended two-strand β ribbons are often used for forming external interaction sites on a protein. Such a ribbon would normally extend from the end of a β sheet or β barrel, but if the interaction site needs to be at one side of the sheet or barrel, then a β bulge at the point of departure can be used to direct the β ribbon out more nearly at right angles. For example, in the immunoglobulins the 47,48;35 bulge in the V_L domain and the 48,49;36 bulge in the V_H domain send out a pair of β ribbons that help complete closure around the V_L–V_H interdomain contact. In prealbumin the Phe-44,Ala-45;Val-32 bulge helps to turn out the extended β ribbon near the 2-fold axis of the dimer that forms the DNA-binding site proposed in Blake and Oatley (1977) (see Fig. 44).

E. Disulfides

Disulfide bridges are, of course, true covalent bonds (between the sulfurs of two cysteine side chains) and are thus considered part of the primary structure of a protein by most definitions. Experimentally they also belong there, since they can be determined as part of, or an extension of, an amino acid sequence determination. However, proteins normally can fold up correctly without or before disulfide formation, and those SS links appear to influence the structure more in the manner of secondary-structural elements, by providing local specificity and stabilization. Therefore, it seems appropriate to consider them here along with the other basic elements making up three-dimensional protein structure.

A modest amount of accurate conformational information is available from small-molecule X-ray structures of various forms of cystine. χ_1 angles are close to $+60°$, and all three dihedral angles internal to the disulfide are close to $\pm90°$. Two mirror-image conformations are observed; in N,N'-diglycyl-L-cystine dihydrate (Yakel and Hughes, 1954) and in L-cystine dihydrobromide (Peterson et al., 1960) or hydrochloride (Steinrauf et al., 1958) all three internal dihedral angles are approximately $-90°$, forming a left-handed spiral, while in hexagonal L-cystine (Oughton and Harrison, 1957, 1959) they are all approximately $+90°$, forming a right-handed spiral. Figure 45 shows the left-handed disulfide of L-cystine dihydrobromide, viewed down the 2-fold axis perpendicular to the S—S bond.

The dihedral angles of disulfides in proteins are very difficult to determine with any accuracy except in refined high-resolution structures. In the first few protein structures to show disulfides at 2 Å resolution, attention was paid mostly to the dihedral angle around the S—S bond (χ_3), since it is the most characteristically interesting

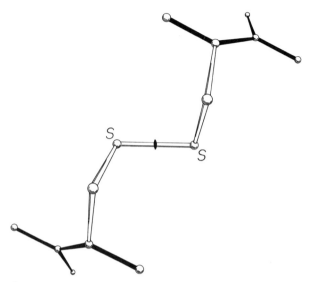

FIG. 45. The structure of L-cystine dihydrobromide, seen down the crystallographic 2-fold axis. The disulfide is in a left-handed spiral conformation.

parameter for cystine, is one of the easier ones to measure, and presumably is correlated with the handedness of the large optical rotation associated with the presence of disulfide bridges. As expected, the χ_3 angles were in the range around 90–100° and were found in both left-handed and right-handed forms (Blake *et al.*, 1967; Wyckoff *et al.*, 1970).

Now that about 70 different disulfides have been seen in proteins and more than 20 of those have been refined at high resolution, it is possible to examine disulfide conformation in more detail, as it occurs in proteins. Many examples resemble the left-handed small-molecule structures extremely closely; Fig. 46 shows the Cys-30–Cys-115 disulfide from egg white lysozyme. The χ_2, χ_3, and χ_2' dihedral angles and the $C\alpha$–$C\alpha'$ distance can be almost exactly superimposed on Fig. 45; the only major difference is in χ_1. All of the small-molecule structures have χ_1 close to 60°. Figure 47 shows the χ_1 values for half-cystines found in proteins. The preferred value is $-60°$ (which puts Sγ trans to the peptide carbonyl), while 60° is quite rare since it produces unfavorable bumps between Sγ and the main chain except with a few specific combinations of χ_2 value and backbone conformation.

If χ_3 is kept between 90° and 100° and χ_2 and χ_2' are varied, the distance between α-carbons across the disulfide can range all the way from about 4 to 9 Å. That corresponds to the extreme range observed

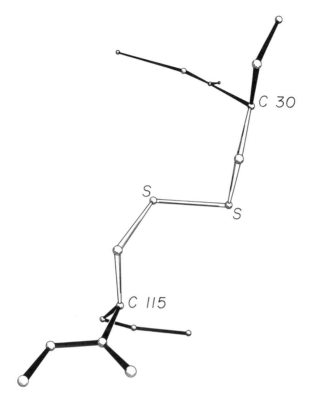

FIG. 46.   A left-handed spiral disulfide from hen egg white lysozyme, viewed from a direction similar to Fig. 45.

refined:

total:

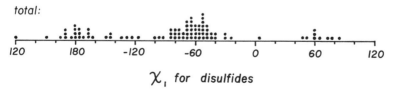

χ_1 for disulfides

FIG. 47.   The χ_1 angles observed for disulfides in protein structures.   The examples from refined, high-resolution structures are shown separately at the top.

for all disulfides. However, the effective range for Cα—Cα' distance is really a good deal narrower than that, since 85% of all the disulfides and 95% of the refined ones fall between 4.4 and 6.8 Å.

Now let us examine the relationships between handedness of χ₃, Cα—Cα distance, and χ₂ values. Among the disulfides for which coordinates were available at 2 Å resolution or better (Deisenhofer and Steigemann, 1975; Imoto *et al.*, 1972; Wyckoff *et al.*, 1970; Quiocho and Lipscomb, 1971; Saul *et al.*, 1978; Epp *et al.*, 1975; Huber *et al.*, 1974; Chambers and Stroud, 1979; Hendrickson and Teeter, 1981; Brookhaven Data Bank, 1980; Feldmann, 1977), there are equal numbers with right-handed and left-handed χ₃. The average Cα—Cα' distance across the left-handed ones is 6.1 Å, exactly what was seen in the small-molecule structures, but for the right-handed ones the average Cα—Cα' distance is 5.2 Å. Clearly the two sets of disulfides as they occur in proteins cannot simply be mirror images of one another.

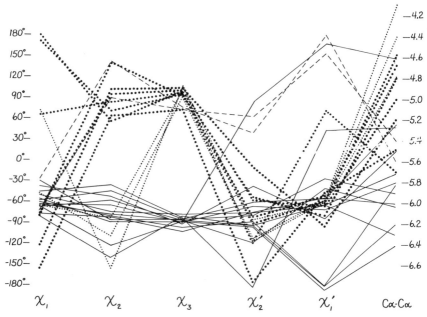

FIG. 48. Plot of all five dihedral angles and Cα-Cα distance for the disulfides from refined, high-resolution (2 Å or better) protein structures. The left-handed disulfides (almost all of which are left-handed spirals) are shown with solid lines, the − + + (right-handed hook) disulfides with large dots, the − + − (short right-handed hook) disulfides with small dots, and the + + + (right-handed spiral) disulfides with dashed lines. The long immunoglobulin disulfides are not included here.

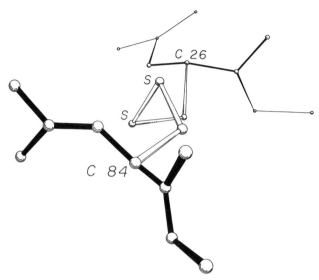

FIG. 49. A left-handed spiral disulfide from ribonuclease S, viewed end-on.

Figure 48 plots all five side chain dihedral angles (χ_1, χ_2, χ_3, χ_2', and χ_1') and the $C\alpha$—$C\alpha'$ distances for the high-resolution disulfides. The overall conformations fall into just two major and two minor categories (plus perhaps one more to be discussed below). Essentially all of the left-handed ones (solid lines in Fig. 48) have approximately $\chi_1 = -60°$, $\chi_2 = -90°$, $\chi_3 = -90°$, $\chi_2' = -90°$, $\chi_1' = -60°$. This can be called the left-handed spiral conformation, and is the same as that seen in Fig. 46. An example from ribonuclease S is shown in Fig. 49, looking down the spiral. A majority of the right-handed disulfides have a conformation of approximately $\chi_1 = -60°$, $\chi_2 = +120°$, $\chi_3 = +90°$, $\chi_2' = -50°$, $\chi_1' = -60°$ (heavy dots in Fig. 48) and a $C\alpha$ separation averaging only 5 Å. This can be called the right-handed hook conformation; a typical example is shown in Fig. 50. Two cases have $- + -$ dihedral angles and a $C\alpha$–$C\alpha$ distance of 4–4.5 Å; these could be called short right-handed hooks. Then there are two cases of right-handed spirals (dashed lines in Fig. 48) which tend to have one χ_1 angle of 180° and are still significantly shorter than the left-handed spirals.

The only cases that were omitted from Fig. 48 are the disulfides that span the β barrels in immunoglobulins (Epp et al., 1975; Saul et al., 1978). They are unusually long, with $C\alpha$ separations of 6.6 to 7.4 Å, which is achieved by having both χ_2 and χ_2' close to 180°. These long disulfides are trans–gauche–trans (180°, ±90°, 180°) in χ_2, χ_3, χ_2',

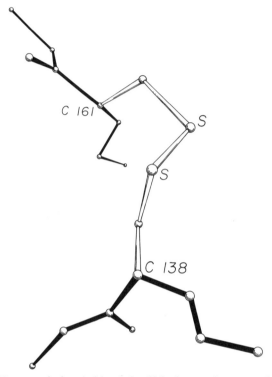

FIG. 50.   A right-handed hook disulfide from carboxypeptidase A.

while both the spiral and the hook conformations described above are gauche–gauche–gauche ($\pm 90°$, $\pm 90°$, $\pm 90°$). The χ_1 angles are generally either 180 or 60° (in contrast to the usually preferred $-60°$) and the χ_3s show no preference for left-handed vs right-handed. Indeed, in the Bence-Jones V_L dimer REI (Epp *et al.*, 1975) the disulfides were found to be a disordered mixture of the right-handed and left-handed forms. Asymmetrical preferences for the handedness of χ_3 presumably must involve either direct or indirect interactions with the backbone; with χ_2 near 180° such interactions are minimized. However, one must also presume that these unusually long disulfides are at least slightly strained and that χ_2 angles in the 60–120° range are more favorable as a general rule. It also may well be true that χ_3 is even more strongly constrained to $\pm 90°$ than is χ_2, but the angle distributions seen in Fig. 48 should not be taken as proof that that is so, because preferred values for χ_3 and not for χ_2 have been built into almost all model-building and refinement routines. Molecular orbital calculations for χ_3 by Pullman and Pullman (1974) give an energy minimum at 100°, but do not rise by more than 0.5 kcal/mol from about 70 to 140°. They did not report calculations for χ_2.

The distinctive differences between the left- and right-handed disulfide conformations have little to do with the χ_3 angle itself. The bumps of the sulfurs with the polypeptide backbone are produced by a combination of χ_1 and χ_2; since it is unfavorable to have χ_2 in the range of $+60$ to $+100°$ when χ_1 has its preferred value near $-60°$, the right-handed disulfides cannot adopt a $+++$ spiral in proteins.

In summary, we can expect that most disulfides will have Cα separations of less than 6.5 Å unless they are stretched across a β barrel or perhaps a short loop. The majority will have either the left-handed spiral conformation or the right-handed hook conformation.

Now let us examine the distribution and position of disulfides in proteins. The simplest consideration is distribution in the sequence (see Fig. 51), which is apparently quite random, except that there must be at least two residues in between connected half-cystines. Even rather conspicuous patterns such as two consecutive half-cystines in separate disulfides turn out, when the distribution is plotted for the solved structures (Fig. 51), to occur at only about the random expected frequency. The sequence distribution of half-cystines is influenced by the statistics of close contacts in the three-dimensional structures, but apparently there are no strong preferences of the cystines that could influence the three-dimensional structure.

Disulfide topology may be considered in terms of the possible patterns of cross-bridges on a floppy string. The cases that occur appear to be a random selection among the possible alternatives, showing no evident preferences for or against any particular features (such as whether nearest-neighbor half-cystines are connected or whether the total connectivity can be drawn in two dimensions). This situation is a very marked contrast to what we found for β sheet topology, where there are a number of quite strong topological preferences. However,

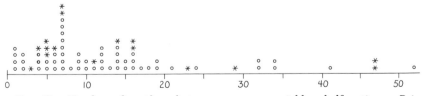

FIG. 51. Number of residues between sequence-neighbor half-cystines. Pairs which are in the same disulfide are shown by asterisks and those which are not by circles.

the topological and sequence randomness of disulfides is what one would expect if their major role is to stabilize close contacts in the final structure but they have no influence on the early stages of the protein-folding process.

We can also examine what types of backbone conformation are found at the ends of disulfides, and here we can see some preferences again. Well over half of the backbone strands are in extended conformation, although a relatively small fraction are actually part of a β sheet. It is not possible for a disulfide to join neighboring strands in a β sheet: any but the closest residues on adjacent strands are too far apart, and a closest pair of residues is slightly too close together. Also, for a close pair on β sheet the $C\alpha$—$C\beta$ bonds of the two are approximately parallel, while they need to be approximately perpendicular for a right-handed hook and antiparallel for either right- or left-handed spirals. Occasionally disulfides join next-nearest-neighbor β strands with only some disruption of the intervening hydrogen-bonding pattern (e.g., Cys-40–Cys-95 in ribonuclease S). An even more common relationship to β sheet is a disulfide joining the continuation of two nearest-neighbor strands after they have separated from the β sheet and can attain a more suitable separation and angle. Typically the disulfide is one or two residues out from the last hydrogen bond on one strand and three or four residues out on the other strand; the β strands are almost always antiparallel rather than parallel. This sort of arrangement is somewhat reminiscent of a β bulge (see Section II,D), and for the case of Cys-65–Cys-72 in ribonuclease S the disulfide actually spans one end of a wide type β bulge.

α-Helix is also quite common as the backbone conformation flanking a disulfide, but there is seldom well-formed helix on both ends of a given disulfide. If one end comes from an α-helix, the other end will usually be an extended chain, or one or more tight turns, or irregular structure past the end of a helix. Presumably the constraints on favorable separation and angle of $C\alpha$—$C\beta$ bonds in disulfides are difficult to satisfy with residues in any of the normal helix packing arrangements (see Section II,A). There are no disulfides at all in any of the helix-bundle structures (see Section III,B). Disulfides do connect a pair of adjacent helices, however, in phospholipase A_2 and in crambin.

In cases in which backbone direction is readily definable (primarily for extended or helical chains), the two chains joined by a disulfide almost always cross at steep angles to one another (60 to 90°).

The third most frequent backbone conformation at disulfide ends is a tight turn. Sometimes it is a succession of turns, or bit of 3_{10}-helix.

FIG. 52. A schematic backbone drawing of insulin, a small structure which is dependent on its disulfides for stability.

Turns seem to be somewhat favored at the hook end of a right-handed hook disulfide.

There is a correlation between the backbone conformations which commonly flank disulfides and the frequency with which disulfides occur in the different types of overall protein structure (see Section III,A for explanation of structure types), although it is unclear which preference is the cause and which the effect. There are very few disulfides in the antiparallel helical bundle proteins and none in proteins based on pure parallel β sheet (except for active-site disulfides such as in glutathione reductase). Antiparallel β sheet, mixed β sheet, and the miscellaneous α proteins have a half-cystine content of 0–5%. Small proteins with low secondary-structure content often have up to 15–20% half-cystine. Figure 52 shows the structure of insulin, one of the small proteins in which disulfides appear to play a major role in the organization and stability of the overall structure.

F. Other Nonrepetitive Structure

When there is a need to divide up protein structure for purposes of description, prediction, spectroscopic characterization, etc., the usual categories have been "helix," "β structure," and "coil." "Coil" is defined in practice as "none of the above." Its major distinguishing feature is that it is nonrepetitive in backbone conformation (although depending on one's definition of β structure, one might or might not include an isolated piece of extended chain as coil). Sometimes coil is referred to as "random coil," or is modeled by properties observed in 6 M guanidine hydrochloride. This seems unfortunate, since the actual portions of crystallographically determined protein structures generally described as coil are not random or disordered in any sense of the word; they are every bit as highly organized and firmly held in place as the repeating secondary structures—they are simply harder to describe. In recent years since recognition of the wide occurrence and importance of tight turns (see Section II,C), they are often separated out as a category of structure, so that now the miscellaneous "coil" category would refer to what is neither helix, β, nor turn.

Turns and bulges are nonrepetitive features which are characterized primarily in terms of backbone conformation and backbone hydrogen bonding. However, much coil structure appears to be very strongly influenced by specific side chain interactions. These have not been very widely analyzed, but a few examples can illustrate the sorts of patterns to be expected. Probably the earliest notice of such a feature is the observation in Kendrew *et al.* (1961) that serine or threonine frequently hydrogen-bonds to a backbone NH exposed at the beginning of an α-helix. Figure 53 illustrates such a conformation. Energy calculations for side chain interactions have, quite understandably, considered only the very local region (e.g., Anfinsen and Scheraga, 1975). Recently there have been some systematic empirical computer surveys of side chain environments, such as in Warme and Morgan (1978) and Crippen and Kuntz (1978), and surveys of side chain conformations, such as Janin *et al.* (1978) and Bhat *et al.* (1979). These do not focus specifically on nonrepetitive structure, but any strong preferences would be especially influential there. So far, however, these studies are still at the initial stages of tabulating raw statistical preferences.

At the other extreme, it is possible to examine individual examples of a single potentially interesting type to see simply what features turn

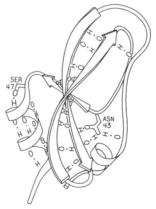

FIG. 53. The main chain hydrogen bonds of basic pancreatic trypsin inhibitor, plus two of the side chains whose hydrogen bonds stabilize the ends of pieces of secondary structure: Ser-47 at the beginning of an α-helix and Asn-43 at the end of a β strand.

TABLE I

*Locations of the Asparagine Residues Are Tabulated
for β Sheets of at Least Three Strands (and Known
Amino Acid Sequence) in the Known Protein
Structures*[a]

	Asparagine position in β sheets		
	At end of strand	On a side strand	In middle of sheet
Antiparallel	30	17	1
Parallel	10	4	3

[a] For all residues in β sheet, approximately 50% are in the middle and 50% at either an end or a side. In contrast, less than 20% of the asparagines in parallel sheet are in the middle, and only 2% of those in antiparallel sheet are in the middle.

up. Asparagine is one such potentially interesting residue, since it combines a side chain that mimics a backbone peptide, along with the conformational constraints of possessing only two side chain variable angles. Asn is more likely than any other nonglycine residue to have ϕ,ψ angles outside the normally allowed regions. Also, it does indeed show a pattern of hydrogen-bonding that is distinct from either glutamine or aspartate. Asparagine is more than twice as likely as any other residue to bond to the first exposed backbone NH or CO at the end of an antiparallel β strand (as in, for example, Fig. 53) and probably for that reason is essentially forbidden from occurring in the interior of antiparallel β sheet although it is very common at or just beyond the edges (see Table I). Asparagine also shows a favored type of hydrogen bond from its side chain CO to the backbone NH of residue $n + 2$. This pattern happens to predispose a type I tight turn at residues $n + 1$ and $n + 2$ (see Fig. 54).

Figure 55a shows a small piece of nonrepetitive structure from pancreatic trypsin inhibitor in which the side chain Oδ of Asn-24 is hydrogen-bonded to both the $n + 2$ and $n + 3$ backbone NH groups. There is a G1 β bulge (see Section II,D) at Gly-28, Leu-29, Asn-24. The corner between the two strands is turned by what could be better described as a five-residue turn with one α-helical hydrogen bond and an Asn stabilizing the loop NH groups, rather than as two successive non-hydrogen-bonded turns (which is how it shows up in any computer search for tight turns). An extremely similar conformation occurs in prealbumin (Blake *et al.*, 1978), with the same G1 bulge and

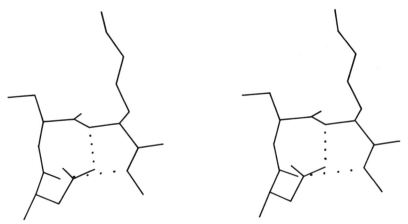

FIG. 54. An asparagine side chain making a hydrogen-bond to the main chain NH of residue $n + 2$, an arrangement which helps stabilize the central peptide of a tight turn. Residues 91–93 from chymotrypsin.

five-residue "α" turn stabilized this time by Asp-18 instead of an asparagine, and with an additional bond to the Nε of Arg-21 (see Fig. 55b).

It will be interesting in the future to see fuller characterization of nonrepetitive structure, especially since it forms many of the more complicated enzyme active sites. A few instances have so far been described in which a particular organization of coil structure can be recognized as providing a particular type of functional site, since it occurs with very similar patterns in more than one example. One of these structures is a loop which binds iron–sulfur clusters in both ferredoxin and high-potential iron protein (Carter, 1977). Another such structure is the central loop portion of the "E–F hands" that form the calcium-binding sites in carp calcium-binding protein (Kretsinger, 1976). The backbone structures curl around in very similar conformations and provide ligands in a definite order to the six octahedral coordination sites around the Ca^{2+}, as illustrated in Fig. 56.

G. Disordered Structure

In contrast to the well-ordered but nonrepetitive coil structures, there are also genuinely disordered regions in proteins, which are either entirely absent on electron density maps or which appear with a much lower and more spread out density than the rest of the protein. The disorder could either be caused by actual motion, on a time scale of anything shorter than about a day, or it could be caused by having multiple alternative conformations taken up by the different mole-

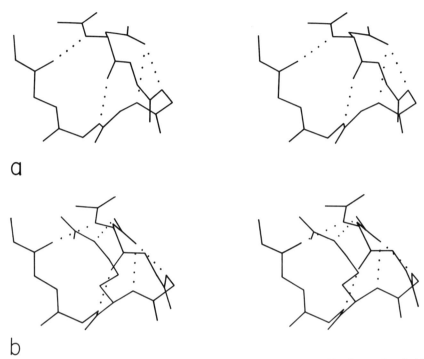

FIG. 55. Two very similar 5-residue turns with a single α-helical hydrogen bond: (a) pancreatic trypsin inhibitor residues 24–28, stabilized by the side chain of Asn-24; (b) prealbumin residues 18–22, stabilized by Asp-18.

cules in the crystal. Well-ordered side chains also move, often very rapidly (see Wüthrich and Wagner, 1978; McCammon *et al.*, 1977), but the movements are brief departures from a single stable conformation.

One simple case of disordered structure involves many of the long charged side chains exposed to solvent, particularly lysines. For example, 16 of the 19 lysines in myoglobin are listed as uncertain past Cδ and 5 of them for all atoms past Cβ (Watson, 1969); for ribonuclease S Wyckoff *et al.* (1970) report 6 of the 10 lysine side chains in zero electron density; in trypsin the ends of 9 of the 13 lysines refined to the maximum allowed temperature factor of 40 (R. Stroud and J. Chambers, personal communication); and in rubredoxin refined at 1.2 Å resolution the average temperature factor for the last 4 atoms in the side chain is 9.2 for one of the four lysines versus 43.6, 74.4, and 79.3 for the others. Figure 57 shows the refined electron density for the well-ordered lysine and for the best of the disordered ones in ru-

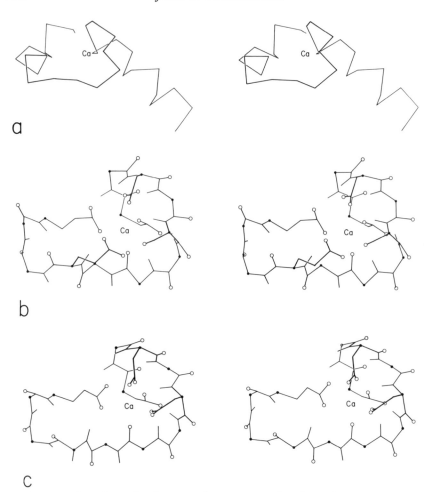

FIG. 56.   The calcium-binding sites from carp muscle calcium-binding protein: (a)
backbone of the entire "E-F hand"; (b) detailed view of the E-F calcium-binding site,
including those side chains which are Ca ligands; (c) detailed view of the C-D
calcium-binding site, rotated to match part b.   Oxygens are shown as open circles and
α-carbons as solid dots.

bredoxin.   Interestingly, arginine side chains do not follow this same
pattern.   In the structures quoted above, 70% of the arginines were
well ordered, as opposed to only 26% of the lysines.   In refinement at
very high resolution it is sometimes possible to express a partially disordered side chain as a mixture of two different specific conformations, as, for instance, isoleucines 7 and 25 in crambin (Hendrickson
and Teeter, 1981).

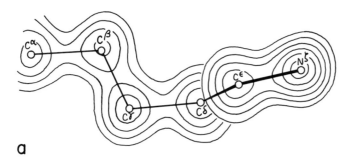

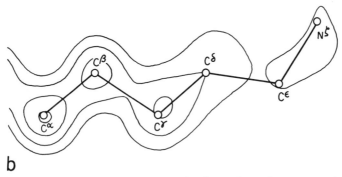

FIG. 57.   Model and electron density in rubredoxin after refinement at 1.2 Å resolution, for (a) the well-ordered lysine, Lys-46 (temperature factor average of 9.2 for the outer four atoms of the side chain); (b) the best of the disordered lysines, Lys-3 (temperature factor average of 43.6 for the outer four side chain atoms).   From Watenpaugh *et al.* (1980), Fig. 12, with permission.

It is fairly common for a few residues at the N-terminus or the C-terminus of the polypeptide chain to be disordered, if they include no large hydrophobic residues.   In some cases it is known that these dangling ends can be cleaved off without any loss of stability or activity in the protein (e.g., Anfinsen *et al.*, 1971).   However, there are certainly other cases in which disordered regions have very definite functional roles.   Sometimes there is disorder in one state and an ordered conformation in another state, with the contrast between the two having a functional role; for example, the intersubunit salt linkages in hemoglobin which provide constraints in the deoxy form and are free and disordered in the oxy form (Perutz, 1970, 1978).   In systems designed for specific proteolytic cleavage, disorder is one way of promoting cleavage at a given loop.   The new chain ends liberated by such cleavage are also very often disordered, as for instance in chymotrypsin (Birktoft and Blow, 1972). In some cases a ligand-binding site may show partial or complete disorder in the absence of the ligand but become well-ordered when the ligand is bound, as for instance the RNA site of tobacco mosaic virus protein (Butler and Klug, 1978).

One of the most intriguing recent examples of disordered structure is in tomato bushy stunt virus (Harrison *et al.*, 1978), where at least 33 N-terminal residues from subunit types A and B, and probably an additional 50 or 60 N-terminal residues from all three subunit types (as judged from the molecular weight), project into the central cavity of the virus particle and are completely invisible in the electron density map, as is the RNA inside. Neutron scattering (Chauvin *et al.*, 1978) shows an inner shell of protein separated from the main coat by a 30-Å shell containing mainly RNA. The most likely presumption is that the N-terminal arms interact with the RNA, probably in a quite definite local conformation, but that they are flexibly hinged and can take up many different orientations relative to the 180 subunits forming the outer shell of the virus particle. The disorder of the arms is a necessary condition for their specific interaction with the RNA, which cannot pack with the icosahedral symmetry of the protein coat subunits.

Although disordered structure is fairly common in the known protein structures, this is undoubtedly one of the cases in which the process of crystallization induces a bias on the results observed. Since extensive disorder makes crystals much harder to obtain, it seems probable that disordered regions are even more prevalent on the proteins that do not crystallize.

H. Water

In a very real sense, the structure of the closely bound water molecules around a protein are a part of the protein structure: they determine conformation of the exposed side chains, stabilize the ends of secondary structures, and occupy positions at active sites where they influence substrate binding and sometimes catalysis. The properties of the bulk water are critical in stabilizing the folded native form of proteins (e.g., Kuntz and Kauzmann, 1974), but it is only the bound water that we will consider to be an actual part of, rather than an influence on, the protein structure.

In high-resolution X-ray structures of proteins it is usual for a small number of solvent molecules to appear fairly clearly as peaks in the electron density map (see Fig. 13). Now that various refinement techniques are being applied to many protein structures, determination of water positions is usually a part of the process. In only a few cases,

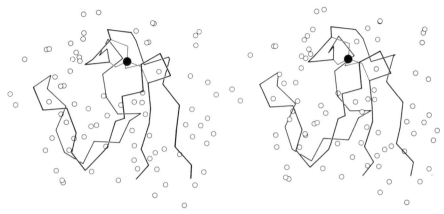

FIG. 58.   Stereo drawing of the rubredoxin backbone with the iron (filled circle) and its cysteine sulfur ligands and all the water molecules (open circles) identified during refinement of the structure at 1.2 Å resolution.   Adapted from Watenpaugh *et al.* (1979), Fig. 11, with permission.

such as the study of rubredoxin in Watenpaugh *et al.* (1978) and the study of actinidin in Baker (1980), has a real attempt been made to locate all of the fairly tightly bound waters and to eliminate spurious peaks.   Figure 58 shows the waters around rubredoxin.   Occupancies as well as positions are refined so that partially ordered as well as tightly bound water can be located.   It is in fact only relatively few waters for which the occupancy approaches 1 (23 of the 130 waters located in rubredoxin had occupancies ≥ 0.9).

Another recent study which provides less direct, but also very detailed, information about the water around a protein is the Monte Carlo calculations performed by Hagler and Moult (1978) for egg lysozyme.   From random starting positions they obtain a very long series of possible sets of water positions for which the statistical properties must obey all the constraints of the energy functions used.   Contour maps can be plotted giving the overall frequency of water location at each point, and they match the refined X-ray electron density contours quite well.   Also, individual sets of positions at single cycles in the simulation can be examined.   The energies of water molecules in various types of locations can be determined for the overall simulation and can be compared with the energy distribution for the bulk water.

The detailed study of water structure around proteins is only just beginning, but a number of conclusions can be drawn from the crystallographic and theoretical work that has already been done.   Isolated water molecules occur trapped inside protein interiors, where they

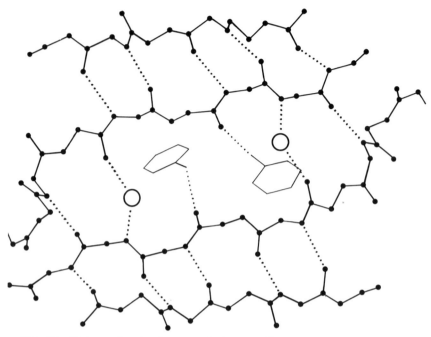

FIG. 59. Water molecules (open circles) in prealbumin, bridging between main chain groups that are too far apart to continue β-type hydrogen-bonding between strands. A hydrogen bond to a tyrosine side chain is also shown.

can fill defects in the side chain packing and usually make some hydrogen bonds to protein atoms. Their energies are rather high, but it is much better to have a water than an empty hole in those locations. The number of such internal waters varies very widely from one protein to another. Both for internal and for surface waters, it is very common that they bond to the first free backbone NH or CO groups at the ends of pieces of secondary structure; for β strands it is common that the last H-bond opens up wider, with a water bridging in between (see Fig. 59).

The most ordered surface waters are those around charged side chains or in surface crevices. Occasionally those crevices can be very deep, such as the active site pocket in carbonic anhydrase, which extends about 15 Å in from the surface, with a network of water molecules (Lindskog et al., 1971). The well-ordered waters at the protein surface are usually part of an approximately tetrahedral (but sometimes planar trigonal) network of hydrogen bonds to the protein and to other waters. An example from rubredoxin is shown in Fig. 60.

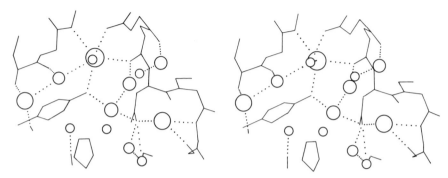

FIG. 60. A stereo view of one of the hydrogen-bonded networks of water molecules at the surface of the rubredoxin molecule [adapted from Watenpaugh *et al.* (1978), Fig. 7, with permission]. The size of the waters is proportional to their occupancy factors, so that the most well-ordered waters are shown largest.

Both crystallographically and also from vapor-pressure measurements of solvent stabilization (Wolfenden, 1978) it appears that water hydrogen bonds more frequently and more strongly to peptide CO groups than NH groups. In rubredoxin, only 24% of the available backbone NH groups are bonded to water and 70% to other protein atoms, while for the CO groups 43% bond to waters, 41% to protein atoms, and another 8% to both (Watenpaugh *et al.*, 1978).

Many of the tightly bound waters have energies substantially lower than the bulk water (Hagler and Moult, 1978). All studies have found that most of the bound waters, and all of the highly ordered ones, are in the first coordination layer, but that they do not by any means cover the whole protein surface. A substantial number of partially ordered waters are found in the second coordination layer (where they hydrogen-bond to the protein only through first-layer water) and essentially none any further out than that, even where there are suitably sized channels between neighboring protein molecules. The degree of motion seen for individual water molecules also increases dramatically as a function of their distance out from the protein.

I. Subunits and Domains

Many protein molecules are composed of more than one subunit, where each subunit is a separate polypeptide chain and can form a stable folded structure by itself. The amino acid sequences can either be identical for each subunit (as in tobacco mosaic virus protein), or similar (as in the α and β chains of hemoglobin), or completely different (as in aspartate transcarbamylase). The assembly of many identical subunits provides a very efficient way of constructing

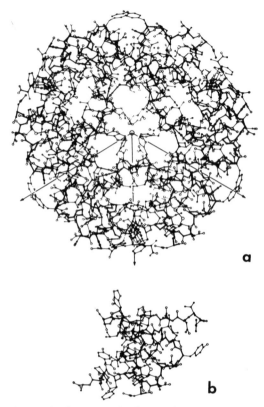

FIG. 61. (a) The insulin hexamer; (b) the insulin monomer. From Blundell *et al.* (1972), with permission.

large structures such as virus coats. Often a multisubunit molecule is more smoothly globular than its component subunits are, as for instance in the insulin hexamer shown in Fig. 61.

The surfaces that form subunit–subunit contacts are very much like parts of a protein interior: detailed fit of generally hydrophobic side chains, occasional charge pairing, and both side chain and backbone hydrogen bonds. Twofold symmetry is the most common relationship between subunits. The 2-fold is often exact and can be part of the actual crystallographic symmetry, as for the prealbumin dimer in Fig. 62. However, in many cases (e.g., Tulinsky *et al.*, 1973; Blundell *et al.*, 1972) individual side chains very close to the approximate 2-fold axis must take up nonequivalent positions in order to avoid overlapping (see Fig. 63). Conformational nonequivalence can extend further away from the axis and produce such effects as different

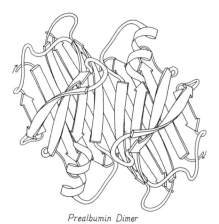

Prealbumin Dimer

FIG. 62. A schematic drawing of the backbone of the prealbumin dimer, viewed down the 2-fold axis. Arrows represent β strands. Two of these dimers combine back-to-back to form the tetramer molecule.

binding constants for ligands (e.g., Hill *et al.*, 1972). Tetrahedral 222 symmetry is also common, either with only one or with all three 2-folds exact (e.g., Adams *et al.*, 1969).

A rather common feature of subunit contacts is β sheet hydrogen-bonding between strands in opposite subunits. Theoretically the relationship could be a pure translation or a 2-fold screw axis with a one-residue translation (for a pair of parallel strands), but all the known cases of intersubunit β sheet bonding turn out to be between equivalent strands related by a local 2-fold axis. For hydrogen-bond formation, the 2-fold must be perpendicular to the β sheet, requiring the two equivalent strands to be antiparallel. Those may be the only two β strands (as in insulin, Fig. 63), or they may be part of antiparallel β sheets (as in prealbumin, Fig. 62), or the rest of the sheets may be parallel (as in alcohol dehydrogenase domain 1).

Similar subunit structures can assemble in quite different ways. The Greek key β barrel of Cu,Zn superoxide dismutase assembles back-to-back across a tight, hydrophobic side chain contact, while the Greek key β barrel of prealbumin joins by continuing the β sheet bonding side-by-side. Even for proteins known to be closely related the subunits may associate in nonhomologous ways, such as the monomers versus tetramers in various hemoglobins, or the atypical contact between chains in the Bence-Jones protein Rhe (Wang *et al.*, 1979). The homologous domains in the immunoglobulin chains associate in three quite different types of pairwise contact: the usual "back-to-back" barrel contact of V_L and V_H (shown in Fig. 101); C_L and

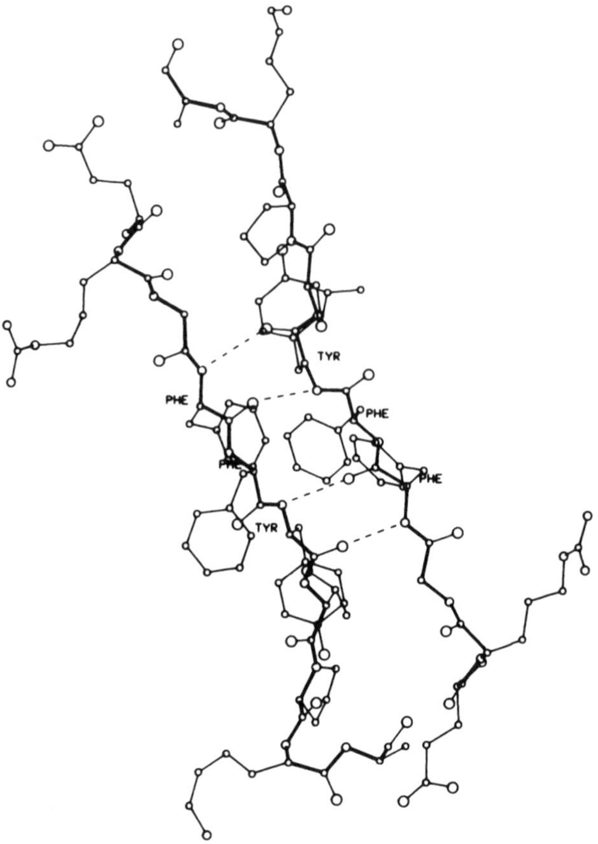

FIG. 63.   Departures from local 2-fold symmetry, especially of side chain positions, in the β strand dimer interaction of insulin.   From Blundell *et al.* (1972), with permission.

C_H1 barrels contact "front-to-front"; and pairs of C_H2 domains are quite widely separated by carbohydrate (Silverton *et al.*, 1977).   Of course, the great majority of homologous proteins have homologous subunit contacts, but it seems that even a quite drastic change in subunit contacts is easier to accommodate than major internal rearrangement.

Twofold contacts are self-homologous—formed by equivalent surfaces from each of the participating subunits, while the occasional 3-fold (e.g., bacteriochlorophyll protein), 4-fold (hemerythrin), or 17-fold (tobacco mosaic virus) contact is heterologous—formed by joining two different surfaces.   An especially interesting type of heterologous contact has been found in hexokinase (Steitz *et al.*, 1976). It was previously assumed that self-associating arrays with a definite number of subunits had to be related by a closed, point-group symmetry operation in order to avoid producing infinite aggregation (e.g., Klotz *et al.*, 1970). Hexokinase, however, has a 156° rotation plus a 13.8 Å translation between subunits, which demonstrated clearly that screw axes are also permissable as long as addition of a third subunit by the same screw operation is blocked by overlap with the first subunit. Such a screw-axis relationship can easily produce very marked nonequivalence between chemically identical subunits.

Subunit contacts need to be relatively extensive and stable if they are to ensure subunit association in the absence of a covalent link. However, in some cases a subunit contact can shift back and forth between two different stable positions, as has been demonstrated for oxy- versus deoxyhemoglobin (Perutz, 1970). Allosteric control can then be exerted by any factors which either affect the local conformation or bind between the subunits. A less elegant but even more extreme example is lamprey hemoglobin, which dissociates altogether in the oxy form (Hendrickson and Love, 1971).

FIG. 64. The tightly associated domains (one shown light and the other dark) of elastase. Figures 64 through 66 use a space-filling representation with a sphere around each α-carbon position; they were photographed from Richard Feldmann's molecular graphics display at the National Institutes of Health.

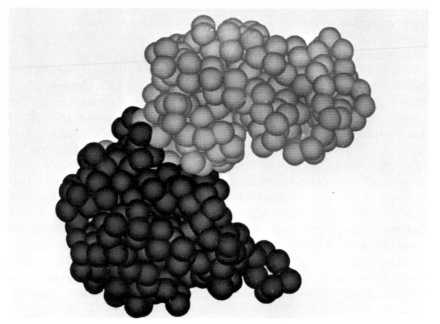

FIG. 65. The "dumbbell" domain organization of phosphoglycerate kinase, with a relatively narrow neck between two well-separated domains.

Subunit motion between two positions is also critical to the assembly of tobacco mosaic virus. In the partially assembled "disks," having two stacked layers of 17 subunits each, the layers are wedged apart toward their inner radius. During assembly of the viral helix, RNA binds between the layers, which then clamp tightly together with $16\frac{1}{3}$ subunits per turn (Bloomer et al., 1978; Butler and Klug, 1978).

Within a single subunit, contiguous portions of the polypeptide chain frequently fold into compact, local, semiindependent units called domains. The separateness of two domains within a subunit varies all the way from independent globular domains joined only by a flexible length of polypeptide chain to domains with extremely tight and extensive contact and a smoothly spherical outside surface for the entire subunit (such as in Fig. 64). An intermediate level of domain separateness is common in the known structures, with an elongated overall subunit shape and a definite neck or cleft between the domains, such as phosphoglycerate kinase shown in Fig. 65.

Another feature frequently seen in both domain and subunit contacts is an "arm" at one end of the chain which crosses over to "embrace" the opposite domain or subunit. Figure 66 shows such "arms" on the domains of papain. "Arms" that cross between domains or subunits almost invariably lie at the surface, but one unusual case in influenza virus hemagglutinin has a piece from a different domain forming the central strand of a five-stranded β sheet (see Fig. 83, where the alien strand is shown by the dotted lines).

The paucity of examples of flexibly hinged domains is almost certainly due to the difficulties of crystallizing such structures. In the immunoglobulins it has long been known from electron microscopic and hydrodynamic evidence that the hinge between the Fab and Fc regions is very flexible. The intact Dob immunoglobulin whose structure has been determined (Silverton *et al.*, 1977) has a substantial deletion in the hinge region which presumably limits its flexibility greatly. Intact immunoglobulins without such a deletion are notoriously difficult to crystallize, and the two cases in which crystallization has succeeded both turned out to have ordered Fab regions and invisible, disordered Fc portions (Colman *et al.*, 1976; Edmundson, 1980). A study of diffracted X-ray intensity as a function of resolution has shown that the Fc disorder is probably a static, statistical disorder among at least four multiple conformations (Marquart *et al.*, 1980).

At the other extreme, with very tightly associated domains, it is rather difficult to make the decision as to how many domains should be said to be present. In naming domains for the present study we have made use wherever possible of experimental evidence about either hinge motions of domains or about their folding or stability as

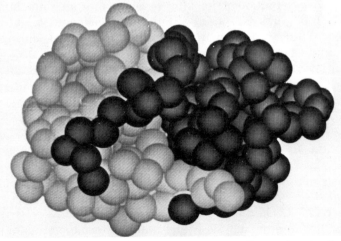

FIG. 66.   The domains of papain, which wrap "arms" around each other.

isolated units. For example, rigid-body hinging has been documented for hexokinase (Bennett and Steitz, 1978), phosphoglycerate
kinase (Pickover *et al.*, 1979), tomato bushy stunt virus (Harrison *et
al.*, 1978), and between immunoglobulin V_L and C_L domains (Schiffer
et al., 1973; Abola *et al*, 1980). On the other hand, the known movements on substrate binding in carboxypeptidase (Quiocho and Lipscomb, 1971), adenylate kinase (Sachsenheimer and Schulz, 1977), and
phosphorylase (Sygusch, *et al.*, 1977) involve only surface movement
of loops. For elastase (Ghelis *et al.*, 1978) it is known that after proteolytic cleavage the domains can fold up into stable isolated units.
But on the other hand, none of the large fragments of staphylococcal
nuclease (Taniuchi and Anfinsen, 1969) or of cytochrome c (Fisher *et
al.*, 1973) can fold up independently. For the majority of proteins,
where such experimental evidence is not available, the decision about
domains was made on the basis of analogy: whether the whole subunit
or its parts more closely resembled other single-domain proteins. It
is possible that some of the larger domains listed here will turn out to
have genuinely independent smaller parts; domain divisions have
been claimed within subtilisin (Wright *et al.*, 1969; Honig *et al.*, 1976)
and the first domain of phosphorylase (Weber *et al.*, 1978). Several
domains of more than 300 residues are now known, however, which
cannot plausibly be subdivided [e.g., triosephosphate isomerase, carboxypeptidase, bacteriochlorophyll protein]. Since large domains
are clearly possible, we have been conservative in assigning divisions.
Only three of the domains consist of two long, noncontiguous segments
(e.g., pyruvate kinase d1); most are a single piece of chain.

The above definition of domains, in which they are thought of as potentially independent, stable folding units analogous to subunits, is
only one of three rather separate concepts of domains in current think-
ing about the subject. The initial recognition of domain divisions as a
general feature in the three-dimensional structures (Wetlaufer, 1973)
was in terms of locally compact globular regions also contiguous in
the sequence. This idea has recently been elaborated and computerized in terms of maximum long-range contacts between segments
with only short-range internal contacts (Crippen, 1978), binary divisions of the sequence that maximally lie on opposite sides of a single
cutting plane (Rose, 1979), or binary sequence divisions that mini-
mize the sum of the separate surface areas (Wodak and Janin, 1980).
Each of these algorithms agrees in many cases with "intuitive" or
"subjective" division into domains, but one of the more striking results of this kind of analysis is that it always produces a hierarchy or
tree of substructures, usually containing two or three levels between

the individual secondary-structure elements and the entire subunit. Domains as usually conceived represent the upper levels of such a hierarchy, while the lower levels may very well represent intermediates in the folding process. One lesson from these studies, however, is that something else is involved in the intuitive concept of a domain besides these purely geometrical criteria of relative compactness. In the current study we have provided that additional criterion by requiring analogy to some structure known to possess stability in isolation.

The third major concept of domains is basically genetic, following quite naturally from the earliest idea of domains: the homologous, internally disulfide-linked regions in the immunoglobulin sequence (Edelman and Gall, 1969). Rossmann (Rossmann et al., 1974) has proposed that the similar nucleotide-binding domains in various dehydrogenases represent genetic segments that have been transferred and combined with differing catalytic domains to produce functionally distinct but partially related enzymes. Recently the discovery of exon (translated and expressed) DNA sequences separated by intron sequences which are clipped out of the mRNA before translation into protein has raised the intriguing possibility that separate exons correspond to structural and/or functional domains recognizable in the proteins and basic to their evolutionary history. In immunoglobulins the exons correspond quite exactly with structural domain divisions, with the hinge region as a separate exon (Sakano et al., 1979). However, in hemoglobin the introns occur not only inside what is usually considered a single domain, but even in the middle of individual helices, although it has been found that the isolated central exon peptide can bind heme (Craik et al., 1980). At the current level of knowledge it is unclear whether exons will provide a clarification of the basis of structural domains, although they are clearly a fundamental breakthrough in our understanding of the evolutionary processes involved.

Domains have proved so fruitful in explaining both the structure and the function of proteins that the concept is certain to survive in one form or another. At some time in the near future it will presumably acquire full scientific respectability with a verifiable definition in terms of either folding units or genetic units or perhaps both.

The domains (as we have defined them) within a subunit very often resemble each other (as discussed in Section IV,B), and those similar domains are frequently related by an approximate 2-fold axis. The 2-fold relationship often occurs even in cases in which it involves considerable inconvenience because the start and end of the chain are on opposite sides of the basic domain structure, so that a long additional

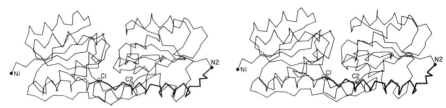

FIG. 67. Stereo α-carbon drawing of the two domains of arabinose-binding protein (viewed perpendicular to the approximate 2-fold axis between domains), with the stretch of chain shown dark which joins the end of the first domain to the beginning of the second one.

loop is required to connect the end of the first domain to the beginning of the second one (see for instance Fig. 67). An even more interesting instance of the pervasiveness of 2-fold domain contacts is in the serine proteases, where the relationship between genetically equivalent portions of the chain is in fact a pure translation. However, the initially heterologous contact seems to have evolved toward 2-fold similarity of the contact surfaces, so that now these proteins give the appearance of having a 2-fold relationship around the midpoint of the sequence (see Fig. 68). This convergent evolutionary process was able to produce an apparent sequence inversion because the topology of the serine protease domains $(+1, +1, +3, -1, -1,$ or $-1, -1, -3,$ $+1, +1)$ is invariant to reversal of N- to C-terminal direction. The kinds of cases illustrated in Figs. 67 and 68 suggest that 2-fold contacts are inherently easier to design well than unmatched, heterologous contacts.

The types of contacts between domains are in general very similar

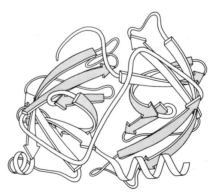

FIG. 68. Schematic backbone drawing of the elastase molecule, showing the similar β barrel structures of the two domains. The outside surfaces of the β barrels are stippled.

to those found between subunits, but there are a few characteristic differences. Continuation of β sheet hydrogen-bonding, which is very common between subunits, is almost unknown between domains (which typically contact through side chains rather than main chain). β barrel domains (see Section III,D) most often associate side-by-side, usually with their barrel axes at a 50 to 90° angle to each other (as in Fig. 68). Doubly wound parallel β sheet domains (see Section III,C) most often associate with the β strands pointing toward each other (as in Fig. 68).

Domains as well as subunits can serve either as moving parts for the functioning protein or as modular bricks to aid in efficient assembly. Undoubtedly the existence of separate domains is important in simplifying the folding process into separable, smaller steps, especially for very large proteins. The commonest domain size is between 100 and 200 residues, but it now appears that there is no strict and definite upper limit on practical folding size: domain sizes vary by an order of magnitude, from 41 residues up to more than 400. The range of domain sizes is somewhat different for each of the major structural categories (see Sections III,B through III,E for definitions): generally less than 100 residues for small disulfide-rich or metal-rich proteins, 80 to 150 for antiparallel α, 100 to 200 for antiparallel β, and 120 to 400 for parallel α/β. The lower limit for each category presumably reflects the smallest stable structure that can be made using that general design pattern. The upper limit may, among other things, reflect the largest domain for each structure type that can fold up efficiently as a single unit.

The other important function of domains is to provide motion. Completely flexible hinging would be impossible between subunits because they would simply fall apart, but it can be done between covalently linked domains. More limited flexibility between domains is often crucial to substrate binding, allosteric control, and assembly of large structures. In hexokinase the two domains hinge together on binding of glucose, enclosing it almost completely (Bennett and Steitz, 1978). Not only does this mechanism provide access to a very tight and hydrophobic specificity site, it has also been hypothesized as necessary in order to discriminate against using water as a substrate and acting as a counterproductive ATPase. It should be remembered, however, that domain motion is not the only way to solve the problem: what hexokinase and alcohol dehydrogenase accomplish by domain hinging, adenylate kinase and lactate dehydrogenase accomplish by large movements of surface loops.

In tomato bushy stunt virus protein, domain hinging helps to solve

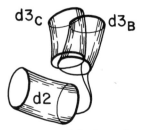

FIG. 69.   The two different positions of the hinge between domains 2 and 3 of tomato bushy stunt virus protein.   Each domain is represented by a cylinder, with domain 2 as the reference and domain 3 shown in the relative positions it takes in type B
subunits and in type C subunits.

the problem of packing 180 identical protein subunits quasiequiva-
lently into the 60-fold icosahedral symmetry of the virus shell (Harrison *et al.*, 1978).   Around the exact 2-fold and the quasi-2-fold axes
of the icosahedron, contact must be made between chemically identical subunits which come together at rather different angles around
the two kinds of 2-fold axes.   Identical contacts are made in both cases
between pairs of protruding d3 domains, but the angle between d2
and d3 domains can change by about 20° (see Fig. 69) so that each type
of d2 domain is placed in the correct orientation to interact around the
5-fold or the quasi-6-fold axes.   Pairs of d2 shell domains form two dif-

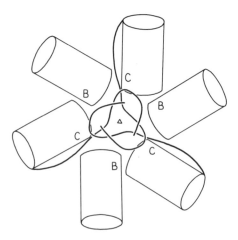

FIG. 70.   Domains 2 (cylinders) and N-terminal tails of B- and C-type subunits
around the quasi-6-fold axis in tomato bushy stunt virus.   The association of the three
C-subunit tails around the quasi-6-fold forms "domain 1" (see Fig. 84).

ferent types of contacts: in one type the two surfaces are tightly packed and the N-terminal arms are disordered, while in the other type of contact the N-terminal arms of "C" subunits fill a wedge-shaped opening between the contact surfaces and then wind around the 3-fold axis (see Figs. 70 and 84).

PROTEIN TAXONOMY

III. Classification of Proteins by Patterns of Tertiary Structure

A. Summary of the Classification System

1. Principles and Methods

Having looked at the characteristics of individual structural features and some of their local combinations, we are now in a position to sort out and classify the major structural patterns, or "folds," that make up entire proteins. This classification will build on earlier work by Rossmann (e.g., Rossmann and Argos, 1976), Richardson (1977), and Levitt and Chothia (1976), but will attempt to combine and extend those systems, as well as including the newer structures now available.

The most useful level at which to categorize protein structures is the domain, since there are many cases of multiple-domain proteins in which each separate domain resembles other entire smaller proteins. We have separated proteins into domains on the basis of whether the pieces could be expected to be stable as independent units or are analogous to other complete structures (see Section II,I for a fuller explanation). Clearly demonstrated homologous families such as trypsin, chymotrypsin, elastase, and the *S. griseus* proteases, or the cytochromes c, c_2, c_{550}, c_{551}, and c_{555}, are treated as single examples. There are between 90 and 105 different domains represented in the current sampling of known protein structures, depending on how one counts the cases of similar domain structures within a given protein. In the schematic drawings of Figs. 72–86 such domains are illustrated separately only if they are at least as different as the range of variation common within the close homology families (and, of course, if suitable coordinates or stereos were available). The domains within each protein are distinguished by numbers in sequence order (e.g., papain d1 and papain d2), except for the immunoglobulins for which we use the standard terminology (V_L, C_H1, etc.) for constant and variable domains.

Structural categories are assigned primarily on the basis of the type

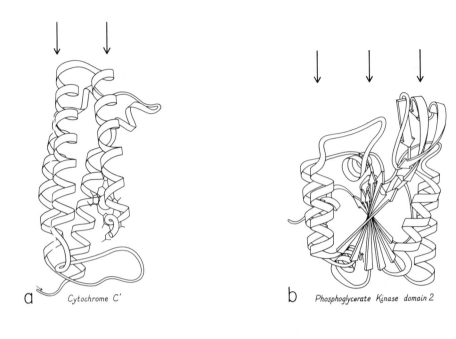

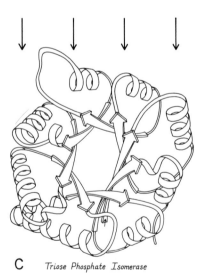

FIG. 71.   Examples of protein domains with different numbers of layers of backbone structure: (a) two-layer cytochrome c'; (b) three-layer phosphoglycerate kinase domain 2; (c) four-layer triosephosphate isomerase.   The arrows above each drawing point to the backbone layers.

and organization of secondary-structure elements, the topology of their connections, and the number of major layers of backbone structure that are present. Since proteins fold to form a protected hydrophobic core of side chains on the interior, the simplest type of stable protein structure consists of polypeptide backbone wrapped more or less uniformly around the outside of a single hydrophobic core. We will describe such a structure as "two layer," because a line from the solvent through the center of the protein and back out again would pass through two principal layers of backbone structure (see Fig. 71a). Over half of the known domain structures have two layers. About a third of the structures have three layers of backbone and two hydrophobic cores; the commonest such type has a central β sheet layer flanked by two helical layers (see Fig. 71b). There are three known four-layer domains (e.g., Fig. 71c) and one five-layer. Isolated loops that curl over the outside are not considered to form a distinct layer. Although there are some ambiguous cases (especially in the very small structures) they are less common than one might expect, presumably because the requirements for rapid folding rule out much tangling or recrossing of the backbone.

The approximately 100 distinctly different domains fall into four broad categories, each of which has several subgroupings. The four broad categories are (I) antiparallel α; (II) parallel α/β; (III) antiparallel β; (IV) small SS-rich or metal-rich. The major determinant in assigning a domain to one of these categories is not just the percentage of a given secondary structure, but whether that type of secondary structure forms the central core and whether its interactions could be the dominant stabilizing ones. The two β categories are the most populous, with 30 to 35 members each; there are about 20 α-helical domains and a dozen of the small proteins. The overall classification scheme is summarized in Section III,A,2. An alphabetical index of the proteins is given in Section III,A,4, with domain assignments, structure subcategories, and literature references.

Obviously this classification is not the only plausible way to categorize protein structures. Indeed, for some of the individual cases there are other descriptions which would be preferable because they emphasize possible relationships to functionally similar proteins. However, the motivation here has been to achieve the most satisfactory compromise for *all* the structures: to fit as many examples into as few groupings as possible, while retaining enough detail to provide meaningful descriptions. Also, the prejudice has been in favor of grouping together domains whose entire structure is approximately the same rather than cases in which a relatively small portion of both structures is more exactly similar.

In order to illustrate this taxonomic system and to facilitate contrasts and comparisons among the structures, schematic backbone drawings have been made of most of the known structures. The drawings are grouped together by categories in Section III,A,3. α-Carbon coordinates were displayed in stereo on Richard Feldmann's computer graphics system at NIH; a suitable view was chosen (consistent for each subcategory of structure), and plotter output was obtained at a consistent scale (approximately 20 Å per inch on the final drawings as reproduced here). The schematic was drawn on top of the plotter output for accuracy, with continual reference to the stereo for the third dimension. Loops, and to some extent β strands, were smoothed for comprehensibility, and shifts of 1 or 2 Å were sometimes necessary in order to avoid ambiguity at crossing points. A uniform set of graphical conventions was adopted (see Section III,A,3 for explanation) in which β strands are shown as arrows, helices as spiral ribbons, and nonrepetitive structure as ropes. Location and extent of β strands and helices are sometimes based on published descriptions and hydrogen-bonding diagrams, but often must be judged from the stereo view itself. Very short β interactions are shown as arrows when they form part of a larger sheet but may be left out if they are isolated. Foreshortening, overlaps, edge appearance, and relative size change are used to provide depth cues.

2. Outline of the Taxonomy

I. Antiparallel α domains
  A. Up-and-down helix bundles
    Myohemerythrin, hemerythrin
    Cytochrome b_{562}
    Cytochrome c'
    Uteroglobin
    Staphylococcal protein A fragment
    Influenza virus hemagglutinin "domain" around 3-fold
    Tobacco mosaic virus protein
    Cytochrome b_5
    Tyrosyl-tRNA synthetase domain 2
    Ferritin (?)
    Purple membrane protein (?)
  B. Greek key helix bundles

Myoglobin, hemoglobin
Thermolysin domain 2
T4 phage lysozyme domain 2
Papain domain 1
Cytochrome c peroxidase domain 1
C. Miscellaneous antiparallel α
Carp muscle calcium-binding protein
Egg lysozyme
Citrate synthase
Catalase domain 2
Cytochrome c peroxidase domain 2
p-Hydroxybenzoate hydroxylase domain 3
II. Parallel α/β domains
A. Singly wound parallel β barrels
Triosephosphate isomerase
Pyruvate kinase domain 1
KDPG aldolase (?)
B. Doubly wound parallel β sheets
1. Classic doubly wound β sheets
Lactate dehydrogenase domain 1
Alcohol dehydrogenase domain 2
Aspartate transcarbamylase catalytic domain 2
Phosphoglycerate kinase domain 1
Tyrosyl-tRNA synthetase domain 1(?)
Phosphorylase domain 2, central three layers
2. Doubly wound variations
Glyceraldehyde-phosphate dehydrogenase domain 1
Phosphorylase domain 1, central three layers
Flavodoxin
Subtilisin
Arabinose-binding protein domains 1 and 2
Dihydrofolate reductase
Adenylate kinase
Rhodanese domains 1 and 2
Glutathione reductase domains 1 and 2
Phosphoglycerate mutase
Phosphoglycerate kinase domain 2
Pyruvate kinase domain 3
Hexokinase domains 1 and 2
Catalase domain 3
Aspartate aminotransferase

Aspartate transcarbamylase catalytic domain 1
Phosphofructokinase domain 1
p-Hydroxybenzoate hydroxylase domain 1
Glucosephosphate isomerase domain 1
Glutathione peroxidase
C. Miscellaneous parallel α/β
  Carboxypeptidase
  Thioredoxin
  Carbonic anhydrase
  Phosphofructokinase domain 2
  Glucosephosphate isomerase domain 2
III. Antiparallel β domains
  A. Up-and-down β barrels
    Papain domain 2
    Soybean trypsin inhibitor
    Catalase domain 1
  B. Greek key β barrels
    1. Simple Greek keys
      Trypsin-like serine proteases domains 1 and 2
      Pyruvate kinase domain 2
      Prealbumin
      Plastocyanin, azurin
      Immunoglobulin, variable and constant domains
      Cu,Zn superoxide dismutase
      Staphylococcal nuclease
    2. "Jellyroll" Greek keys
      Tomato bushy stunt virus protein domains 2 and 3
      Southern bean mosaic virus protein
      Concanavalin A
      Influenza virus hemagglutinin HA1
      γ-Crystallin domains 1 and 2
  C. Multiple, partial, and other β barrels
    Acid proteases domains 1 and 2
    Alcohol dehydrogenase domain 1
    Pancreatic ribonuclease
  D. Open-face β sandwiches
    T4 lysozyme domain 1
    Aspartate transcarbamylase regulatory domains 1 and 2
    Streptomyces subtilisin inhibitor
    Glutathione reductase domain 3
    Thermolysin domain 1

Glyceraldehyde-phosphate dehydrogenase domain 2
Bacteriochlorophyll protein
p-Hydroxybenzoate hydroxylase domain 2
Influenza virus hemagglutinin HA2
L7/L12 ribosomal protein
E. Miscellaneous antiparallel β
Gene 5 protein, *E. coli*
Lactate dehydrogenase domain 2
Tomato bushy stunt virus protein "domain" 1
IV. Small disulfide-rich or metal-rich domains
A. SS-rich
1. Toxin-agglutinin fold
Erabutoxin, cobra neurotoxin
Wheat germ agglutinin domains 1, 2, 3, and 4
2. Other SS-rich
Pancreatic trypsin inhibitor
Insulin
Phospholipase A_2
Crambin
B. Metal-rich
1. Up-and-down ligand cages
Rubredoxin
Cytochrome b_5
Cytochrome c
Cytochrome c_3
2. Greek key ligand cages
Ferredoxin
High-potential iron protein

3. Schematic Drawings of the Protein Domains by Structure Type

Detailed discussions of the categories and subgroupings are given in Sections III,B through III,E. The scale of these drawings (Figs. 72–86) is approximately 20 Å to the inch. β strands are shown as arrows with thickness, helices as spiral ribbons, and nonrepetitive structure as ropes. Disulfides are shown as "lightning bolts." Circles represent metals, and some prosthetic groups are shown as atomic skeletons, but not for all cases in which they are known to be present. A question mark in the label means that backbone connectivity is uncertain in some places. Where needed for clarity, the N-terminus of the domain is indicated by a small arrow; for a few two-chain domains the C-terminus is indicated as well.

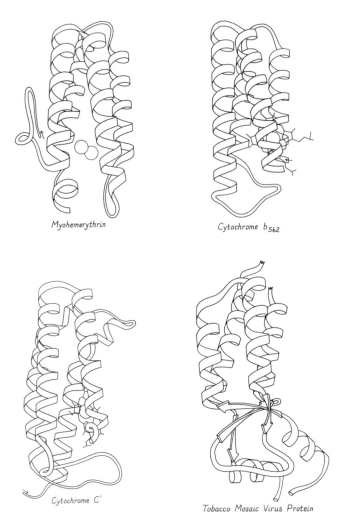

Myohemerythrin

Cytochrome b₅₆₂

Cytochrome C'

Tobacco Mosaic Virus Protein

FIG. 72.   Antiparallel α: up-and-down helix bundles.

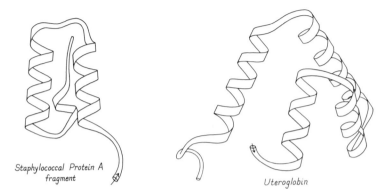

Staphylococcal Protein A
fragment

Uteroglobin

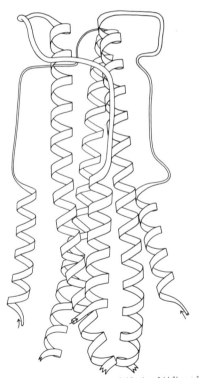

Influenza Virus Haemagglutinin HA2, threefold "domain"

FIG. 72 (continued)

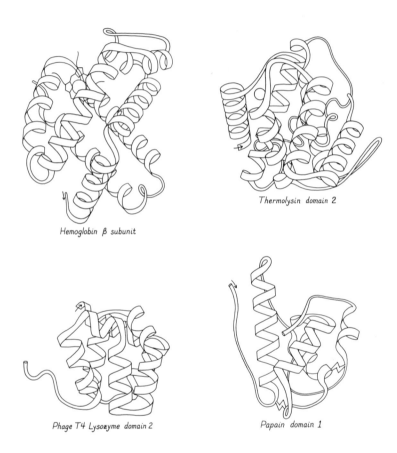

Hemoglobin β subunit

Thermolysin domain 2

Phage T4 Lysozyme domain 2

Papain domain 1

FIG. 73.   Antiparallel α: Greek key helix bundles.

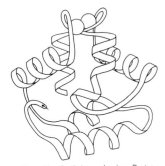

Carp Muscle Calcium-binding Protein

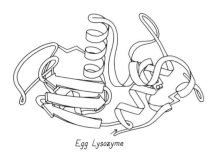

Egg Lysozyme

FIG. 74. Antiparallel α: miscellaneous.

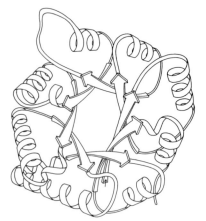

Triose Phosphate Isomerase

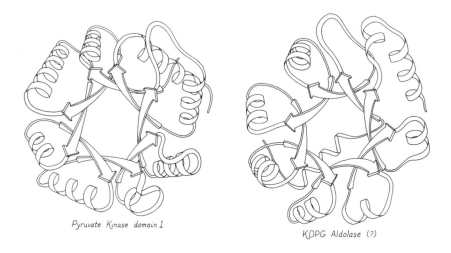

Pyruvate Kinase domain I

KDPG Aldolase (?)

FIG. 75.   Parallel α/β: singly wound parallel β barrels.

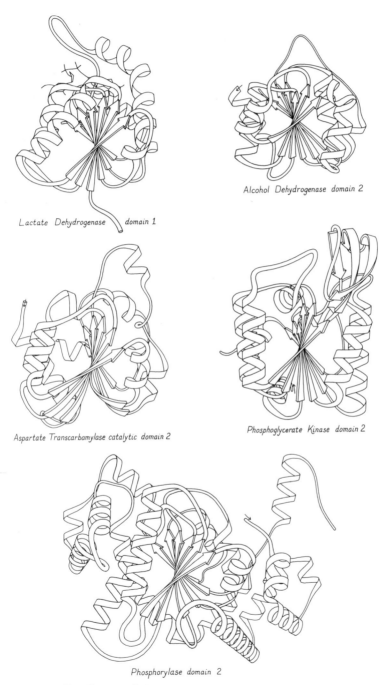

Lactate Dehydrogenase domain 1

Alcohol Dehydrogenase domain 2

Aspartate Transcarbamylase catalytic domain 2

Phosphoglycerate Kinase domain 2

Phosphorylase domain 2

FIG. 76. Parallel α/β: classic doubly wound β sheets.

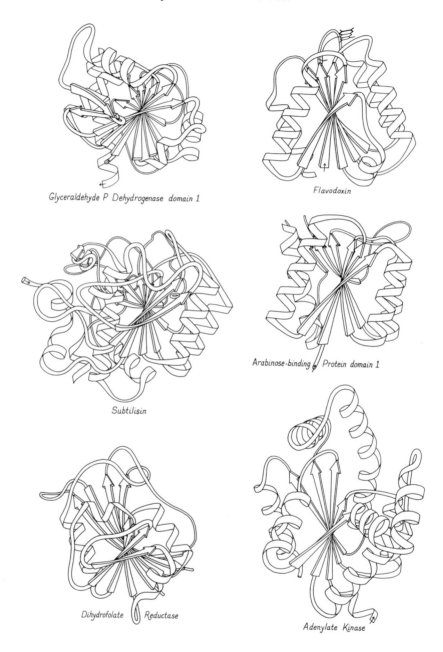

Glyceraldehyde P Dehydrogenase domain 1

Flavodoxin

Subtilisin

Arabinose-binding Protein domain 1

Dihydrofolate Reductase

Adenylate Kinase

FIG. 77. Parallel α/β: doubly wound parallel β sheets.

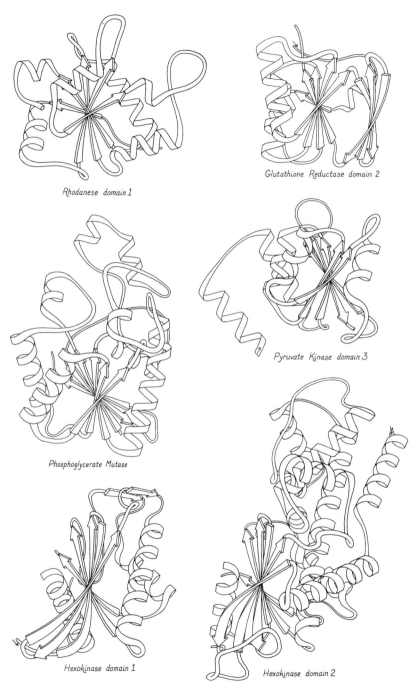

Rhodanese domain 1

Glutathione Reductase domain 2

Pyruvate Kinase domain 3

Phosphoglycerate Mutase

Hexokinase domain 1

Hexokinase domain 2

FIG. 77 (continued)

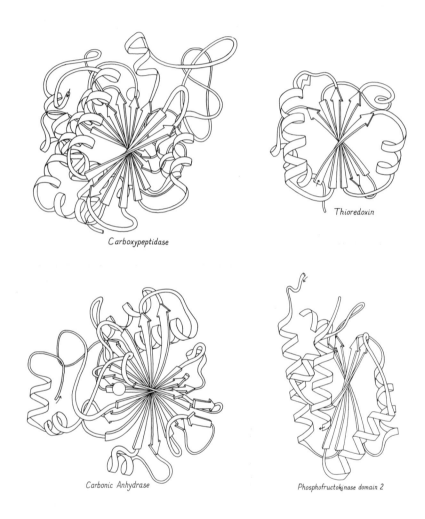

Carboxypeptidase

Thioredoxin

Carbonic Anhydrase

Phosphofructokinase domain 2

FIG. 78. Parallel α/β: miscellaneous.

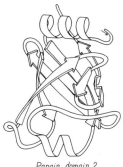

Papain domain 2

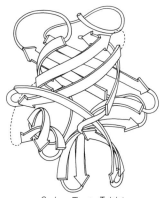

Soybean Trypsin Inhibitor

FIG. 79. Antiparallel β: up-and-down β barrels.

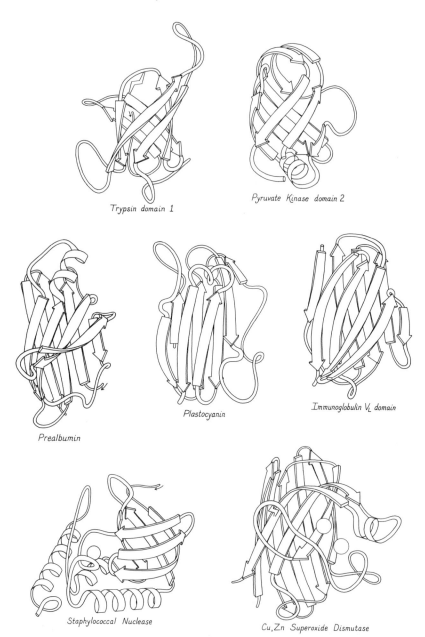

Trypsin domain 1

Pyruvate Kinase domain 2

Prealbumin

Plastocyanin

Immunoglobulin V_L domain

Staphylococcal Nuclease

Cu,Zn Superoxide Dismutase

FIG. 80.   Antiparallel β: Greek key β barrels.

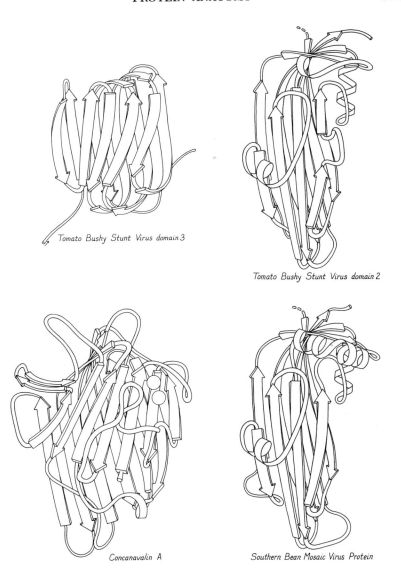

Tomato Bushy Stunt Virus domain 3

Tomato Bushy Stunt Virus domain 2

Concanavalin A

Southern Bean Mosaic Virus Protein

FIG. 81. Antiparallel β: "jellyroll" Greek key β barrels.

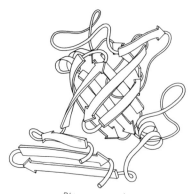

Rhizopuspepsin domain 1

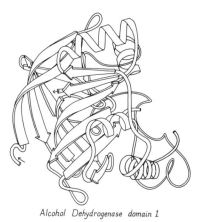

Alcohol Dehydrogenase domain 1

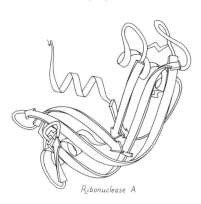

Ribonuclease A

FIG. 82.   Antiparallel β: other, multiple, and partial barrels.

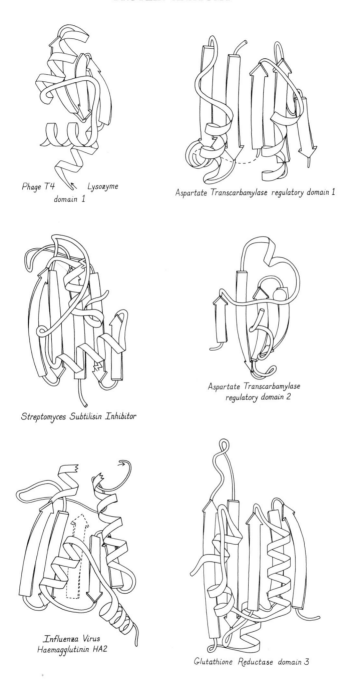

Phage T4 Lysozyme
domain 1

Aspartate Transcarbamylase regulatory domain 1

Streptomyces Subtilisin Inhibitor

Aspartate Transcarbamylase
regulatory domain 2

Influenza Virus
Haemagglutinin HA2

Glutathione Reductase domain 3

FIG. 83.   Antiparallel β: open-face sandwich β sheets.

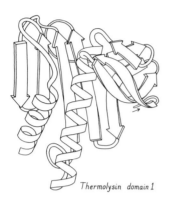

Thermolysin domain 1

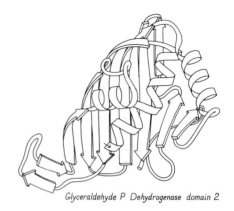

Glyceraldehyde P Dehydrogenase domain 2

Bacteriochlorophyll Protein

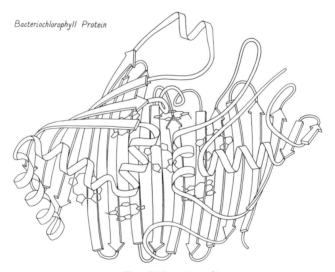

FIG. 83 (continued)

Gene 5 Protein

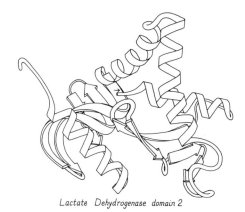

Lactate Dehydrogenase domain 2

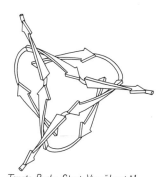

Tomato Bushy Stunt Virus "domain" 1

FIG. 84. Antiparallel β: miscellaneous.

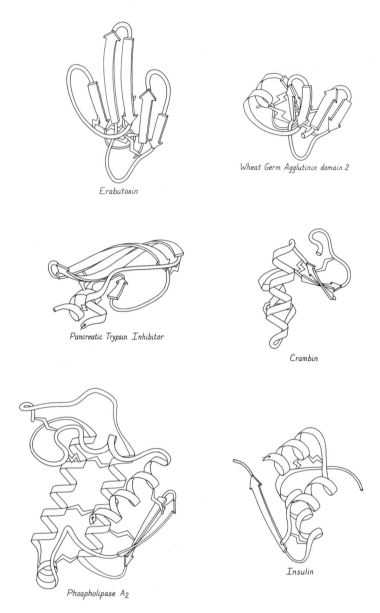

Erabutoxin

Wheat Germ Agglutinin domain 2

Pancreatic Trypsin Inhibitor

Crambin

Phospholipase A$_2$

Insulin

FIG. 85. Small disulfide-rich.

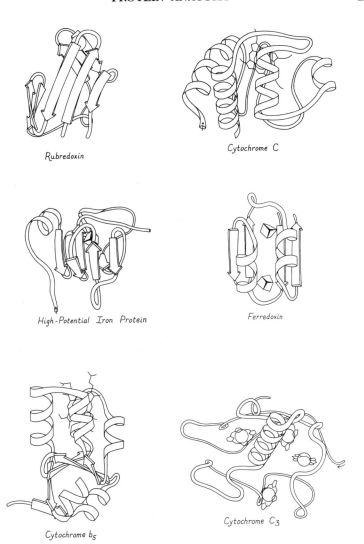

Rubredoxin

Cytochrome C

High-Potential Iron Protein

Ferredoxin

Cytochrome b₅

Cytochrome C₃

FIG. 86. Small metal-rich.

4. Index of Proteins

Acid proteases, *see* Rhizopuspepsin

Actinidin (Baker, 1980), *see* Papain

Adenylate kinase (Schulz *et al.*, 1974a)
    Doubly wound parallel β sheet (Fig. 77)

Agglutinin, wheat germ (Wright, 1977)
    Domains 1, 2, 3, and 4: small SS-rich (Fig. 85)

Alcohol dehydrogenase, liver (Eklund *et al.*, 1976)
    Domain 1: multiple β barrel (Fig. 82)
    Domain 2: classic doubly wound β sheet (Fig. 76)

Aldolase, 2-keto-3-deoxy-6-phosphogluconate (Mavridis and Tulin-sky, 1976; Richardson, 1979)
    Singly wound parallel β barrel? (Fig. 75)

Arabinose-binding protein (Quiocho *et al.*, 1977)
    Domains 1 and 2: doubly wound parallel β sheet (Fig. 77)

Aspartate aminotransferase (Ford *et al.*, 1980)
    Doubly wound parallel β sheet

Aspartate carbamoyltransferase, (Monaco *et al.*, 1978)
    Regulatory domain 1: open-face β sandwich (Fig. 83)
    Regulatory domain 2: open-face β sandwich (Fig. 83)
    Catalytic domain 1: doubly wound parallel β sheet
    Catalytic domain 2: classic doubly wound β sheet (Fig. 76)

Aspartate transaminase: *see* Aspartate aminotransferase

Aspartate transcarbamylase (Monaco *et al.*, 1978), *see* Aspartate carbamoyltransferase

Azurin (Adman *et al.*, 1978), *see* Plastocyanin

Bacteriochlorophyll protein (B. W. Matthews *et al.*, 1979)
    Open-face β sandwich (Fig. 83)

Bacteriorhodopsin, *see* Purple membrane protein

Bence-Jones protein, *see* Immunoglobulin

Calcium-binding protein, carp muscle (Kretsinger and Nockolds, 1973)
    Miscellaneous antiparallel α (Fig. 74)

Carbonate dehydratase, *see* Carbonic anhydrase

Carbonic anhydrase C (Lindskog *et al.*, 1971)
    Miscellaneous parallel α/β (Fig. 78)

Carboxypeptidase A (Quiocho and Lipscomb, 1971)
    Miscellaneous parallel α/β (Fig. 78)

Catalase (Vainshtein *et al.*, 1980)
    Domain 1: up-and-down β barrel
    Domain 2: miscellaneous antiparallel α
    Domain 3: doubly wound parallel β sheet

Chymotrypsin (Birktoft and Blow, 1972), *see* Trypsin
Citrate synthase (Wiegand *et al.*, 1979)
    Miscellaneous antiparallel α
Concanavalin A (Reeke *et al.*, 1975)
    Jellyroll Greek key β barrel (Fig. 81)
Crambin (Hendrickson and Teeter, 1981)
    Small SS-rich (Fig. 85)
γ-Crystallin (Blundell *et al.*, 1981)
    Domains 1 and 2: jellyroll Greek key β barrel
Cytochrome b_5 (Mathews *et al.*, 1972)
    Small metal-rich (Fig. 86)
Cytochrome b_{562} (Mathews *et al.*, 1979)
    Up-and-down helix bundle (Fig. 72)
Cytochrome c (Swanson *et al.*, 1977)
    Small metal-rich (Fig. 86)
Cytochrome c' (Weber *et al.*, 1980)
    Up-and-down helix bundle (Fig. 72)
Cytochrome c_2 (Salemme *et al.*, 1973), *see* Cytochrome c
Cytochrome c_3 (Haser *et al.*, 1979)
    Small metal-rich (Fig. 86)
Cytochrome c_{550} (Timkovich and Dickerson, 1973), *see* Cytochrome c
Cytochrome c_{551} (Almassy and Dickerson, 1978), *see* Cytochrome c
Cytochrome c_{555} (Korszun and Salemme, 1977), *see* Cytochrome c
Cytochrome c peroxidase (Poulos *et al.*, 1980)
    Domain 1: Greek key helix bundle
    Domain 2: miscellaneous antiparallel α
Dehydrogenases, *see* Alcohol, Glyceraldehyde phosphate, Malate, or Lactate
Dihydrofolate reductase (Matthews *et al.*, 1977)
    Doubly wound parallel β sheet (Fig. 77)
Elastase (Sawyer *et al.*, 1978), *see* Trypsin
Erabutoxin (Low *et al.*, 1976)
    Small SS-rich (Fig. 85)
Ferredoxin (Adman *et al.*, 1973)
    Small metal-rich (Fig. 86)
Ferritin (Banyard *et al.*, 1978; Clegg *et al.*, 1980)
    Up-and-down helix bundle
Flavodoxin (Burnett *et al.*, 1974)
    Doubly wound parallel β sheet (Fig. 77)
Gene 5 protein, fd phage (McPherson *et al.*, 1979)
    Miscellaneous antiparallel β

Glucosephosphate isomerase (Shaw and Muirhead, 1977)
    Domains 1 and 2: miscellaneous parallel α/β
Glutathione peroxidase (Ladenstein et al., 1979)
    Doubly wound parallel β sheet
Glutathione reductase (Schulz et al., 1978)
    Domains 1 and 2: doubly wound parallel β sheet (Fig. 77)
    Domain 3: open-face β sandwich (Fig. 83)
Glyceraldehyde-phosphate dehydrogenase (Buehner et al., 1974)
    Domain 1: doubly wound parallel β sheet (Fig. 77)
    Domain 2: open-face β sandwich (Fig. 83)
Glycogen phosphorylase (Sprang and Fletterick, 1979)
    Domain 1: doubly wound parallel β sheet, five-layer
    Domain 2: classic doubly wound β sheet, five-layer (Fig. 76)
Hemagglutinin, influenza virus (Wilson et al., 1981)
    HA1: jellyroll Greek key β barrel
    HA2: open-face β sandwich
    HA2 around 3-fold: miscellaneous helix cluster
Hemerythrin (Stenkamp et al., 1978), see Myohemerythrin
Hemoglobin (Ladner et al., 1977)
    Greek key helix bundle (Fig. 73)
Hexokinase (Steitz et al., 1976)
    Domains 1 and 2: doubly wound parallel β sheet (Fig. 77)
High-potential iron protein (Carter et al., 1974)
    Small metal-rich (Fig. 86)
p-Hydroxybenzoate hydroxylase (4-hydroxybenzoate 3-mono-
    oxygenase) Wierenga et al., 1979)
    Domain 1: doubly wound parallel β sheet
    Domain 2: open-face β sandwich
    Domain 3: miscellaneous antiparallel α
Immunoglobulin (Epp et al., 1974; Silverton et al., 1977)
    Variable and constant domains: Greek key β barrel (Fig. 80)
Insulin (Blundell et al., 1972)
    Small SS-rich (Fig. 85)
Kinases, see Adenylate kinase, Hexokinase, Phosphoglycerate
    kinase, Phosphofructokinase, or Pyruvate kinase
Lactate dehydrogenase (Adams et al., 1970)
    Domain 1: classic doubly wound β sheet (Fig. 76)
    Domain 2: miscellaneous antiparallel β (Fig. 84)
Lysozyme, hen egg white (Imoto et al., 1972)
    Miscellaneous antiparallel α (Fig. 73)
Lysozyme, T4 phage (Matthews and Remington, 1974)
    Domain 1: open-face β sandwich (Fig. 83)

Domain 2: Greek key helix bundle (Fig. 73)
L7/L12 ribosomal protein (Leijonmarck *et al.*, 1980)
  Open-face β sandwich
Malate dehydrogenase (Hill *et al.*, 1972), *see* Lactate dehydrogenase
Myoglobin (Watson, 1969), *see* Hemoglobin
Myohemerythrin (Hendrickson and Ward, 1977)
  Up-and-down helix bundle (Fig. 72)
Neurotoxin
  Cobra (Walkinshaw *et al.*, 1980), *see* Erabutoxin
  Sea snake (Tsernoglou and Petsko, 1977), *see* Erabutoxin
Nuclease, staphylococcal (or micrococcal) (Arnone *et al.*, 1971)
  Greek key β barrel (Fig. 80)
Papain (Drenth *et al.*, 1974)
  Domain 1: Greek key helix bundle (Fig. 73)
  Domain 2: up-and-down β barrel (Fig. 79)
Parvalbumin, *see* Calcium-binding protein
Pepsin (Andreeva and Gustchina, 1979), *see* Rhizopuspepsin
Plastocyanin (Colman *et al.*, 1978)
  Greek key β barrel (Fig. 80)
Phosphoglycerate kinase (Banks *et al.*, 1979)
  Domain 1: doubly wound parallel β sheet
  Domain 2: classic doubly wound β sheet (Fig. 76)
Phosphoglycerate mutase (Campbell *et al.*, 1974)
  Doubly wound parallel β sheet (Fig. 77)
Phospholipase A$_2$ (Djikstra *et al.*, 1978)
  Small SS-rich (Fig. 85)
Phosphorylase, *see* Glycogen phosphorylase
Prealbumin (Blake *et al.*, 1978)
  Greek key β barrel (Fig. 80)
Protein A fragment, staphylococcal (Deisenhofer *et al.*, 1978)
  Up-and-down helix bundle (Fig. 72)
Purple membrane protein (Henderson and Unwin, 1975)
  Either up-and-down or Greek key helix bundle
Pyruvate kinase (Stuart *et al.*, 1979)
  Domain 1: singly wound parallel β barrel (Fig. 75)
  Domain 2: Greek key β barrel (Fig. 80)
  Domain 3: doubly wound parallel β sheet (Fig. 77)
Rhodanese (Ploegman *et al.*, 1978)
  Domains 1 and 2: doubly wound parallel β sheet (Fig. 77)
Rhizopuspepsin (Subramanian *et al.*, 1977)
  Domains 1 and 2: other β barrel (Fig. 82)

Ribonuclease, bovine pancreatic (Wyckoff *et al.*, 1970)
Partial β barrel (Fig. 82)
Rubredoxin (Watenpaugh *et al.*, 1979)
Small metal-rich (Fig. 86)
Serine proteases, *see* Trypsin, Chymotrypsin, Elastase, *Streptomyces griseus* proteases A and B, or Subtilisin
Southern bean mosaic virus protein (Abad-Zapatero *et al.*, 1980)
Jellyroll Greek key β barrel (Fig. 81)
Streptomyces griseus protease A (Brayer *et al.*, 1978), *see* Trypsin
Streptomyces griseus protease B (Delbaere *et al.*, 1975), *see* Trypsin
Subtilisin (Wright *et al.*, 1969)
Doubly wound parallel β sheet (Fig. 77)
Subtilisin inhibitor, *Streptomyces* (Mitsui *et al.*, 1979)
Open-face β sandwich (Fig. 83)
Sulfhydryl proteases, *see* Actinidin, Papain
Superoxide dismutase, Cu,Zn (Richardson *et al.*, 1975)
Greek key β barrel (Fig. 80)
Thermolysin (Colman *et al.*, 1972)
Domain 1: open-face β sandwich (Fig. 83)
Domain 2: Greek key helix bundle (Fig. 83)
Thioredoxin, *E. coli* (Holmgren *et al.*, 1975)
Miscellaneous parallel α/β (Fig. 78)
Thiosulfate sulfurtransferase, *see* Rhodanese
Tobacco mosaic virus protein (Bloomer *et al.*, 1978)
Up-and-down helix bundle (Fig. 72)
Tomato bushy stunt virus protein (Harrison *et al.*, 1978)
"Domain" 1: miscellaneous antiparallel β (Fig. 84)
Domains 2 and 3: jellyroll Greek key β barrel (Fig. 81)
Triosephosphate isomerase (Banner *et al.*, 1975)
Singly wound parallel β barrel (Fig. 75)
tRNA synthetase, tyrosyl (Irwin *et al.*, 1976; D. M. Blow, personal communication)
Domain 1: classic doubly wound β sheet
Domain 2: up-and-down helix bundle
Trypsin (Stroud *et al.*, 1974)
Domains 1 and 2: Greek key β barrel (Fig. 80)
Trypsin inhibitor, pancreatic (Deisenhofer and Steigemann, 1975)
Small SS-rich (Fig. 85)
Trypsin inhibitor, soybean (Sweet *et al.*, 1974)
Up-and-down β barrel (Fig. 79)
Uteroglobin (Mornon *et al.*, 1980)
Up-and-down helix bundle (Fig. 72)

Viral coat proteins, *see* Southern bean mosaic virus, Tobacco mosaic virus, or Tomato bushy stunt virus

B. *Antiparallel α Domains*

The first major grouping of structures contains domains that are essentially all α-helical. Since there is relatively little other structure besides the helices, the simplest ways of connecting them involve predominantly antiparallel helix interactions, and that is in fact what is observed for these proteins. This category corresponds to Levitt and Chothia's all-α category, but it has more members both because of a number of new structures and because of helical domains in proteins they classified as α + β (such as thermolysin).

Figures 72 through 74 show schematic diagrams of the antiparallel α domains, grouped into subcategories. Almost all of them are two-layer structures. The simplest and commonest subgroup looks like a bundle of sticks: usually four helices bundled in a cylinder with simple + 1 connections. Most of the helices are quite close to exactly antiparallel, with typically a left-handed superhelical twist of less than 15° relative to the common axis of the bundle. These structures were first described as a group in Argos *et al.* (1977). Figure 87 illustrates myohemerythrin as an example of this structure type, showing an α-carbon stereo, a schematic drawing, and a topology diagram.

The simple up-and-down helix bundle structures include the hemerythrins (myohemerythrin and the hemerythrin subunits), cytochrome b_{562}, cytochrome c′, uteroglobin, tobacco mosaic virus protein, staphylococcal protein A fragment, and probably the ferritin subunits. Tyrosine-tRNA synthetase domain 2 has quite a similar organization, but the last helix tilts away from the bundle (Blow *et al.*, 1977). The uteroglobin subunit also has its fourth helix out to one side, but in the dimer molecule (Fig. 88) those final helices each complete a compact four-helix bundle with the rest of the opposite subunit. In cytochrome c′ there is a similar but less extreme arrangement in which the first helix lies at a greater angle to the bundle axis and forms the tightest part of the dimer contact.

Tobacco mosaic virus protein has a small, highly twisted antiparallel β sheet at the base of the helix bundle, with two more helices underneath the sheet (see Fig. 72). Cytochrome b_5 looks remarkably similar (see Fig. 105), but the helices are much shorter. That structure could have been classified as an up-and-down helix bundle, but we have placed it in the small metal-rich proteins because its helix bundle is very small and distorted and the heme interactions appear more important than the direct helix contacts.

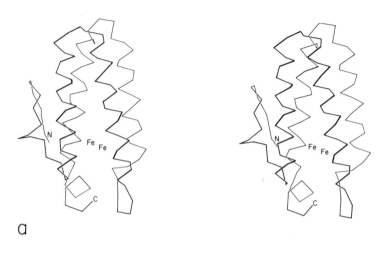

a

b   *Myohemerythrin*

FIG. 87.   Myohemerythrin as an example of an up-and-down helix bundle.   (a) α-
Carbon stereo; (b) schematic drawing of the backbone structure, from the same viewpoint as in a.

All but one of the above structures have four helices in the bundle,
with $+1, +1, +1$ connections.   For the up and down topology on a cylinder, handedness can be defined by whether the chain turns to the
right or to the left at the end of the first structure element (whether it is
a helix or a β strand).   With an even number of helices, reversing N to
C direction of the chain also reverses handedness of the topology; for
an odd number of helices or strands handedness is invariant to chain
reversal.   For $+1$, $+1$, $+1$ topologies in general, handedness is not

Uteroglobin dimer

FIG. 88. The dimer association of uteroglobin, with one subunit shown shaded and one open.

a very robust criterion of similarity, since it reverses on addition or deletion of one of the structure elements at the N-terminus but not at the C-terminus, so that a given five-helix structure could have evolved from either handedness of four-helix structure. Hemerythrin, cytochrome b_{562}, cytochrome c', uteroglobin, and tobacco mosaic virus protein are all right-handed, while cytochrome b_5, tyrosine-tRNA synthetase, and staphylococcal protein A fragment are left-handed.

The connectivity is not known for the seven-helix bundle of purple membrane protein (Henderson and Unwin, 1975), but on the basis of its resemblance to other antiparallel α proteins the most likely topologies would be either up-and-down or Greek key (see below). An analysis based on the sequence and the relative electron-densities of the helices (Engelman *et al.*, 1980) considers a left-handed up-and-down topology as the most probable model.

Many of the up-and-down helix bundle proteins form large multisubunit arrays. Hemerythrin is an octomer, with the end of one helix bundle butting against the side of the next one around the 4-fold axis (Ward *et al.*, 1975). The 24 ferritin subunits form a hollow spherical shell with the helix bundles approximately tangential to the shell and the subunit interactions around the 3-fold and 4-fold axes rather like the interactions in hemerythrin. Tobacco mosaic virus protein, on the other hand, forms a tightly packed long helix of subunits; the α-helical bundles are aligned radially, with RNA bound at their inner ends. Purple membrane protein spans the membrane, forming a two-

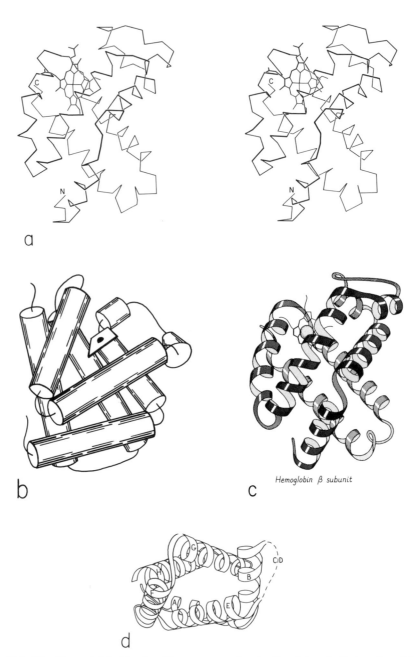

a

b

c

Hemoglobin β subunit

d

FIG. 89. Hemoglobin (β subunit) as an example of a Greek key helix bundle. (a) α-Carbon stereo; (b) schematic drawing of the backbone as two perpendicular layers of α-helices (shown here as cylinders); (c) schematic drawing of the backbone as a Greek key helix bundle (from the same viewpoint as in a); (d) schematic end-on view of the hemoglobin helix bundle, to show that it is a slightly flattened cylinder in cross section (the C-D loop is shown dashed because it would cover part of the cylinder).

dimensional crystalline array with the helix bundles perpendicular to the membrane and parallel to each other around the 3-fold axis.

One of the most important and interesting antiparallel α structures is the globin fold, which has been found in the three-dimensional structures of a large group of related proteins including myoglobin and the hemoglobins of various mammals, glycera, lamprey, insect, and even legume root nodules. The globin fold is a good example of how there may be several alternative useful ways of describing a given structure. To someone studying hemoglobin function the relevant level of description includes all the structural detail that can be made comprehensible, or perhaps generalized to include what is common to all the globin structures. On the other hand, if one is concerned, as we are here, with obtaining a memorably simple description of the whole structure and relating it to other protein structures, then the issue is deciding which features are most important to include in the simplification and with which if any other proteins it can meaningfully be compared. Classifying the globins as all-α proteins is obviously true and useful, but Levitt and Chothia's (1976) scheme of representing the globin topology does not suggest similarities to any of the other all-α proteins, even when the more recent structures are included. Argos and Rossmann (1979) have suggested an interesting similarity of structure around the heme pocket for the globins, cytochrome b_5, and cytochrome c_{551}. Their description is probably the most relevant one for trying to understand how heme-binding pockets are organized, but it does not seem suitable as a general structural description since the omitted halves of the three structures are all extremely different and do not form separable domains.

Figure 89 illustrates two different tries at simplified representation of the globin structure. For reference, Fig. 89a shows the hemoglobin β chain in stereo. Figure 89b shows the globin structure schematically as two layers of helices with the elements in one layer approximately perpendicular to those in the other layer; this can be contrasted with a possible description of the up-and-down helix bundles as two layers with their elements approximately parallel to each other. The perpendicular layers provide a rather successful simple schema for the globin structure, but unfortunately there are no other proteins that can be adequately described as two perpendicular layers of helices. Also, specification of the topology in this scheme is cumbersome, since the chain skips back and forth between layers.

Figure 89c schematizes the globin structure as a twisted cylinder of helices, analogous to the antiparallel β barrels to be discussed in Section III,D. The up-and-down helix bundle structures are of course also readily described as cylinders, so that this schema makes the

majority of antiparallel α structures directly analogous to the majority
of antiparallel β structures. Their topologies can be conveniently
specified by the simple nomenclature listing connection types (see
Section II,B). The major irregularity of the globin fold when considered as a cylinder is that one element (the A and B helices) bends
sharply to close the cylinder; this feature is also seen in five- and
six-stranded β barrels such as trypsin. But perhaps the most satisfying
feature of schematizing the globin fold on a cylinder is that it can then
be grouped with other structures (thermolysin d2, T4 phage lysozyme
d2, papain d1, and cytochrome c peroxidase d2) which also show the
"Greek key" (see Richardson, 1977) topology of $+3, -1, -1$. Papain
domain 1 also shows the diagnostic feature of Greek key structures by
containing a non-nearest-neighbor connection which skips across the
end of the cylinder; however, most of its helices are short and they
form a rather irregular bundle. Papain domain 1 contains two disulfides; we will find repeatedly that increasing disulfide content goes
along with decreasing regularity of both secondary and tertiary structure.

These four structures then form the second major subgrouping of
antiparallel α domains, which we will call Greek key helix bundles
(see Fig. 73). The helix elements lie on an approximate cylinder (see
Fig. 89d for an end view), with 0 to 45° right-handed twist relative to
the cylinder axis; they are connected with a Greek key topology which
can have either a counterclockwise (globins) or a clockwise (thermolysin d2 and T4 lysozyme d2) swirl when viewed from the outside.

The remaining structures in this category (carp Ca-binding protein,
egg lysozyme, citrate synthase, catalase d2, and p-hydroxybenzoate
hydroxylase d3) are miscellaneous helical domains. However, there
is good evidence from sequences and from functional resemblances
(Kretsinger, 1976) that carp Ca-binding protein exemplifies a whole
group of proteins that are constructed of "E-F hands" (see Section
II,F) and that regulate or are regulated by changes in Ca^{2+} concentration. Citrate synthetase may be the first example of a group of larger
helical domains with three layers. Irregular helical structures with a
moderate number of disulfides can be classified either here or as small
SS-rich. We have classified egg lysozyme here (with only 4 disulfides
in 129 residues), while phospholipase A_2 (with 7 disulfides in 123 residues) is classified with the small SS-rich proteins.

C. Parallel α/β Domains

The largest grouping of structures contains domains organized
around a parallel or mixed β sheet, the connections for which form
structure (usually helical) protecting both sides of the sheet, with the

helices also predominantly parallel to each other. Of course, each helix and its neighboring β strand are antiparallel to one another, but this structure category is called parallel α/β because both the β sheet interactions and the α-helix interactions are internally parallel. The parallel α/β category is the same as Levitt and Chothia's α/β proteins. Figures 75–78 show schematic drawings of this group of structures. It is interesting to note that there seems no a priori reason not to have parallel all-α structures or parallel all-β structures formed of two helix layers or of two parallel β sheets, yet such structures are not found. All of the domains with parallel organization have both a β sheet of at least four or five strands and at least three or four α-helices. Almost all have at least three layers.

The first subgrouping under the parallel α/β category contains two of the largest but simplest domain structures that have yet been found. They are the eight-stranded parallel β barrels of triosephosphate isomerase (see Fig. 90) and pyruvate kinase domain d1, both of which are connected in $+1x,+1x$ topology all the way around. (In structures with both β sheet and also several helices it is convenient to use just the β strands for designating the topology.) The connections are α-helices, which form a larger cylinder of parallel helices concentric with the β barrel. The structural elements of both α and β cylinders have a pronounced right-handed twist around the cylinder axis. Connections between the parallel β strands must lie on the outside of the barrel since the interior is filled by the packed hydrophobic side chains. If all of the crossover connections must be right-handed and no knots are allowed, then the chain must wind consistently around the barrel in one direction, and the $+1x,+1x,+1x$ topology is not only the simplest but essentially the only possible topology for such a structure (Richardson, 1977), since all other topologies are knotted and unfoldable. We call this structure the singly wound parallel β barrel, since successive crossover connections are wound on the barrel progressing in a single direction with no reversal or backtracking. Figure 91a is a highly schematized representation of the "singly wound" structure, viewed from one end of the barrel.

The largest subgrouping within the parallel α/β category contains structures with a central twisted wall of parallel or mixed β sheet, protected on both sides by its crossover connections (most of which are helical). This is called the doubly wound parallel β sheet, because with right-handed crossovers the simplest way of protecting both sides of the sheet is to start near the middle and wind toward one edge, then return to the middle and wind to the other edge. Figure 91b is a highly schematized representation of the "doubly wound" structure, viewed from one end of the sheet (compare with Fig. 91a).

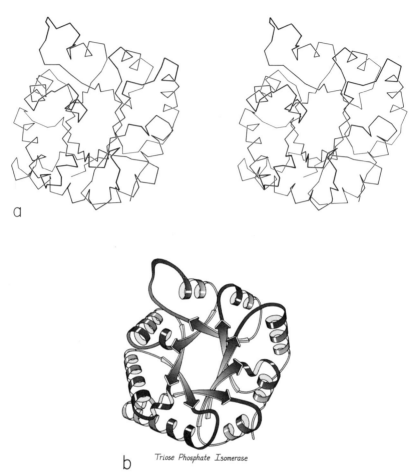

b   *Triose Phosphate Isomerase*

FIG. 90.   Triosephosphate isomerase as an example of a singly wound parallel β barrel.   (a) α-Carbon stereo, viewed from one end of the barrel; (b) backbone schematic, viewed as in a; (c) α-carbon stereo, viewed from the side of the barrel; (d) backbone schematic, viewed as in c; (e) topology diagram showing the + 1x right-handed connections between the β strands.

The singly wound barrel has four major layers of backbone structure and the doubly wound sheet has three major layers (with two separate hydrophobic cores); most other domain structures have only two major backbone layers with a single hydrophobic core, and are on the average considerably smaller.

The doubly wound structures were first recognized as a category by Rossmann in comparing flavodoxin with lactate dehydrogenase d1. As more and more protein structures were solved which fell into this

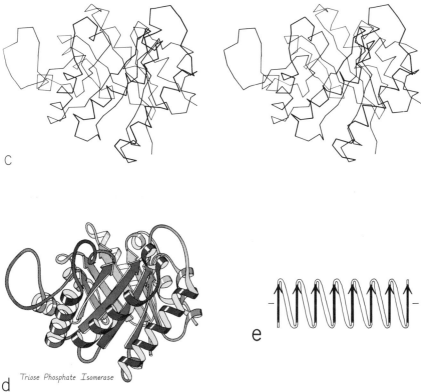

c

d *Triose Phosphate Isomerase*

e

FIG. 90 (continued)

category, the relationships between them have been described and
debated at considerable length. The initial descriptions were in
terms of the $\beta-\alpha-\beta-\alpha-\beta$ unit as a supersecondary structure (Rao and
Rossmann, 1973). Quite soon the emphasis shifted to the functional
properties of the nucleotide-binding site which most of them share,
and to the probable evolutionary relationships between these
"nucleotide-binding domains" (Schulz and Schirmer, 1974; Rossmann *et al.*, 1974). By now the consensus appears to be that some of
the most similar of these structures must certainly be related to each
other, while at least some of the most dissimilar examples surely
cannot be related (Rossmann and Argos, 1976; Matthews *et al.*, 1977;
Levine *et al.*, 1978; McLachlan, 1979a).

We will group these domains into five gradually loosening levels of
topological similarity, without attempting to make any definite decision as to where the dividing line lies between divergent and convergent examples.

Figure 92 shows stereo and schematic drawings of lactate dehy-

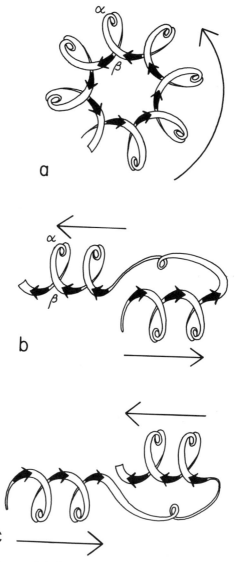

FIG. 91. Highly simplified sketches (viewed from the C-terminal end of the β strands) of (a) a singly wound parallel β barrel; (b) a classic doubly wound β sheet; (c) a reverse doubly wound β sheet. Thin arrows next to the diagrams show the direction in which the chain is progressing from strand to strand in the sheet.

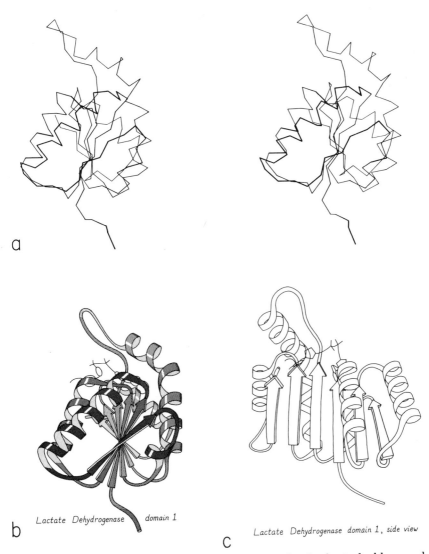

a

b   *Lactate Dehydrogenase*   *domain 1*

c

Lactate Dehydrogenase domain 1, side view

FIG. 92. Lactate dehydrogenase domain 1 as an example of a classic doubly wound parallel β sheet. (a) α-Carbon stereo, viewed from one edge of the sheet; (b) backbone schematic, viewed as in a; (c) backbone schematic, viewed from one face of the sheet.

drogenase domain 1, which is the classic example of a nucleotide-binding domain or doubly wound sheet.

The darkest inner box in Fig. 93 includes those "classic" doubly wound parallel sheets that exactly match the topology of lactate dehydrogenase d1. Phosphorylase domain 2 is a five-layer structure in

Lactate DH$_{d1}$, Alcohol DH$_{d2}$, P Glycerate K$_{d2}$, ATCase cat.$_{d2}$, Phosphorylase$_{d2}$

Glyceraldehyde P DH$_{d1}$

Flavodoxin

Subtilisin

Arabinose-binding$_{d1}$

Dihydrofolate Reductase

Adenylate K

Rhodanese$_{d1}$

Glutathione Reductase$_{d2}$

P Glycerate Mutase

Pyruvate K$_{d3}$

HexoK$_{d1}$

HexoK$_{d2}$

Carboxypeptidase

Thioredoxin

Carbonic Anhydrase

P Fructo K$_{d2}$

FIG. 93.   Topology diagrams for the doubly wound and miscellaneous α/β domains illustrated in Figs. 76 through 78. Arrows represent the β strands; thin connections lie behind the β sheet and fat ones above it. The darkest upper box surrounds the classic doubly wound sheets; successively lighter solid boxes include domains that are progressively less like the classic topology; the dotted box encloses the miscellaneous α/β structures.   K = kinase; P = phospho; DH = dehydrogenase; ATCase = aspartate transcarbamylase.

which the central three layers are a classic doubly wound sheet and the outer helical layers are formed by the two ends of the chain.   The next box includes examples in which deleting one or two strands

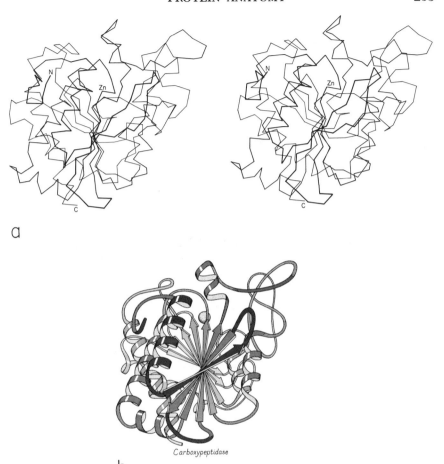

a

b

Carboxypeptidase

FIG. 94.   Carboxypeptidase A as an example of a miscellaneous α/β structure.   (a)
α-Carbon stereo, viewed from one edge of the mixed β sheet; (b) backbone schematic,
viewed as in a.

either at an end of the chain or at an edge of the β sheet will produce a
five- or six-strand doubly wound sheet, while in the next box such deletions yield four doubly wound strands.   In the outermost solid box it
is necessary to omit strands interior to both the sheet and the sequence in order to get four strands of doubly wound β sheet.   The
structures inside the dotted box can yield no more than three such
strands and cannot really be described as doubly wound; they share
with the rest of this large subgrouping only the general organization of
a central wall of parallel or mixed β sheet protected on both sides by
its connections (see Fig. 94 for carboxypeptidase as an example).   As

one progresses outward from the classic to the most peripheral cases, the number of antiparallel strand pairs mixed in with the parallel gradually increases. Aside from the "classic" examples in the inner box, there are several other exact duplicates of doubly wound topologies between different proteins: phosphorylase d1 versus glyceraldehyde-phosphate dehydrogenase d1; aspartate transcarbamylase catalytic d1 versus rhodanese d1,d2; catalase d3 versus flavodoxin; and p-hydroxybenzoate hydroxylase d1 versus glutathione reductase d1.

As one progresses from classic to peripheral doubly wound sheets, the number of domains that bind nucleotides also decreases. A favorable site for binding dinucleotides (or in a few cases, mononucleotides) is associated with this general category of structure and to a large extent with the doubly wound topology. The dinucleotides are all bound in approximately equivalent positions at the C-terminal end of the β sheet strands, within one strand of the central position where the winding switches direction (see Fig. 91b). Nucleotides are also bound at the C-termini of β strands in the singly wound barrels. In most of these cases, each nucleotide is associated with a "mononucleotide-binding fold" of three β strands and two helices with $+1x,+1x$ topology; combination of two of these folds around a local 2-fold axis produces the classic doubly wound sheet. In some cases, however (such as hexokinase or dihydrofolate reductase), the topology is quite significantly different. Also there seems to be another quite different type of nucleotide-binding site such as the active site in staphylococcal nuclease (Arnone et al., 1971) or the AMP site in phosphorylase (Sygusch et al., 1977); both of these sites rely mainly on arginines for binding the nucleotide phosphates.

One quite surprising and intriguing feature of this group of structures is that it contains extremely few examples of the "reverse doubly wound" topology (see Fig. 91c), a different but equally plausible pattern related to the doubly wound sheet by reversing the N- to C-terminal direction of the chain or by switching relative positions of the two halves of the β sheet. Of the four reverse examples (found in PGK d2, GPDH d1, glucosephosphate isomerase d1, and phosphorylase d1) none forms a nucleotide-binding site, all belong to a sheet that also has a normal doubly wound section, and none includes more than four strands. Those cases do demonstrate that such a topology is stable and can fold, but there must be some strong reason why it is so rare. Some simple explanations of this regularity would be either that most of the nucleotide-binding domains are related, or that they must fold strictly from the N-terminus, or that the requirements for forming a nucleotide-binding site are restrictive enough to constrain the usual doubly wound topology. None of these explanations is completely satisfying, however, because a number of domains are known that cannot fold strictly from the N-terminus, because the relative placement of features forming the nucleotide sites is only rather approximate (e.g., see D. A. Matthews *et al.*, 1979), and because the rearrangement necessary to produce a reverse doubly wound sheet seems much less drastic than many of the rearrangements that must be proposed if all of these proteins are related.

D. Antiparallel β Domains

The next major grouping consists of domains that are organized around an antiparallel β sheet. They are as numerous as the parallel α/β structures, and their topology and classification have been discussed before (see Levitt and Chothia, 1976; Sternberg and Thornton, 1977a,b; Richardson, 1977). This category is the most varied in terms of size and organizational patterns. Figures 79 through 84 show backbone schematics for the antiparallel β domains, grouped into subcategories.

Most of the antiparallel β domains have their β sheets wrapped around into a cylinder, or barrel, shape. None of the antiparallel barrels has as symmetrical or as continuously hydrogen-bonded a cylindrical sheet as the singly wound parallel β barrels of triosephosphate isomerase and pyruvate kinase d1; however, antiparallel barrels are very much more common. Because of gaps in the hydrogen-bonding, some of these structures have been described as two β sheets facing each other (e.g., Schiffer *et al.*, 1973; Blake *et al.*, 1978; Harrison *et al.*, 1978). Our reasons for treating them all as barrels are that the gap positions are sometimes different in domains that are probably related, and that the barrel description yields very much simpler and more unified topologies.

Barrels seem to prefer pure parallel or antiparallel β structure even more strongly than does β sheet in general. All the known singly wound barrels are pure parallel. An antiparallel barrel with an odd number of strands is constrained to have one parallel interaction, but no other parallel strand pairs occurs within antiparallel barrels except in the acid proteases. Also, even-stranded barrels are much more common than odd-stranded ones.

The first type of antiparallel β barrel, in analogy with the first type of helix bundle, has simple up and down $+1, +1, +1$ connections all around. Although it is relatively unusual for a barrel to be composed entirely of up-and-down strands, many of the larger barrels and sheets have four- to six-stranded sections of simple up-and-down topology

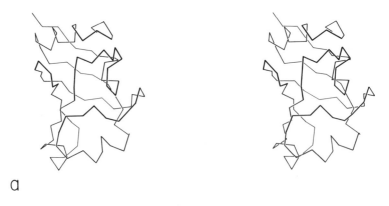

a

b   *Papain   domain   2*

FIG. 95.   Papain domain 2 as an example of an up-and-down antiparallel β barrel.   (a) α-Carbon stereo, viewed from one side of the barrel; (b) backbone schematic, viewed as in a.

embedded within them.   There are three examples of pure up-and-down β barrels: soybean trypsin inhibitor, papain d1, and catalase d1. Figure 95 shows a stereo and a schematic drawing of papain d1.   Soybean trypsin inhibitor has long excursions at the ends of three of the β strand pairs, forming separate, twisted β ribbons; there is a strong internal 3-fold symmetry which includes these ribbons as well as the strand pairs in the barrel (McLachlan, 1979c).   Catalase d1 is an eight-stranded up-and-down barrel with less extreme loop excursions. Rubredoxin could be considered as a very irregular and incomplete up-and-down barrel in which β-type hydrogen bonds are formed between only about half of the strand pairs (see Fig. 76).   It is very small and compact, and is presumably stabilized partly by the network of Cys ligands to the iron; therefore we have placed it in the small metal-rich category.

Soybean trypsin inhibitor, papain d1, and rubredoxin have identical topologies: six strands of $+1, +1, +1, \ldots$ proceeding to the left around the barrel if the chain termini are at the bottom. However, handedness is not nearly as meaningful a property for up-and-down topologies as it is for Greek keys, since up-and-down handedness can change on addition or delection of a single strand.

The commonest subgroup of antiparallel β barrel structures has a Greek key topology, with $-3, +1, +1, -3$ connections or a close variant. The first Greek key barrel structures were compared in Richardson *et al.* (1976), and they and the up-and-down barrels were described as categories in Richardson (1977). Figure 96 illustrates Cu,Zn superoxide dismutase as an example of a Greek key β barrel.

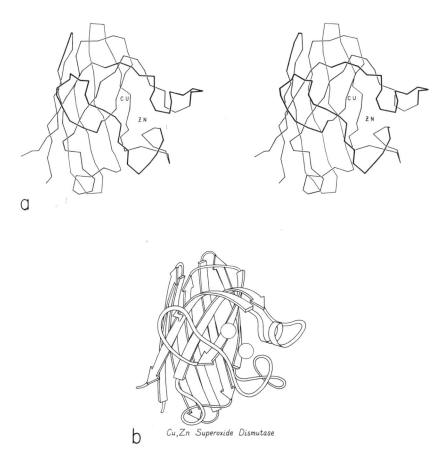

FIG. 96. Cu,Zn superoxide dismutase as an example of a Greek key antiparallel β barrel. (a) α-Carbon stereo, viewed from one side of the barrel; (b) backbone schematic, viewed as in a.

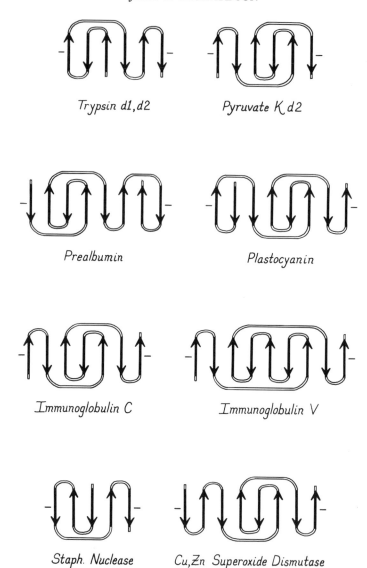

FIG. 97.   Topology diagrams of the Greek key antiparallel β barrels. The dashes on either side of a topology diagram indicate that the barrel was opened up at that point and laid out flat; all barrels are shown viewed from the outside.

There are 13 Greek key barrels in our sample, and 12 of them (all except staphylococcal nuclease) have the same handedness: viewed from the outside, the Greek key pattern forms a counterclockwise swirl (see Fig. 97). The four barrels shown in Figure 81 have a more

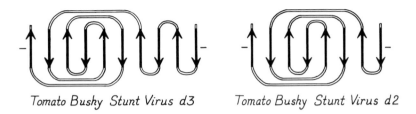

Tomato Bushy Stunt Virus d3   Tomato Bushy Stunt Virus d2

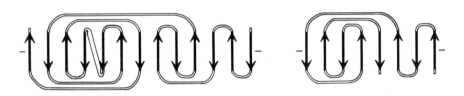

Concanavalin A   γ Crystallin

FIG. 98. Topology diagrams of the "jellyroll" Greek key β barrels.

complicated "jellyroll" topology with an extra swirl in the Greek key (this pattern was also common on Greek vases); the "jellyroll" Greek key topologies are shown in Fig. 98. The jellyroll pattern is produced by having a pair of connections, rather than just one connection, crossing each end of the barrel.

The Greek key barrels have between 5 and 13 strands, but in all cases they enclose approximately the same cross-sectional area (see Section II,B). The cross sections are somewhat elliptical, with more flattening the more strands there are. For 8- to 10-stranded barrels, it is noticeable that the direction of the long axis of the cross-section twists from one end of the barrel to the other by close to 90° (see Fig. 99).

± 3 connections are not particularly common outside of the barrels, so that the prevalence of Greek key topologies is not due simply to chance combination of the connection types that make it up. There are two different ways of analyzing the Greek key which could perhaps explain both its frequent occurrence and its strongly preferred handedness. The first approach is to consider the stability of the final barrel, given its size, shape, and twist. Figure 99 shows that the Greek key pattern provides neat, efficient connections across the top and bottom of the barrel, lying next to the ± 1 connections. In tomato bushy stunt virus d3 there is actually some β-type hydrogen-bonding between the + 1, − 3, and + 5 connecting loops. In combination with the twist of the strands and of the barrel cross section, a counterclockwise Greek key (as shown) produces ±3 connections

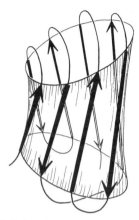

FIG. 99. A highly simplified sketch of the slightly flattened cylinder of a β barrel, showing how the direction of flattening twists from top to bottom.

that are approximately perpendicular to each other on opposite ends of the barrel and that can both cross along a short axis of the cross section. A clockwise Greek key would place the ±3 connections in a weaker position approximately parallel to each other, and one of them would be along the long axis. This argument could not account for the handedness of the partial Greek keys with −3,+1,+1 topology (such as staphylococcal nuclease and chymotrypsin) where there is a ±3 connection at only one end of the barrel.

The other possible explanation hypothesizes an effect during the protein folding process, very similar to the one proposed to explain crossover handedness (see Section II,B). All Greek keys, even the "jellyrolls," necessarily have a folding point halfway along the chain from which two paired strands can be followed back next to each other as they curl around the structure. Given the prevalence of Greek key patterns in the known structures, it seems very likely that the polypeptide chain can fold up by first folding in half and forming a long, two-stranded β ribbon, and then curling up the ribbon to produce the further β sheet interactions. This sort of process is illustrated in Fig. 100. Since the initial ribbon would presumably have a strong right-handed twist (see Section II,B), it would impart a right-handed twist to the curling direction and always end up with a counterclockwise Greek key. Besides the β barrels, there are other pieces of protein structure that suggest this sort of process, such as the long β ribbons in lactate dehydrogenase d2 (see Fig. 74). This kind of folding hypothesis has been utilized by Ptitsyn and Finkelstein (1980) to obtain rather successful predictions of β strand contacts and topologies.

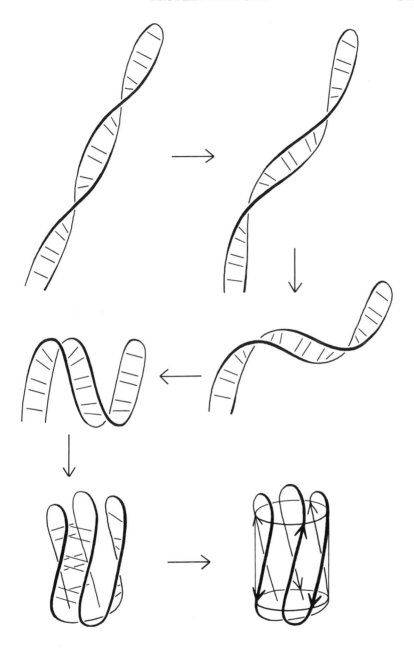

FIG. 100. A hypothetical folding scheme for Greek key β barrels which could explain why essentially all of the Greek key and jellyroll β barrels have the same handedness of topology.

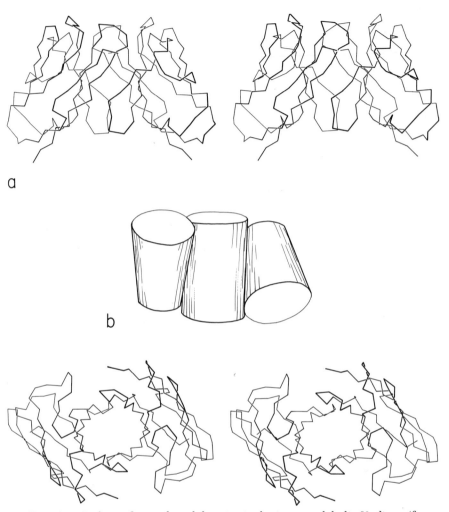

a

b

FIG. 101.   Packing of two β barrel domains in the immunoglobulin V_L dimer (from Bence-Jones REI): (a) α-carbon stereo, viewed from the sides of the barrels; (b) simplified schematic of the barrels as cylinders, viewed as in a; (c) α-carbon stereo, viewed from one end of the barrels.   The contact between the two domains forms a third barrel in the center.

Partial, multiple, and other barrels have been grouped together as another subgroup within the antiparallel β category (see Fig. 82).   Ribonuclease contains a four-stranded antiparallel β sheet that looks like a five-stranded barrel with one strand missing.   Alcohol dehydrogenase d1 includes a five-stranded antiparallel barrel (with a topology of +1, +3x, −2, +1) and another partial five-stranded barrel.

Back-to-back β barrels that share one wall occur in the variable half of immunoglobulin Fab structures (except for Rhe: see Wang *et al.*, 1979), where V_L and V_H are each antiparallel β barrels and the contact between them forms an even more regular eight-stranded barrel with four strands contributed from each domain (see Fig. 101). The three barrels pack against each other with a right-handed superhelical twist, and the angle between the axes of adjoining barrels is the same as the angle between opposite strands in one of the barrels. The two domains of the acid proteases have complicated, very similar mixed β sheets that could be described either as a six-stranded barrel with side sheets or as several interlocking β sheets. When more examples are available, it will probably be possible to find patterns to the ways in which small subsidiary β sheets can interlock into the edges of larger sheets (such as in the acid proteases or thermolysin d1), but for now no attempt has been made to classify them.

The next subcategory of antiparallel β domains each has a single, more or less twisted β sheet, either pure antiparallel or predominantly so, but not closing around to form a barrel. They are shown in Fig. 83, and Fig. 102 shows glyceraldehyde-phosphate dehydrogenase as an example. Their common feature is a layer of helices and loops which covers only one side of the sheet, so that they are two-layer structures. Many β barrels have been described as "sandwiches," with two slices of β sheet "bread" and a "filling" of hydrophobic side chains; based on that analogy these structures would be "open-face sandwiches," with a single slice of β sheet "bread" and a "topping" of helices and loops. The open-face β sandwiches could rival a Danish buffet for variety on a theme: they range from 3 to 15 strands, with a wide assortment of topologies, curvatures, and placement of helices and loops. Bacteriochlorophyll protein, the largest of them, encloses between the β sheet layer and the helical layer a core of seven bacteriochlorophyll molecules, tightly packed in an orderly but quite asymmetrical array.

The remaining three antiparallel β structures form a miscellaneous category (see Fig. 84). Lactate dehydrogenase d2 and gene 5 protein each has several two-stranded antiparallel β ribbons, but they do not coalesce into any readily described overall pattern. The N-terminal domain of tomato bushy stunt virus protein has a unique β structure in which equivalent pieces of chain from three different subunits wrap around a 3-fold axis to form what has been called a "β annulus" (Harrison *et al.*, 1978). Each of the three chains contributes a short strand segment to each of three three-stranded, interlocking β sheets. This "domain" provides one of the subunit contacts that hold the virus

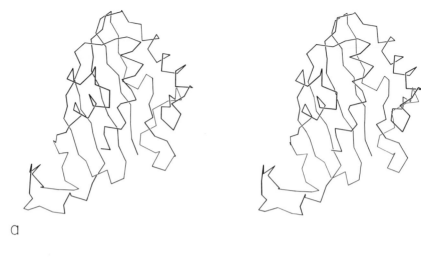

a

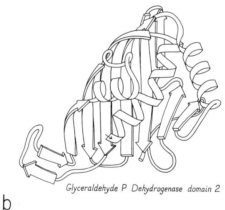

Glyceraldehyde P Dehydrogenase domain 2

b

FIG. 102.   Glyceraldehyde-phosphate dehydrogenase domain 2 as an example of an open-face sandwich antiparallel β sheet.   (a) α-Carbon stereo, viewed from the buried side of the sheet; (b) backbone schematic, viewed as in a.

shell together.   However, only one-third of the 180 subunits contribute to the β annuli; for the other quasiequivalent subunits, the N-terminal part of the chain is disordered with respect to the virus shell.

E. Small Disulfide-Rich or Metal-Rich Domains

The last major category (shown in Figs. 85 and 86) consists of small (usually less than 100 residues) domains whose structures seem to be strongly influenced by their high content either of disulfide bonds (S)

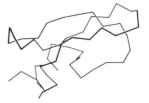

a

b   *Pancreatic Trypsin Inhibitor*

FIG. 103. Basic pancreatic trypsin inhibitor as an example of a small disulfide-rich structure. (a) α-Carbon stereo; (b) backbone schematic, viewed as in a, with disulfides shown as zig-zags. Figure 2 shows an all-atom stereo of this protein with side chains.

or of metal ligands (M). These S–M proteins often look like distorted versions of other, more regular, proteins. The disulfide-rich ones include many toxin and enzyme-inhibitor structures. For most of the disulfide-rich proteins it is known experimentally that they are completely unstable if the disulfides are broken (in contrast to larger disulfide-containing proteins, for which disulfides merely provide additional stabilization for an already-determined structure). Figure 103 shows pancreatic trypsin inhibitor as an example of a disulfide-rich protein, and Fig. 104 shows cytochrome c as an example of a metal-rich protein. Most of the S–M proteins are single-domain and monomeric: only wheat germ agglutinin has multiple domains, and only insulin has multiple subunits in the molecule.

The only subgroup of similar structures within the S–M proteins is the toxin-agglutinin folds of the snake neurotoxins and the domains of wheat germ agglutinin (see Fig. 85). They are made up of extended-chain loops with an almost identical topology of $-1, +3, -1, +2x$ rather like a series of half-hitch knots (the β structure is extremely minimal in wheat germ agglutinin) strongly linked by a core of four disulfides, three of which are equivalent (see Drenth *et al.*, 1980). High-potential iron protein and ferredoxin share a local loop structure that binds the iron–sulfur cluster, but otherwise are different.

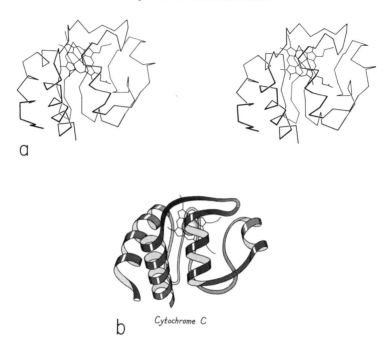

Cytochrome C

FIG. 104. Cytochrome c as an example of a small metal-rich protein. (a) α-Carbon stereo, with heme; (b) backbone schematic, viewed as in a. The backbone forms an approximate up and down cylinder with the heme tucked into the center, but the elements forming the cylinder are a mixture of helices and extended strands.

Most of the metal-rich proteins form approximately cylindrical two-layer structures with either an up and down (rubredoxin, cytochrome c) or a Greek key (ferredoxin) topology, but in which the elements forming the cylinder are a mixture of helices, β strands, and more or less extended portions of the backbone. Cytochrome c_3 is perhaps the ultimate example of an S–M protein, with four hemes in just over a hundred residues, and essentially no secondary structure at all except for one helix.

One way of considering these proteins is as distorted versions of the other structural types. Most S–M proteins can fairly clearly be

FIG. 105. Examples of small disulfide-rich or metal-rich proteins (shown on the right side) compared with their more regular counterparts in other structural categories (shown at the left). (a) Tobacco mosaic virus protein, an up-and-down helix bundle; (b) cytochrome b_5, a distorted up-and-down helix bundle; (c) trypsin domain 1, a Greek key antiparallel β barrel; (d) high-potential iron protein, a distorted Greek key β barrel; (e) glutathione reductase domain 3, an open-face sandwich β sheet; (f) ferredoxin, a distorted open-face sandwich β sheet.

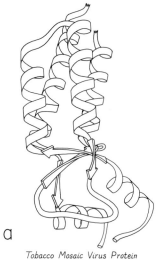

a

Tobacco Mosaic Virus Protein

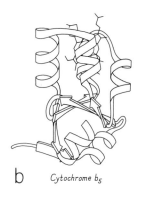

b   *Cytochrome b₅*

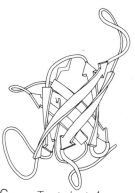

c   *Trypsin domain 1*

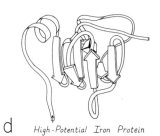

d   *High-Potential Iron Protein*

f   *Ferredoxin*

e

Glutathione Reductase domain 3

grouped as either distorted helix clusters (phospholipase, cytochrome c, cytochrome b_5), distorted β barrels (rubredoxin, high-potential iron protein), or distorted open-face sandwiches (erabutoxin, wheat germ agglutinin, pancreatic trypsin inhibitor, or ferredoxin). Figure 105 shows an example of each of these relationships. In fact, one reasonable taxonomy would do away with this fourth major category altogether and place all the S–M proteins as irregular examples of either an α or a β category. We have not chosen that approach, however, because several of the structures (crambin, insulin, and cytochrome c_3) are rather difficult to place in one of the other categories, and also because these small proteins influenced by nonpolypeptide interactions appear to share important features, especially in terms of the probable complexity of their folding process (see Section IV,C).

Another suggestive fact is that there are no small, irregular structures related to the parallel α/β category. Perhaps this reflects the fact that domains organized around parallel β sheet are necessarily fairly large and seem to be dependent on large, buried, and quite regular β structure for their stability. There are in fact no hemes (in spite of all the helices) or iron–sulfur clusters in parallel α/β proteins, and no disulfides except for the single active-site disulfides of thioredoxin, glutathione peroxidase, and glutathione reductase.

IV. Discussion

A. Implications for Noncrystallographic Determination of Protein Structure

Aside from the direct techniques of X-ray or electron diffraction, the major possible routes to knowledge of three-dimensional protein structure are prediction from the amino acid sequence and analysis of spectroscopic measurements such as circular dichroism, laser Raman spectroscopy, and nuclear magnetic resonance. With the large data base now available of known three-dimensional protein structures, all of these approaches are making considerable progress, and it seems possible that within a few years some combination of noncrystallographic techniques may be capable of correctly determining new protein structures. Because the problem is inherently quite difficult, it will undoubtedly be essential to make the best possible use of all hints available from the known structures.

The most important general point to be emphasized is that it is now both possible and essential to frame and test hypotheses about exactly which structural features a given technique is really measuring or predicting. Structure surveys like the current one can help in choosing proteins that will provide critical tests of such hypotheses, both by locating proteins that vary most in the parameters under consideration, and also in helping to control for the effects of differences in other major structural parameters. It seems inherently unlikely that any spectroscopic feature is a direct measure of percentage α, β, turn, and coil as defined in any of the usual ways. But since such percentages are certainly not the only useful way to describe protein structure, it should be fruitful to combine theoretical analysis (where possible) with careful empirical tests in order to determine the set of descriptions most applicable for a given technique. Let us consider several examples of what could be attempted with this approach.

Methods for predicting secondary structure from amino acid sequence could presumably benefit from considering parallel and antiparallel β sheet separately, since the two types have rather different single and pairwise residue preferences. An overall classification scheme could help in choosing a large and characteristic sample. However, there is the difficulty of dealing with mixed β sheets. For a given set of parameters that successfully distinguished pure parallel from pure antiparallel sheets, it would be possible for instance to test whether the characteristics of strands in mixed β sheets depended mainly on their local hydrogen-bonding type or depended mainly on whether the overall sheet organization was "antiparallel β" type or "parallel α/β" type (for example, prealbumin is an "antiparallel" mixed sheet and carboxypeptidase is a "parallel" mixed sheet). This sort of question should be asked also for the amide I bands in infrared (see Miyazawa and Blout, 1961; Chirgadze and Nevskaya, 1976) and Ramam spectra (see Krimm and Abe, 1972; Yu et al., 1975) that are thought to be sensitive to the differences between parallel and antiparallel β sheet. It would be especially useful if it turned out that some features were sensitive to local and some to overall structure. In general, the parallel α/β structures have been grossly underrepresented in spectroscopic studies of protein conformation, because they do not occur in the small proteins that made up most of the early X-ray structure determinations. Now that α/β proteins have been shown to be extremely common, this sampling bias can be corrected.

The C—S and S—S stretch vibrations of disulfides (Edsall et al., 1950) can be observed in the Raman spectra of proteins, but their interpretation is still somewhat controversial (see, for example, Klis and Siemion, 1978; Spiro and Gaber, 1977). Using series of model compounds, Van Wart et al. (1973) have related S—S stretch frequency to the χ_3 (Cβ—S—S—Cβ') dihedral angle, while Sugeta et al. (1972, 1973) have related the S—S frequency to the χ_2 (Cα—Cβ—S—S) dihedral angle and C—S stretch frequency to χ_1 angle; these latter correlations have been further modified by Van Wart and Scheraga (1976). The relationship of spectrum to conformation seems to be quite complex in proteins, where constraints at either end of the disulfide would tend to increase coupling between the modes. The S—S stretch may be sensitive to the relative sign as well as the absolute value of χ_2, and therefore may reflect the difference between the spiral and the hook conformations (see Section II,E). It should be possible to determine characteristic spectra for the three common disulfide conformations found in proteins (the left-handed spiral, the right-handed hook, and the extended form in immunoglobulins) by choosing accurately refined proteins with a single or a dominant disulfide conformation (e.g., immunoglobulins, carboxypeptidase, egg lysozyme, and pancreatic trypsin inhibitor).

Very low-frequency vibrations have been observed in proteins (e.g., Brown *et al.*, 1972; Genzel *et al.*, 1976), which must involve concerted motion of rather large portions of the structure. By choosing a suitable set of proteins to measure (preferably in solution), it should be possible to decide approximately what structural modes are involved. Candidates include helix torsion, coupled changes of peptide orientation in β strands, and perhaps relative motions of entire domains or subunits. These hypotheses should be tested, because the low-frequency vibrations probably reflect large-scale structural properties that would be very useful to know.

In using circular dichroism to estimate percentages of the various secondary structures in a protein (e.g., Saxena and Wetlaufer, 1971; Grosse *et al.*, 1974), helix can be judged more reliably than other features, as is usually true for almost any method including prediction (e.g., Maxfield and Scheraga, 1976). This is presumably because α-helices are relatively uniform in both local and longer range patterns, while β structure is widely variable in hydrogen-bonding pattern, regularity, twist, exposure, and overall shape. There is at least a real possibility that differences in shape and organization of β structure are reflected in the circular dichroism spectrum; that possibility should be tested, because it would be even more useful to be able to categorize a structure as a doubly wound sheet or an antiparallel β barrel than to say it had 35% β structure, even supposing that we could reliably do the latter.

Successful examples of the sort of correlations postulated above would add additional independent pieces of information for use in a combined strategy of noncrystallographic protein structure determination. Empirical regularities such as the handedness of crossover connections (see Section II,B) can help in narrowing down the possibilities. Another need is to decide whether, and at what point, a protein is divided into domains. The more tenuously connected domain pairs can often be recognized by such techniques as electron microscopy, viscosity, low-angle scattering, or proteolysis, and it might prove possible to recognize domain-connection regions in the sequence. Knowledge of a set of common overall structure types (such as the major subgroupings in our classification scheme) can provide prototypes with which to match the distribution of predicted secondary structures and the characteristics suggested by various spectroscopic measures. For a given protein, combination of all these methods in an overall strategy that can deal with their probabilistic nature and disparate information content may some day be able to produce a fairly small number of alternative structures, one of which (by some process such as energy minimization) would converge to what could be recognized as the correct native structure.

Even an infallible method of structure prediction would not make protein crystallography obsolete; detailed prior knowledge of the globin structure has not removed the necessity or interest of high-resolution X-ray structures for other species, mutants, and ligand forms of hemoglobin. What it would do is to take away a great deal of the fun and excitement of discovering new structures by protein crystallography; but that is not too large a price for the kind of increased understanding that is likely to accompany even the most ad hoc of successful structure prediction methods.

B. Implications for Protein Evolution

One important reason for classifying proteins is simply to make the structures more memorable. The system proposed above can help to do that, especially for those cases which fall into one of the more narrowly defined subgroups. However, we also want to know to what extent this classification is a true taxonomy: that is, whether or not it expresses underlying genetic relationships. In addition, among so many structure examples, almost any major rule governing either protein evolution or protein folding would have predictable statistical consequences on the pattern of structural resemblances to be expected. Therefore, it is worthwhile examining the distribution of features that is actually found, because it may suggest various conclusions about how proteins evolve and fold.

One significant feature evident in the known structures is the frequency with which domain pairs within a given protein are found to match each other closely in structure. It is known from amino acid se-

TABLE II

Internal Similarity or Dissimilarity of Domains within Multidomain Proteins[a]

Similar domain pairs	Different domain pairs
Phosphorylase d1, d2	Papain d1, d2
Phosphoglycerate kinase d1, d2	Tyrosyl-tRNA synthetase d1, d2
Aspartate carbomoyltransferase catalytic	Thermolysin d1, d2
d1, d2	T4 phage lysozyme d1, d2
Arabinose-binding protein d1, d2	Glucosephosphate isomerase d1, d2
Phosphofructokinase d1, d2	Pyruvate kinase d1, d2
Rhodanese d1, d2	Pyruvate kinase d2, d3
Hexokinase d1, d2	Lactate dehydrogenase d1, d2
Glutathione reductase d1, d2	Alcohol dehydrogenase d1, d2
Tomato bushy stunt virus d2, d3	Glyceraldehyde-phosphate dehydrogenase
Chymotrypsin d1, d2	d1, d2
γ-Crystallin d1, d2	Glutathione reductase d2, d3
Immunoglobulin C, V	Influenza virus hemagglutinin HA1, HA2
Immunoglobulin C1, C2	p-Hydroxybenzoate hydroxylase
Immunoglobulin C2, C3	(4-hydroxybenzoate 4-monooxygenose)
Acid protease d1, d2	d1, d2
Wheat germ agglutinin d1, d2	p-Hydroxybenzoate hydroxylase d2, d3
Wheat germ agglutinin d1 and d2,	Catalase d1, d2
d3 and d4	Catalase d2, d3

[a] For proteins with more than two domains, each potential duplication is listed separately: e.g., a minimum of two duplications would be necessary to produce either a three-domain or a four-domain structure. Members of the pairs in the left-hand column both fall within the same structural subcategory and have fairly similar topologies; such pairs are perhaps the result of internal gene duplications. Members of pairs in the right-hand column almost all fall into different major categories of tertiary structure (e.g., one all-helical and one antiparallel β); presumably they could not have been produced by internal gene duplication.

quences (e.g., Dayhoff and Barker, 1972) that internal gene duplication can occur in proteins. For recent or highly conserved duplications with closely related sequences the duplication event can be conclusively demonstrated. However, study of sequences cannot tell us how widespread and frequent gene duplication has been in the evolution of proteins because it cannot detect old duplications whose sequences have had time to diverge beyond recognizable homology. There are 26 multidomain proteins in our sample, which would have required the introduction of new domains 35 different times; they are listed in Table II. In slightly over half (17) of those cases, the structure of the new and old domains is basically the same (Fig. 106 shows the two domains of rhodanese as an example); in two cases (cytochrome c peroxidase and aspartate transcarbamylase regulatory chain) the level of similarity is ambiguous; while in the other 16 cases

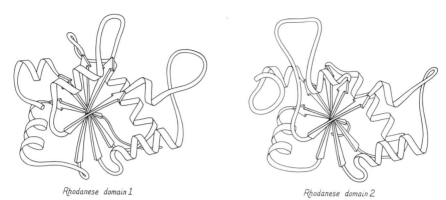

Rhodanese domain 1   *Rhodanese domain 2*

FIG. 106.   Rhodanese domains 1 and 2 as an example of a protein with two domains which resemble each other extremely closely.

the structures are totally different and presumably could not be the result of internal gene duplication (e.g., Fig. 107).   Many of the 17 similar cases involve rather unusual structures, such as the complex mixed sheets of the acid proteases, the five-layer domains of phosphorylase, or the mixed doubly wound sheets of hexokinase (Fig. 108).

In only 4 of the 17 similar domain pairs is it possible to find a domain in some other protein that matches the structure of one of the pair as well as they match each other (two of those four cases are classic-topology doubly wound sheets).   Only for the immunoglobulins (and probably for wheat germ agglutinin, if its sequence were known) is there any significant sequence homology detectable between the similar domain pairs.   Purely by chance one would expect vaguely similar structures (within the same major subgroup) perhaps one time out of ten, and detailed resemblance of relatively unusual structures only about one time in 50 or 100.   It is unlikely, therefore, that more than one or two of the 17 similar domain pairs happened by chance.   Only one or two of the pairs could be the products of convergent evolution, because in the other cases the two domains of the pair have quite different functions.   Therefore it seems fairly certain that almost all of the similar domain pairs are indeed the result of internal gene duplications.   We are left then with the rather interesting conclusion that about half the time multiple domains are produced by gene duplication.

It is also possible that the large and relatively complex domain structures we find today were initially produced by gene duplication from smaller substructures; many of these cases have been analyzed by McLachlan (e.g., McLachlan *et al.*, 1980; McLachlan, 1979b,c).

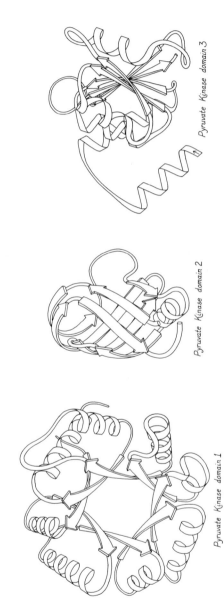

Pyruvate Kinase domain 1

Pyruvate Kinase domain 2

Pyruvate Kinase domain 3

FIG. 107. Pyruvate kinase domains 1, 2, and 3 as an example of a protein whose domains show no structural resemblance whatsoever.

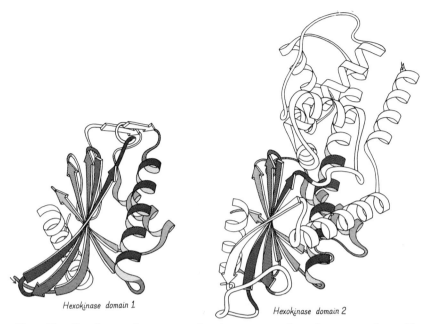

Hexokinase domain 1

Hexokinase domain 2

FIG. 108.   Hexokinase domains 1 and 2: the proteins whose domains are least alike of all the cases that may represent gene duplications.   The equivalent portions of the two domains are shown shaded.

There is very strong evidence from sequence as well as structure that this occurred in ferredoxin (Adman *et al.*, 1973) and probably in the carp calcium-binding protein (Kretsinger, 1972).   These substructures would not have been very stable by themselves, but they could perhaps have survived under less rigorously competitive conditions by associating as identical subunits.   (There must have been a stage early in the evolution of life when there were few proteases to degrade a temporarily unfolded protein; also, marginally stable proteins become a selective disadvantage only when other organisms develop more stable ones.)   It is very difficult to tell from the present structures whether or not this process commonly occurred.   Most such possible substructures are so simple that they are very likely to occur often no matter how the domains originated, which means that the internal symmetry of the doubly wound, singly wound, up-and-down α, and up-and-down β structures does not prove that they originated by duplication.   Also, such duplications would have to have been extremely ancient genetic events and there could have been much alteration since then, so the fact that most other structures show no internal symmetries does not prove that they did *not* originate by such a process. For example, the uteroglobin subunit (see Fig. 88) is not internally symmetric because its fourth helix exchanges position with the one on the other subunit of the dimer; it is very unclear, however, that uteroglobin is any less likely to be the product of an internal duplication than, for instance, myohemerythrin.

The next evolutionary question is how many of the different proteins are related to one another. To what extent are the various proteins in one of the structure subgroups all related? To what extent are proteins within a functional category related? In the end, this question comes down to asking how difficult it is to originate completely new proteins: are there very many, or relatively few, independent evolutionary lines among the proteins? At one extreme we would expect to see a fairly small number of distinguishable general structure types, and all members of one functional category of proteins would usually be found within the same structural class. If there were improbable structures represented, as there might well be, there would be only a few different such structure types but each would include several similar protein examples. At the other extreme, we would expect a fairly random distribution of functional types within the various structural categories, with many different improbable structures represented by only a single example of each. Neither of these extreme situations quite applies to the observed distribution of protein structures, but in general it conforms much more closely to the multiple-origin, random model. The simplest, most probable structure types are extremely common, while the more peculiar, complicated patterns each show up only once or twice.

In a very broad overview of the structural categories one can state several statistical correlations with type of function. Hemes are almost always bound by helices, but never in parallel α/β structures. Relatively complex enzymatic functions, especially those involving allosteric control, are occasionally antiparallel β but most often parallel α/β. Binding and receptor proteins are most often antiparallel β, while the proteins that bind in those receptor sites (i.e., hormones, toxins, and enzyme inhibitors) are most apt to be small disulfide-rich structures. However, there are exceptions to all of the above generalizations (such as cytochrome c_3 as a nonhelical heme protein or citrate synthase as a helical enzyme), and when one focuses on the really significant level of detail within the active site then the correlation with overall tertiary structure disappears altogether. For almost all of the dozen identifiable groups of functionally similar proteins that are represented by at least two known protein structures, there are at least

TABLE III

Correlation between Functional Descriptions of Proteins and Their Overall Tertiary Structures[a]

Dehydrogenases	Kinases
⎰ Lactate, malate dehydrogenase	⎡ Hexokinase
⎱ Liver alcohol dehydrogenase	⎢ Pyruvate kinase
⎩ Glyceraldehyde-phosphate dehydrogenase	⎨ Adenylate kinase
Proteases	⎢ Phosphoglycerate kinase
Trypsin, chymotrypsin, elastase, etc.	⎣ Phosphofructokinase
Subtilisin	Protease inhibitors
Papain, actinidin	Pancreatic trypsin inhibitor
Thermolysin	Soybean trypsin inhibitor
Carboxypeptidase	*Streptomyces* subtilisin inhibitor
Acid proteases	Nucleases
Isomerases	Pancreatic ribonuclease
Triosephosphate isomerase	Staphylococcal nuclease
Glucosephosphate isomerase	Peroxidases
Cytochromes	Glutathione peroxidase
Cytochromes c, c_2, c_{550}, c_{551}, c_{555}	Cytochrome c peroxidase
Cytochrome b_5	Oxygen carriers
⎰ Cytochrome b_{562}	Myoglobin, hemoglobin
⎱ Cytochrome c'	Myohemerythrin, hemerythrin
Redox Fe-S proteins	Hormone-binding proteins
High-potential iron protein	Uteroglobin
Ferredoxin	Prealbumin
Viral coat proteins	Lectins
⎰ Tomato bushy stunt virus protein	⎰ Concanavalin A
⎱ Southern bean mosaic virus protein	⎱ Influenza virus hemagglutinin
Tobacco mosaic virus protein	Wheat germ agglutinin

[a] Within each functional grouping, proteins known to be homologous are listed on a single line, and proteins that fall within the same tertiary-structure subcategory (for at least one of their domains) are bracketed. In spite of the detailed resemblances commonly found within active sites, the great majority of examples shown no similarity of overall structure.

two and sometimes four or five totally different tertiary-structure types which share that function, as shown in Table III. Probably the most dramatic case is the proteolytic enzymes: although the trypsin-like serine proteases form a structurally related group, the proteases as a whole are represented by six widely different structures, including two textbook examples of convergent evolution: subtilisin versus the trypsin family and thermolysin versus carboxypeptidase. Even in the cases in which only two protein structures are known from a general functional category (such as lectins, nucleases, peroxidases, oxygen carriers, etc.) those two structures are quite different. It seems possible, then, that active sites are easier to alter or to redevelop independently than one would have thought, compared with the total time scale involved in the evolution of proteins.

The really puzzling fact, however, is that there is one glaring exception to the above pattern: the nucleotide-binding domains, especially the dehydrogenases and kinases. Within that functional group, and within the parallel α/β structures, the distribution is exactly what one would expect from the model in which large groups of proteins share an evolutionary origin (for at least one of their domains). Such a pattern could also be explained by especially stringent selective pressures, although there is no evidence at all that the requirements for nucleotide-binding sites are any more restrictive than for any other function (see Section III,C). Or the pattern could result from a combination of moderate selective pressure and some as-yet-unknown restrictions on folding within this general structure category. Whatever the explanation, it must somehow account for the fact that nucleotide-binding proteins (or, perhaps, enzymes in the glycolytic pathway) appear to be different in some fairly fundamental way from any other functional category sampled so far.

Another general approach that has commonly been taken to the problem of evaluating relationships between proteins is calculation of the minimum root-mean-square difference between superimposed α-carbon positions for the similar parts of the structures. This was initially done by Rossmann (Rao and Rossmann, 1973) to compare nucleotide-binding domains and other structures (e.g., Rossmann and Argos, 1976). These Cα comparisons in general corroborate and quantitate similarities found by inspection, and sometimes have uncovered relationships no one had previously noticed. Considerable progress has been made recently on ways of evaluating the significance level of Cα superpositions (Remington and Matthews, 1980; McLachlan, 1979b; Schulz, 1980). Two logically distinct problems are involved: the first problem is evaluating the significance of a given similarity relative to the probability of its "chance" or "random" occurrence; the second problem is estimating the likelihood that a given significant resemblance was produced by divergent rather than convergent evolution. In practice the two problems are attacked together, because no one is as interested in the more obvious (and highly significant) similarities that all proteins share simply as a result of globularity, covalent bonding, and preferred backbone conformations. Ideally one wants the "control" or reference comparisons to incorporate all nonhistorical constraints that apply to protein structure in general: requirements of overall stability, side chain packing, efficient folding, and all the other factors we do not yet know. Then any closer degree of resemblance can be assumed to be due to an historical evolutionary relationship (with a calculable confidence level).

The apparent objectivity of quantitative comparison methods obscures the fact that we do not yet know nearly enough about either the genetic processes or the stability and folding requirements to be sure the estimated probabilities of relationship are correct within an order of magnitude. Most comparison methods cannot readily allow for insertions and deletions; we know that they are an important factor that should be included, but even if the computational difficulties can be overcome, we simply do not have any idea of the relative likelihood of, for instance, one long versus two short insertions or of whether an insertion that makes a wide spatial excursion is any less likely than one which stays close. Because there are fewer degrees of freedom, spatial equivalence between helices must be less significant per residue than between β strands than between nonrepetitive structure, but we cannot quantitate this effect. Functional resemblance is certainly a strong argument for the significance of a resemblance, but it cannot make a case for divergent rather than convergent evolution. Perhaps the most fundamental difficulty is that it is an a priori assumption, not an empirically determined fact, that closeness of spatial coordinates is a suitable measure of evolutionary distance.

At the same time, we need to know more about the genetic mechanisms that may be influential in protein evolution, since our current paradigms are almost certainly too simple. We need to understand more about the practical consequences of exon–intron organization on the gene and whether it generally correlates with domain divisions or with smaller internal units. It would be useful to know how unusual is the circularly permuted amino acid sequence of favin versus concanavalin A (Cunningham *et al.*, 1979). And we might consider the possibility, for instance, that the helical portion of the larger domain of hexokinase (see Fig. 108) could "bud off" as an independent protein that would have an historical evolutionary relationship to the doubly wound sheet portion of hexokinase but no structural or sequence resemblance whatsoever. In the worst case, it could be that evaluating probable evolutionary relationships in terms of structural resemblance is not generally possible, because the constraints of stability and folding requirements might turn out to be more demanding than the limitations on rapid evolutionary change. However, one must start out with the more optimistic attitude that a sufficiently varied and open-minded program of structure comparisons will teach us a great deal both about the folding constraints and also about the evolutionary history of proteins.

C. Implications for Protein Folding

It has been evident for some time both that a random search through all conformations could not possibly explain protein folding (Levinthal, 1968) and also that the structures themselves show evidence of systematic local folding patterns. The consistent presence of domains in the larger proteins strongly suggests that they are folding units (Gratzer and Beaven, 1969; Wetlaufer, 1973), and for some proteins it is known experimentally that an isolated domain can fold spontaneously (e.g., Ghelis *et al.*, 1978). A domain usually is made up from a single continuous portion of the backbone; however, the idea of separately folding domains, which then associate to form the intact protein, gains additional support from the frequency with which there occurs a short "tail" or "arm" at one end of a domain sequence which folds over to wrap against the outside of a neighboring domain. Figure 66 shows the structure of papain, which is a classic example of domain-clasping arms. Presumably, the placement of such arms is one of the last events in protein folding, which helps bind together the preformed domains.

It has frequently been pointed out (Wetlaufer, 1973; Ptitsyn and Rashin, 1975; Richardson, 1975; Levitt and Chothia, 1976) that the very high occurrence of associations between secondary-structure elements that are adjacent in the sequence is almost certainly a result of the fact that such nearest-neighbor elements are far more likely to come together during folding. This sort of regularity implies that at least some features of the final protein structure are under fairly strong control by the kinetic requirements of the folding process.

Additional sorts of regularities seen in our general classification of structures allow one to generalize the above idea still further. The prevalence of a few simple patterns of overall topology, and especially such features as the right-handedness of crossover connections and the frequency and handedness of Greek keys, strongly suggest the hypothesis that medium-sized as well as strictly local sections of polypeptide backbone have correlated conformations and tend to fold up as a concerted, interacting unit. One of the most interesting supports for this idea is the difference in statistical distribution of topologies that is seen between antiparallel α, antiparallel β, and parallel α/β structures. The parallel α/β structures are greatly influenced by the relatively long-range regularity of crossover handedness, which together with protection for both sides of the sheet produces the doubly wound α/β structure. In contrast to that situation, three-helix units with the first helix parallel to the third one show no handedness preference whatsoever. Although the possible topologies are exactly equivalent for antiparallel α bundles and for antiparallel β barrels, the frequency with which the various possibilities occur is very different for the two cases. Greek key topologies are about four times as common relative to up-and-down topologies for β structures as compared with α ones; the helical Greek keys occur with either handedness, while 12 of the 13 Greek key β barrels are counterclockwise. Pair associations in the β and α/β structures unambiguously show quite long-range correlations. Such correlations are most easily understood if fairly long portions of the polypeptide chain tend to fold as concerted units, such as the coiling up of a twisted, two-stranded β ribbon shown in Fig. 100. The distribution of features seen in helical proteins is ambiguous: it does not rule out the possibility of long-range concerted folding units, but it does not provide any particular support for such an idea. The observed helical structures could be explained by a simple model in which each new helix-pair association is independent of the topology of earlier pairs.

It has long been assumed that among the fluctuating conformational states early in the protein folding process, local elements of secondary structure are formed for a significant portion of the time; evidence comes from the experimentally observed behavior of synthetic polypeptides (e.g., Yaron *et al.*, 1971), from theoretical calculations of locally determined stability (e.g., Ralston and DeCoen, 1974), and even from the degree of success achieved by secondary-structure predictions based only on single-residue, pair, or triplet sequences (e.g., Chou and Fasman, 1974). Particularly favorable such local regions of structure can act as nucleation sites to start and guide the folding process. Many proposed schemes of folding nucleation single out just one type of structure that seems especially suited for forming the first nuclei. The chief candidates that have been proposed as folding nuclei are α-helices, either alone (Levinthal, 1966; Anfinsen, 1972; and Lim, 1978) or in combination with β strands (Nagano, 1974); pairs of β strands brought together by a tight turn (Lewis *et al.*, 1971; Crawford *et al.*, 1973) or as long double ribbons (Ptitsyn and Finkelstein, 1980); and hydrophobic clusters (Matheson and Scheraga, 1978). The proposals for helical nuclei postulate that in predominantly β proteins the helices in the nucleation structures later unfold into extended strands. However, backbone connectivity has its maximum influence early in the folding process, so that topological patterns in the final structure are very sensitive to the order and mechanism of folding, as we have seen before in the contrast between the orderly topology of β strands and the random topology of disulfide connections. Therefore, if nucleation sites are basically similar for all types of structures, that similarity should show up in the overall topological patterns. Instead, as we have seen above, each of the broad types of structure shows characteristically different patterns of pair associations, coiled features, and handedness. Nucleation by hydrophobic clusters is harder to judge from the appearance of the final, folded structures. In proteins with strong long-range regularities of secondary structure it seems very unlikely that the earliest stages of folding are controlled entirely by hydrophobic associations, but there might be pure nucleation by hydrophobic clusters in the more irregular structures.

Judging from the types of regularities seen in the final structures, it seems likely that the typical folding nuclei are different for each of the three largest categories of structure: presumably those nuclei are individual helices and pairs of helices for the antiparallel α structures, $\beta-\alpha-\beta$ loops for the parallel α/β group, and two-stranded β ribbons for the antiparallel β structures. The small S–M proteins presumably either nucleate by helices or β ribbons which may be partially lost later, or else by hydrophobic clusters. This diversity of folding nuclei would fit fairly well with Rose's "lincs and hinges" model (Rose *et al.*, 1976) except that different types of lincs are not equivalent, and only for the antiparallel α case could they be considered as joined by completely flexible hinges. Tanaka and Scheraga (1977) have also proposed a model with diverse nuclei that are determined by near-diagonal regions of local interaction on the diagonal contact plot which fold by steps rather similar to the ones proposed below, except that forming contacts in rigorous order of increasing separation in the sequence does not permit explanation of any topological regularities larger than pairwise.

One last suggestive feature that is seen in the known protein structures is the frequency with which they "almost match" some prototypical structure. As an example of this sort of deviation, plastocyanin (Colman *et al.*, 1978) is an antiparallel β barrel with seven well-formed β strands and an eighth strand which makes only one or two β-type hydrogen bonds, includes a short helix and an irregular excursion, and is slightly displaced from the position for an eighth β strand. If just the seven good β strands are counted as part of the barrel it has an unusual and complicated topology, but if the irregular eighth strand is included the structure is a Greek key barrel of the usual handedness. It may well be that plastocyanin folds as a more regular eight-stranded barrel but effectively loses the β structure in that eighth strand during the final process of adjustment to optimize fit for all the side chains. The significance of the eight-stranded Greek key structure for plastocyanin is reinforced by the fact that the Greek key structure is clearly present in the related protein azurin, with well-formed β structure for that same eighth strand (Adman et $al.$, 1978). There are many other examples of such "approximate" pieces of structure although there is not always a convenient related protein to confirm the assignment. Such features could be explained if proteins first fold to form a maximum amount of regular secondary structure but then may lose some portions of the secondary structure in the final stage of adjusting all interactions for maximum stability. This sort of unfolding and loss of regularity at the final stages has been suggested before on varied sorts of evidence, both for helices (Carter et $al.$, 1974; Lim, 1978) and for β strands (Ko et $al.$, 1977; Richardson et $al.$, 1978). The entire category of S–M proteins is presumably an exaggerated case of this sort of process, in which the amount of adjustment needed to accommodate disulfides or metals into these small proteins is often enough to disrupt the secondary structure almost beyond recognition.

By putting together all of the ideas discussed above, we can propose a speculative general scheme of protein folding as suggested by the properties of the final structures.

The proposed folding process involves four stages, which could be expected to be at least partially separated in time but are not rigorously sequential. Figure 109 illustrates the stages of folding as they might apply to each of the major structure categories. The first stage is the classic one of forming, in a probabilistic and fluctuating sense, individual elements of α-helix, extended strand, or tight turns, and of combining two or three of those elements into the first folding nuclei. This does not involve backbone conformations different from those that would be present in a rigorously random coil; it simply involves a difference in the statistics of their distribution in favor of more correlation between the conformations of adjacent residues. Helices have the advantage of hydrogen-bond formation and of cooperativity, and the helices undoubtedly are more regular and can persist for much longer times than isolated, or perhaps even than paired, β strands. However, extended strands have the advantage that a much broader range of conformational angles is capable of taking part in β structure, and it could well be that extended strands capable of further interaction are present for about as large a fraction of the time as are individual helices. Once a pair of helices, β–α–β loop, a two-stranded β ribbon, or a large hydrophobic cluster has formed, it

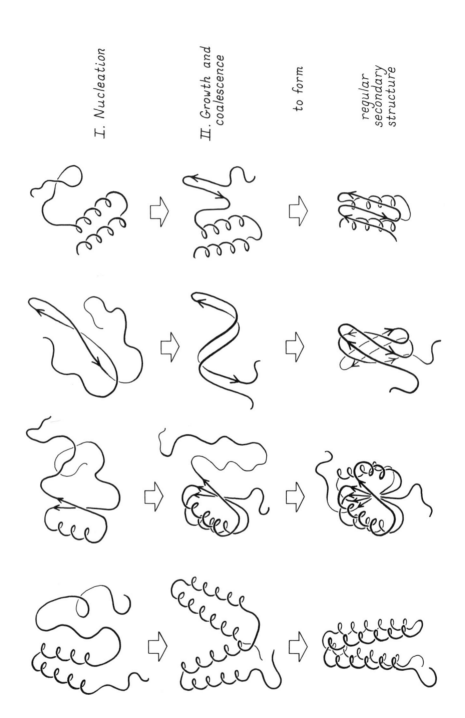

I. Nucleation

II. Growth and coalescence

to form

regular secondary structure

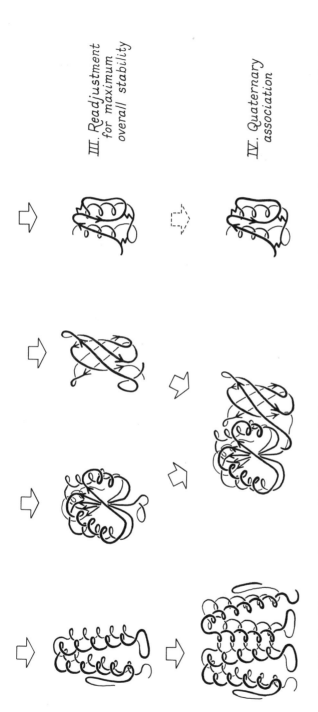

III. *Readjustment for maximum overall stability*

IV. *Quaternary association*

Fig. 109. Possible successive steps in the protein folding process as they might apply to a typical example of each of the four major categories of structure. See text for fuller explanation.

would presumably have enough stability to act as a nucleation site for further folding. At least for large domains, it seems unlikely that there is a unique initial folding nucleus, since the relative stability and probability of occurrence would often be similar among, for instance, several possible $\beta-\alpha-\beta$ loops. Indeed, the most common topologies are the ones that would permit the most alternative folding pathways (Richardson, 1977).

The second stage of folding is the growth and coalescence of secondary-structure elements two or three at a time to form successively larger substructures. The characteristic associations formed at this stage depend on the type and order of secondary-structure elements in the sequence. All-helical structures may associate fairly independently, one nearest-neighbor pair at a time. It is proposed that $\beta-\alpha-\beta$ structures fold concertedly by throwing up loops. Antiparallel β structures probably form two-stranded ribbons from nearest-neighbor strands separated by turns; they can then add on strands or pairs of strands to either side of an initial ribbon, or they can coil up a very long ribbon into a Greek key. At the end of this second stage all of the major regular structures are in place, sometimes in a more complete or more regular form than in the final native structure.

The third stage is a process of many readjustments to settle down into a comfortable, stable overall structure. At this stage disulfides are joined in their final native pairing, metals and prosthetic groups are bound, β bulges are formed, and cis–trans isomerization of prolines occurs if necessary (see Brandts *et al.*, 1975, 1977). Side chain conformations are adjusted to provide optimal fit, and some main chain conformations are also adjusted. Occasionally this might produce additional secondary-structure interactions, but it is much more likely to disrupt some of the preexisting secondary structure; main-chain hydrogen-bonding lost at this stage is more than compensated by side chain interactions. This third, readjustment, stage of folding would normally be expected to be very much slower than any of the other steps. For a one-domain, single subunit protein the folding process would then be complete (unless proteolytic cleavages or some other modifications are needed). It may be that the kind of major reshuffling seen during the folding of pancreatic trypsin inhibitor (Creighton, 1977) can be considered as an especially pronounced example of these final readjustments, although the fact that the incorrect intermediates are not very compact suggests that they may represent a rather different process that can happen in addition to the steps considered here. In general, the final structures of the small S–M proteins suggest that they undergo more extensive rearrangement than other proteins.

The fourth stage of folding is the association of domains (and/or subunits). Sometimes association might start at the end of the second stage, but in general it would probably happen only after readjustments within domains were fairly complete. Domains primarily associate as rigid bodies, but there are usually adjustments of side chains at the contact surface, and "arms" that clasp opposite domains cannot fold into their final conformation until this last stage. Association of subunits is equivalent to association of domains, except for the difference in kinetics produced by the covalent attachment.

The most characteristic features of this proposed folding scheme are the proposal of different kinds of nucleation for the different major structure types, the postulation of some rather large-scale concerted folding units, and the prediction of folding intermediates with somewhat greater amounts of the same sort of secondary structure found in the final native protein. The last effect might turn out to be most pronounced in those proteins with very irregular secondary structures.

In the final analysis protein folding will be really understood only with the aid of much more extensive, direct experimental evidence. Speculative hypotheses can be useful, however, in suggesting potentially fruitful questions for experimental investigation. Probably the most important idea suggested by the above schema is that there are likely to be considerable systematic differences in the kinetics of folding between the various major structural categories of proteins.

ACKNOWLEDGMENTS

I am especially grateful to David Richardson for, among other things, the meticulous technical photography; to Richard Feldmann for extensive use of his molecular display system; and to Chris Anfinsen for suggesting that this article be written.

I am also greatly indebted to the following people, who provided unpublished coordinates or other information: Pat Argos, Frances Bernstein, Colin Blake, David Blow, Tom Blundell, Rick Bott, Carl Brändén, John Chambers, David Davies, Phil Evans, Alyosha Finkelstein, Bob Fletterick, Hans Freeman, Irving Geis, Pauline Harrison, Steve Harrison, Wayne Hendrickson, Isabella Karle, Aaron Klug, Joe Kraut, Michael Leibman, Anders Liljas, Bill Lipscomb, Martha Ludwig, Scott Mathews, Brian Matthews, Alex McPherson, Hilary Muirhead, George Némethy, Eduardo Padlan, Oleg Ptitsyn, Michael Rossmann, Ray Salemme, Charlotte Schellman, George Schulz, Christine Slingsby, Tom Steitz, Michael Sternberg, Bob Stroud, Martha Teeter, Janet Thornton, Al Tulinsky, B. K. Vainshtein, Pat Weber, Don Wetlaufer, Don Wiley, Alex Wlodawer, and Christine Wright.

This work was supported by NIH grant GM-15000.

REFERENCES[3]

Abad-Zapatero, C., Abdel-Meguid, S. S., Johnson, J. E., Leslie, A. G. W., Rayment, I., Rossmann, M. G., Suck, D., and Tsukihara, T. (1980). *Nature, (London)* **286**, 33–39.

Abola, E. E., Ely, K. R., and Edmundson, A. B. (1980). *Biochemistry* **19**, 432–439.

Adams, M. J., Haas, D. J., Jeffery, B. A., McPherson, A., Jr., Mermall, H. L., Rossmann, M. G., Schevitz, R. W., and Wonacott, A. J. (1969). *J. Mol. Biol.* **41**, 159–188.

Adams, M. J., Ford, G. C., Koekoek, R., Lentz, P. J., Jr., McPherson, A., Jr., Rossmann, M. G., Smiley, I. E., Schevitz, R. W., and Wonacott, A. J. (1970). *Nature (London)* **227**, 1098–1103.

Adman, E. T., Sieker, L. C., and Jensen, L. H. (1973). *J. Biol. Chem.* **248**, 3987–3996.

Adman, E. T., Stenkamp, R. E., Sieker, L. C., and Jensen, L. H. (1978). *J. Mol. Biol.* **123**, 35–47.

Almassy, R. J., and Dickerson, R. E. (1978). *Proc. Natl. Acad. Sci. U.S.A.* **75**, 2674–2678.

Anderson, C. M., McDonald, R. C., and Steitz, T. A. (1978). *J. Mol. Biol.* **123**, 1–13.

Anderson, C. M., Zucker, F. H., and Steitz, T. A. (1979). *Science* **204**, 375–380.

Andreeva, N. S., and Gustchina, A. E. (1979). *Biochem. Biophys. Res. Commun.* **87**, 32–42.

Anfinsen, C. B. (1972). *Biochem. J.* **128**, 737–749.

Anfinsen, C. B., and Scheraga, H. A. (1975). *Adv. Protein Chem.* **29**, 205–300.

Anfinsen, C. B., Cuatrecasas, P., and Taniuchi, H. (1971). *In* "The Enzymes" (P. D. Boyer, ed.), Vol. 4, 3rd Ed. Academic Press, New York.

Argos, P., and Rossmann, M. G. (1979). *Biochemistry* **18**, 4951–4960.

Argos, P., Rossmann, M. G., and Johnson, J. E. (1977). *Biochem. Biophys. Res. Commun.* **75**, 83–86.

Arnone, A., Bier, C. J., Cotton, F. A., Day, V. W., Hazen, E. E., Jr., Richardson, D. C., Richardson, J. S., and Yonath, A. (1971). *J. Biol. Chem.* **246**, 2302–2316.

Astbury, W. T. (1933). *Trans. Faraday Soc.* **29**, 193.

Astbury, W. T., and Bell, F. O. (1941). *Nature (London)* **147**, 696–699.

Baker, E. N. (1980). *J. Mol. Biol.* **141**, 441–484.

Banks, R. D., Blake, C. C. F., Evans, P. R., Haser, R., Rice, D. W., Hardy, G. W., Merrett, M., and Phillips, A. W. (1979). *Nature (London)* **279**, 773–777.

Banner, D. W., Bloomer, A. C., Petsko, G. A., Phillips, D. C., Pogson, C. I., and Wilson, I. A. (1975). *Nature (London)* **255**, 609–614.

Banyard, S. H., Stammers, D. K., and Harrison, P. M. (1978). *Nature (London)* **271**, 282–284.

Bennett, W. S., Jr., and Steitz, T. A. (1978). *Proc. Natl. Acad. Sci. U.S.A.* **75**, 4848–4852.

Bhat, T. N., Sasisekharan, V., and Vijayan, M. (1979). *Int. J. Peptide Protein Res.* **13**, 170–184.

Birktoft, J. J., and Blow, D. M. (1972). *J. Mol. Biol.* **68**, 187–240.

Blake, C. C. F., and Oatley, S. J. (1977). *Nature (London)* **268**, 115–120.

Blake, C. C. F., Mair, G. A., North, A. C. T., Phillips, D. C., and Sarma, V. R. (1967). *Proc. R. Soc. Ser. B* **167**, 365–377.

Blake, C. C. F., Geisow, M. J., Oatley, S. J., Rérat, B., and Rérat, C. (1978). *J. Mol. Biol.* **121**, 339–356.

---

[3] To find references for a given protein, see Section III,A,4.

Bloomer, A. C., Champness, J. N., Bricogne, G., Staden, R., and Klug, A. (1978). *Nature (London)* **276**, 362–368.

Blow, D. M., Birktoff, J. J., and Hartley, B. S. (1969). *Nature (London)* **221**, 337–340.

Blow, D. M., Irwin, M. J., and Nyborg, J. (1977). *Biochem. Biophys. Res. Commun.* **76**, 728–734.

Blundell, T. L., Dodson, G. G., Hodgkin, D. C., and Mercola, D. A. (1972). *Adv. Protein Chem.* **26**, 279–402.

Blundell, T., Lindley, P., Miller, L., Moss, D., Slingsby, C., Tickle, I., Turnell, B., and Wistow, G. (1981). *Nature (London)* **289**, 771–777.

Brahms, S., and Brahms, J. (1980). *J. Mol. Biol.* **138**, 149–178.

Brandts, J. F., Halvorsen, H. R., and Brennan, M. (1975). *Biochemistry* **14**, 4953–4963.

Brandts, J. F., Brennan, M., and Lin, L. N. (1977). *Proc. Natl. Acad. Sci. U.S.A.* **74**, 4178–4181.

Brayer, G. D., Delbaere, L. T. J., and James, M. N. G. (1978). *J. Mol. Biol.* **124**, 261–283.

Brown, K. G., Erfurth, S. C., Small, E. W., and Peticolas, W. L. (1972). *Proc. Natl. Acad. Sci. U.S.A.* **69**, 1467–1469.

Buehner, M., Ford, G. C., Moras, D., Olsen, K. W., and Rossmann, M. G. (1974). *J. Mol. Biol.* **90**, 25–49.

Burnett, R. M., Darling, G. D., Kendall, D. S., LeQuesne, M. E., Mayhew, S. G., Smith, W. W., and Ludwig, M. (1974). *J. Biol. Chem.* **249**, 4383–4392.

Butler, P. J. G., and Klug, A. (1978). *Sci. Am.* **239**, (5), 62–69.

Campbell, J. W., Watson, H. C., and Hodgson, G. I. (1974). *Nature (London)* **250**, 301–303.

Carter, C. W. (1977). *J. Biol. Chem.* **252**, 7802–7811.

Carter, C. W., Jr., Kraut, J., Freer, S. T., Xuong, N. H., Alden, R. A., and Bartsch, R. G. (1974). *J. Biol. Chem.* **249**, 4212–4225.

Chambers, J. L., and Stroud, R. M. (1979). *Acta Crystallogr.* **B35**, 1861–1874.

Champness, J. N., Bloomer, A. C., Bricogne, G., Butler, P. J. G., and Klug, A. (1976). *Nature (London)* **259**, 20–24.

Chandrasekaran, R., and Ramachandran, G. N. (1970). *Int. J. Protein Res.* **2**, 223–233.

Chauvin, C., Witz, J., and Jacrot, B. (1978). *J. Mol. Biol.* **124**, 641–651.

Chirgadze, Yu. N., and Nevskaya, N. A. (1976). *Biopolymers* **15**, 627–636.

Chothia, C. (1973). *J. Mol. Biol.* **75**, 295–302.

Chothia, C., Levitt, M., and Richardson, D. (1977). *Proc. Natl. Acad. Sci. U.S.A.* **74**, 4130–4134.

Chou, P. Y., and Fasman, G. D. (1974). *Biochemistry* **13**, 211–245.

Chou, P. Y., and Fasman, G. D. (1977). *J. Mol. Biol.* **115**, 135–175.

Clegg, G. A., Stansfield, R. F. D., Bourne, P. E., and Harrison, P. M. (1980). *Nature (London)* **288**, 298–300.

Cohen, F. E., Sternberg, M. J. E., and Taylor, W. R. (1980). *Nature (London)* **285**, 378–382.

Colman, P. M., Jansonius, J. N., and Matthews, B. W. (1972). *J. Mol. Biol.* **70**, 701–724.

Colman, P. M., Deisenhofer, J., Huber, R., and Palm, W. (1976). *J. Mol. Biol.* **100**, 257–282.

Colman, P. M., Freeman, H. C., Guss, J. M., Murata, M., Norris, V. A., Ramshaw, J. A. M., and Venkatappa, M. P. (1978). *Nature (London)* **272**, 319–324.

Cook, D. A. (1967). *J. Mol. Biol.* **29**, 167.

Craik, C. S., Buchman, S. R., and Beychok, S. (1980). *Proc. Natl. Acad. Sci. U.S.A.* **77**, 1384–1388.

Crawford, J. L., Lipscomb, W. N., and Schellman, C. G. (1973). *Proc. Natl. Acad. Sci. U.S.A.* **70**, 538–542.

Creighton, T. E. (1977). *J. Mol. Biol.* **113**, 275–341.

Crick, F. H. C. (1953). *Acta Crystallogr.* **6**, 689–697.

Crippen, G. M. (1978). *J. Mol. Biol.* **126**, 315–332.

Crippen, G. M., and Kuntz, I. D. (1978). *Int. J. Peptide Protein Res.* **12**, 47–56.

Cunningham, B. A., Hemperly, J. J., Hopp, T. P., and Edelman, G. M. (1979). *Proc. Natl. Acad. Sci. U.S.A.* **76**, 3218–3222.

Davies, D. R. (1964). *J. Mol. Biol.* **9**, 605–609.

Dayhoff, M. O., and Barker, W. C. (1972). "Atlas of Protein Sequence and Structure," Vol. 5, pp. 41–45. National Biomedical Research Foundation, Silver Spring, Maryland.

Deisenhofer, J., and Steigemann, W. (1975). *Acta Crystallogr.* **B31**, 238–250.

Deisenhofer, J., Jones, T. A., and Huber, R. (1978). *Hoppe-Seylers Z. Physiol. Chem.* **359**, 975–985.

Delbaere, L. T. J., Hutcheon, W. L. B., James, M. N. G., and Thiessen, W. E. (1975). *Nature (London)* **257**, 758–763.

Dickerson, R. E., and Geis, I. (1969). "Structure and Action of Proteins," Chap. 2. Harper, New York.

Dijkstra, B. W., Drenth, J., Kalk, K. H., and Vandermaelen, P. J. (1978). *J. Mol. Biol.* **124**, 53–60.

Drenth, J., Jansonius, J. N., Koekoek, R., and Wolthers, B. G. (1971). *Adv. Protein Chem.* **25**, 79–115.

Drenth, J., Low, B., Richardson, J. S., and Wright, C. S. (1980). *J. Biol. Chem.* **255**, 2652–2655.

Edelman, G. M., and Gall, W. E. (1969). *Annu. Rev. Biochem.* **38**, 415–466.

Edmundson, A. B. (1981). *Proc. Int. Symp. Biomol. Struct. Conform. Function Evol., Madras, Indian,* in press.

Edsall, J. T., Otvos, J. W., and Rich, A. (1950). *J. Am. Chem. Soc.* **72**, 474–477.

Efimov, A. V. (1977). *Dokl. Akad. Nauk S.S.S.R.* **235**, 699–702.

Efimov, A. V. (1979). *J. Mol. Biol.* **134**, 23–40.

Eklund, H., and Brändén, C. I. (1979). *J. Biol. Chem.* **254**, 3458–3461.

Eklund, H., Nordström, B., Zeppezauer, E., Söderlund, G., Ohlsson, I., Boiwe, T., Söderberg, B.-O., Tapia, O., Brändén, C.-I., and Åkeson, Å. (1976). *J. Mol. Biol.* **102**, 27–59.

Engelman, D. M., Henderson, R., McLachlan, A. D., and Wallace, B. A. (1980). *Proc. Natl. Acad. Sci. U.S.A.* **77**, 2023–2027.

Epp, O., Colman, P., Fehlhammer, H., Bode, W., Schiffer, M., Huber, R., and Palm, W. (1974). *Eur. J. Biochem.* **45**, 513–524.

Epp, O., Lattman, E. E., Schiffer, M., Huber, R., and Palm, W. (1975). *Biochemistry* **14**, 4943–4952.

Feldmann, R. J., (1977). "Atlas of Macromolecular Structure on Microfiche." Tracor-Jitco, Rockville, Maryland.

Finkelstein, A. V., and Ptitsyn, O. B. (1976). *J. Mol. Biol.* **103**, 15–24.

Fisher, W. R., Taniuchi, H., and Anfinsen, C. B. (1973). *J. Biol. Chem.* **248**, 3188–3195.

Ford, G. C., Eichele, G., and Jansonius, J. N. (1980). *Proc. Natl. Acad. Sci. U.S.A.* **77**, 2559–2563.

Genzel, L., Keilman, F., Martin, T. P., Winterling, G., Yacoby, Y., Fröhlich, H., and Makinen, M. W. (1976). *Biopolymers* **15**, 219–225.

Ghelis, C., Tempete-Gaillourdet, M., and Yon, J. M. (1978). *Biochem. Biophys. Res. Commun.* **84**, 31–36.

Gratzer, W. B., and Beaven, G. H. (1969). *J. Biol. Chem.* **244**, 6675–9.

Grosse, D. M., Malur, J., Meiske, W., and Repke, K. R. H. (1974). *Biochim. Biophys. Acta* **359**, 33–46.

Guzzo, A. V. (1965). *Biophys. J.* **5**, 809.

Hagler, A. T., and Moult, J. (1978). *Nature (London)* **272**, 222–226.

Harrison, S. C., Olson, A. J., Schutt, C. E., Winkler, F. K., and Bricogne, G. (1978). *Nature (London)* **276**, 368–373.

Haser, R., Pierrot, M., Frey, M., Payan, F., Astier, J. P., Bruschi, M., and LeGall, J. (1979). *Nature (London)* **282**, 806–810.

Henderson, R., and Unwin, P. N. T. (1975). *Nature (London)* **257**, 28–32.

Hendrickson, W. A., and Love, W. E. (1971). *Nature (London)* **232**, 197–203.

Hendrickson, W. A., and Teeter, M. M. (1981). *Nature (London)*, in press.

Hendrickson, W. A., and Ward, K. B. (1977). *J. Biol. Chem.* **252**, 3012–3018.

Hendrickson, W. A., Klippenstein, G. L., and Ward, K. B. (1975). *Proc. Natl. Acad. Sci. U.S.A.* **72**, 2160–2164.

Hill, E., Tsernoglou, D., Webb, L., and Banaszak, L. J. (1972). *J. Mol. Biol.* **72**, 577–591.

Hol, W. G. J., van Duignen, P. T., and Berendsen, H. J. C. (1978). *Nature (London)* **273**, 443–446.

Holbrook, S. R., Sussmann, J. L., Warrant, R. W., and Kim, S. H. (1978). *J. Mol. Biol.* **123**, 631–660.

Holmgren, A., Söderberg, B.-O., Eklund, H., and Brändèn, C.-I. (1975). *Proc. Natl. Acad. Sci. U.S.A.* **72**, 2305–2309.

Honig, B., Ray, A., and Levinthal, C. (1976). *Proc. Natl. Acad. Sci. U.S.A.* **73**, 1974–1978.

Huber, R., and Steigemann, W. (1974). *FEBS Lett.* **48**, 235–237.

Huber, R., Kukla, D., Ruhlmann, A., and Steigemann, W. (1971). *Cold Spring Harbor Symp. Quant. Biol.* **36**, 141–148.

Huber, R., Kukla, D., Bode, W., Schwager, P., Bartels, K., Deisenhofer, J., and Steigemann, W. (1974). *J. Mol. Biol.* **89**, 73–101.

Huber, R., Deisenhofer, J., Colman, P. M., Matsushima, M., and Palm, W. (1976). *Nature (London)* **264**, 415–420.

Imoto, T., Johnson, L. M., North, A. C. T., Phillips, D. C., and Rupley, J. A. (1972). *In* "The Enzymes" (P. D. Boyer, ed.), Vol. 7, pp. 665–868. Academic Press, New York.

Irwin, M. J., Nyborg, J., Reid, B. R., and Blow, D. M. (1976). *J. Mol. Biol.* **105**, 577–586.

Isogai, Y., Nemethy, G., Rackovsky, S., Leach, S. J., and Scheraga, H. A. (1980). *Biopolymers* **19**, 1183–1210.

IUPAC-IUB Commission on Biochemical Nomenclature (1970). *J. Biol. Chem.* **245**, 6489–6497.

Janin, J., Wodak, S., Levitt, M., and Maigret, B. (1978). *J. Mol. Biol.* **125**, 357–386.

Kauzmann, W. (1959). *Adv. Protein Chem.* **14**, 1–64.

Kendrew, J. C., Bodo, G., Dintzis, H. M., Parrish, R. G., Wyckoff, H., and Phillips, D. C. (1958). *Nature (London)* **181**, 662–666.

Kendrew, J. C., Dickerson, R. E., Strandberg, B. E., Hart, R. G., Davies, D. R., Phillips, D. C., and Shore, V. C. (1960). *Nature (London)* **185**, 422–427.

Kendrew, J. C., Watson, H. C., Strandberg, B. E., Dickerson, R. E., Phillips, D. C., and Shore, V. C. (1961). *Nature (London)* **190**, 666–670.

Klis, W. A., and Siemion, I. Z. (1978). *Int. J. Peptide Protein Res.* **12**, 103–113.

Klotz, I. M., Langerman, N. R., and Darnall, D. W. (1970). *Annu. Rev. Biochem.* **39**, 25–62.

Ko, B. P. N., Yazgan, A., Yeagle, P. L., Lottich, S. C., and Henkens, R. W. (1977). *Biochemistry* **16**, 1720–1725.

Korszun, Z. R., and Salemme, F. R. (1977). *Proc. Natl. Acad. Sci. U.S.A.* **74**, 5244–5247.

Kotelchuck, D., and Scheraga, H. A. (1969). *Proc. Natl. Acad. Sci. U.S.A.* **62**, 14.

Kretsinger, R. H. (1972). *Nature (London) New Biol.* **240**, 85–88.

Kretsinger, R. H. (1976). *Annu. Rev. Biochem.* **45**, 239–266.

Kretsinger, R. H., and Nockolds, C. E. (1973). *J. Biol. Chem.* **248**, 3313–3326.

Krimm, S., and Abe, Y. (1972). *Proc. Natl. Acad. Sci. U.S.A.* **69**, 2788–2792.

Kuntz, I. D. (1972). *J. Am. Chem. Soc.* **94**, 4009–4012.

Kuntz, I. D., and Kaufmann, W. (1974). *Adv. Protein Chem.* **28**, 239–345.

Kuntz, I. D., Crippen, G. M., Kollmann, P. A., and Kimelman, D. (1976). *J. Mol. Biol.* **106**, 983–994.

Ladenstein, R., Epp, O., Bartels, K., Jones, A., Huber, R., and Wendel, A. (1979). *J. Mol. Biol.* **134**, 199–218.

Ladner, R. C., Heidner, E. J., and Perutz, M. F. (1977). *J. Mol. Biol.* **114**, 385–414.

Lee, B., and Richards, F. M. (1971). *J. Mol. Biol.* **55**, 379–400.

Leijonmarck, M., Pettersson, I., and Liljas, A. (1980). *Proc. Aharon Katzir-Katchalsky Conf.* in press.

Lesk, A. M., and Chothia, C. (1980). *J. Mol. Biol.* **136**, 225–270.

Levine, M., Muirhead, H., Stammers, D. K., and Stuart, D. I. (1978). *Nature (London)* **271**, 626–630.

Levinthal, C. (1966). *Sci. Am.* **215** (6), 42–52.

Levinthal, C. (1968). *J. Chim. Phys.* **65**, 44–45.

Levitt, M. (1977). *Biochemistry* **17**, 4277–4285.

Levitt, M., and Chothia, C. (1976). *Nature (London)* **261**, 552–558.

Levitt, M., and Greer, J. (1977). *J. Mol. Biol.* **114**, 181–239.

Lewis, P. N., Go, N., Go, M., Kotelchuk, D., and Scheraga, H. A. (1970). *Proc. Natl. Acad. Sci. U.S.A.* **65**, 810.

Lewis, P. N., Momany, F. A., and Scheraga, H. A. (1971). *Proc. Natl. Acad. Sci. U.S.A.* **68**, 2293–2297.

Lewis, P. N., Momany, F. A., and Scheraga, H. A. (1973). *Biochim. Biophys. Acta* **303**, 211–229.

Lifson, S., and Sander, C. (1979). *Nature (London)* **282**, 109–111.

Lifson, S., and Sander, C. (1980a). *J. Mol. Biol.* **139**, 627–639.

Lifson, S., and Sander, C. (1980b). In "Protein Folding" (R. Jaenicke, ed.), pp. 289–316. Elsevier, Amsterdam.

Lim, V. I. (1974). *J. Mol. Biol.* **88**, 857–894.

Lim, V. I. (1978). *FEBS Lett.* **89**, 10–14.

Lindskog, S., Henderson, L. E., Kannan, K. K., Liljas, A., Nyman, P. O., and Strandberg,

B. (1971). *In* "The Enzymes" (P. D. Boyer, ed.), Vol. 5, pp. 608–622. Academic Press, New York.

Low, B. W., Preston, H. S., Sato, A., Rosen, L. S., Searl, J. E., Rudko, A. D., and Richardson, J. S. (1976). *Proc. Natl. Acad. Sci. U.S.A.* **73**, 2991–2994.

McCammon, J. A., Gelin, B. R., and Karplus, M. (1977). *Nature (London)* **267**, 585–590.

McLachlan, A. D. (1979a). *Eur. J. Biochem.* **100**, 181–187.

McLachlan, A. D. (1979b). *J. Mol. Biol.* **128**, 49–79.

McLachlan, A. D. (1979c). *J. Mol. Biol.* **133**, 557–563.

McLachlan, A. D., Bloomer, A. C., and Butler, P. J. G. (1980). *J. Mol. Biol.* **136**, 203–224.

McPherson, A., Jurnak, F. A., Wang, A. H. J., Molineux, I., and Rich, A. (1979). *J. Mol. Biol.* **134**, 379–400.

Maigret, B., Pullmann, B., and Perahia, D. (1971). *J. Theor. Biol.* **31**, 269–285.

Mandel, N., Mandel, G., Trus, B., Rosenberg, J., Carlson, G., and Dickerson, R. E. (1977). *J. Biol. Chem.* **252**, 4619–4636.

Marquart, M., Deisenhofer, J., Huber, R., and Palm, W. (1980). *J. Mol. Biol.* **141**, 369–391.

Matheson, R. R., Jr., and Scheraga, H. A. (1978). *Macromolecules* **11**, 819.

Mathews, F. S., Levine, M., and Argos, P. (1972). *J. Mol. Biol.* **64**, 449–464.

Mathews, F. S., Bethge, P. H., and Czerwinski, E. W. (1979). *J. Biol. Chem.* **254**, 1699–1706.

Matthews, B. W. (1972). *Macromolecules* **5**, 818–819.

Matthews, B. W. (1975). *Biochim. Biophys. Acta* **405**, 442–451.

Matthews, B. W., and Remington, S. J. (1974). *Proc. Natl. Acad. Sci. U.S.A.* **71**, 4178–4182.

Matthews, B. W., Fenna, R. E., Bolognesi, M. C., Schmid, M. F., and Olson, J. M. (1979). *J. Mol. Biol.* **131**, 259–285.

Matthews, D. A., Alden, R. A., Bolin, J. T., Freer, S. T., Hamlin, R., Xuong, N., Kraut, J., Poe, M., Williams, M., and Hoogsteen, K. (1977). *Science* **197**, 452–455.

Matthews, D. A., Alden, R. A., Freer, S. T., Xuong, N.-H., and Kraut, J. (1979). *J. Biol. Chem.* **254**, 4144–4151.

Mavridis, I. M., and Tulinsky, A. (1976). *Biochemistry* **15**, 4410–4417.

Maxfield, F. R., and Scheraga, H. A. (1976). *Biochemistry* **15**, 5138–5153.

Mitsui, Y., Satow, Y., Watanabe, Y., and Iitaka, Y. (1979). *J. Mol. Biol.* **131**, 697–724.

Miyazawa, T., and Blout, E. R. (1961). *J. Am. Chem. Soc.* **83**, 712–719.

Momany, F. A., McGuire, R. F., Burgess, A. W., and Scheraga, H. A. (1975). *J. Phys. Chem.* **79**, 2361–2381.

Monaco, H. L., Crawford, J. L., and Lipscomb, W. N. (1978). *Proc. Natl. Acad. Sci. U.S.A.* **75**, 5276–5280.

Mornon, J. P., Fridlansky, F., Bally, R., and Milgrom, E. (1980). *J. Mol. Biol.* **137**, 415–429.

Nagano, K. (1974). *J. Mol. Biol.* **84**, 337–372.

Nagano, K. (1977a). *J. Mol. Biol.* **109**, 235–250.

Nagano, K. (1977b). *J. Mol. Biol.* **109**, 251–274.

Némethy, G., and Printz, M. P. (1972). *Macromolecules* **5**, 755–758.

Némethy, G., Phillips, D. C., Leach, S. J., and Scheraga, H. A. (1967). *Nature (London)* **214**, 363–365.

Oughton, B. M., and Harrison, P. M. (1957). *Acta Crystallogr.* **10**, 478–80.

Oughton, B. M., and Harrison, P. M. (1959). *Acta Crystallogr.* **12**, 396–404.

Pain, R. H., and Robson, B. (1970). *Nature (London)* **227**, 62–63.

Pauling, L., and Corey, R. B. (1951). *Proc. Natl. Acad. Sci. U.S.A.* **37**, 729–40.

Pauling, L., Corey, R. B., and Branson, H. R. (1951). *Proc. Natl. Acad. Sci. U.S.A.* **37**, 205–211.

Perutz, M. F. (1949). *Proc. R. Soc. London Ser. A* **195**, 474–499.

Perutz, M. F. (1951). *Nature (London)* **167**, 1053–1054.

Perutz, M. F. (1964). *Sci. Am.* **211** (5), 64–76.

Perutz, M. F. (1970). *Nature (London)* **228**, 726–739.

Perutz, M. F. (1978). *Sci. Am.* **239** (6), 92–125.

Peterson, J., Steinrauf, L. K., and Jensen, L. H. (1960). *Acta Crystallogr.* **13**, 104–109.

Pickover, C. A., McKay, D. B., Engelman, D. M., and Steitz, T. A. (1979). *J. Biol. Chem.* **254**, 11323–11329.

Ploegman, J. H., Drent, G., Kalk, K. H., and Hol, W. G. J. (1978). *J. Mol. Biol.* **123**, 557–594.

Poulos, T. L., Freer, S. T., Alden, R. A., Edwards, S. L., Skogland, U., Takio, K., Eriksson, B., Xuong, N.-H., Yonetani, T., and Kraut, J. (1980). *J. Biol. Chem.* **255**, 575–580.

Prothero, J. W. (1966). *Biophys. J.* **6**, 367–370.

Ptitsyn, O. B. (1969). *J. Mol. Biol.* **42**, 501.

Ptitsyn, O. B., and Finkelstein, A. V. (1980). *In* "Protein Folding" (R. Jaenicke, ed.), pp. 101–115. Elsevier, Amsterdam.

Ptitsyn, O. B., and Rashin, A. A. (1975). *Biophys. Chem.* **3**, 1–20.

Pullman, B., and Pullman, A. (1974). *Adv. Protein Chem.* **28**, 347–526.

Quiocho, F. A., and Lipscomb, W. N. (1971). *Adv. Protein Chem.* **25**, 1–77.

Quiocho, F. A., Gilliland, G. L., and Phillips, G. N., Jr. (1977). *J. Biol. Chem.* **252**, 5142–5149.

Raghavendra, K., and Sasisekharan, V. (1979). *Int. J. Peptide Protein Res.* **14**, 326–338.

Ralston, E., and DeCoen, J. L. (1974). *J. Mol. Biol.* **83**, 393–420.

Ramachandran, G. N. (1974). *In* "Peptides, Polypeptides, & Proteins" (E. R. Blout, F. A. Bovey, M. Goodman, and N. Lotan, eds.), pp. 14–34. Wiley (Interscience), New York.

Ramachandran, G. N., and Sasisekharan (1968). *Adv. Protein Chem.* **23**, 284–438.

Ramachandran, G. N., Ramakrishnan, C., and Sasisekharan, V. (1963). *J. Mol. Biol.* **7**, 95–99.

Rao, S. T., and Rossmann, M. G. (1973). *J. Mol. Biol.* **76**, 241–256.

Reeke, G. N., Becker, J. W., and Edelman, G. M. (1975). *J. Biol. Chem.* **250**, 1525–1547.

Remington, S. J., and Matthews, B. W. (1978). *Proc. Natl. Acad. Sci. U.S.A.* **75**, 2180–2184.

Remington, S. J., and Matthews, B. W. (1980). *J. Mol. Biol.* **140**, 77–99.

Richards, F. M. (1977). *Annu. Rev. Biophys. Bioeng.* **6**, 151–176.

Richardson, J. S. (1976). *Proc. Natl. Acad. Sci. U.S.A.* **73**, 2619–2623.

Richardson, J. S. (1977). *Nature (London)* **268**, 495–500.

Richardson, J. S. (1979). *Biochem. Biophys. Res. Comm.* **90**, 285–290.

Richardson, J. S., Thomas, K. E., Rubin, B. H., and Richardson, D. C. (1975). *Proc. Natl. Acad. Sci. U.S.A.* **72**, 1349–1353.

Richardson, J. S., Richardson, D. C., Thomas, K. A., Silverton, E. W., and Davies, D. R. (1976). *J. Mol. Biol.* **102**, 221–235.

Richardson, J. S., Getzoff, E. D., and Richardson, D. C. (1978). *Proc. Natl. Acad. Sci. U.S.A.* **75**, 2574–2578.

Richmond, T. J., and Richards, F. M. (1978). *J. Mol. Biol.* **119**, 537–555.

Rose, G. D. (1978). *Nature (London)* **272**, 586–590.

Rose, G. D. (1979). *J. Mol. Biol.* **134**, 447–470.

Rose, G. D., and Seltzer, J. P. (1977). *J. Mol. Biol.* **113**, 153–164.

Rose, G. D., Winters, R. H., and Wetlaufer, D. B. (1976). *FEBS Lett.* **63**, 10–16.

Rossmann, M. G., and Argos P. (1976). *J. Mol. Biol.* **105**, 75–96.

Rossmann, M. G., Moras, D., and Olsen, K. W. (1974). *Nature (London)* **250**, 194–199.

Sakano, H., Rogers, J. H., Hüppi, K., Brack, C., Traunecker, A., Maki, R., Wall, R., and Tonegawa, S. (1979). *Nature (London)* **277**, 627–633.

Salemme, F. R., Freer, S. T., Xuong, N. H., Alden, R. A., and Kraut, J. (1973). *J. Biol. Chem.* **248**, 3910–3921.

Saul, F. A., Amzel, L. M., and Poljak, R. J. (1978). *J. Biol. Chem.* **253**, 585–597.

Sawyer, L., Shotton, D. M., Campbell, J. W., Wendell, P. L., Muirhead, H., Watson, H. C., Diamond, R., and Ladner, R. C. (1978). *J. Mol. Biol.* **118**, 137–208.

Saxena, V. P., and Wetlaufer, D. B. (1971). *Proc. Natl. Acad. Sci. U.S.A.* **68**, 969–972.

Schellman, C. (1980). *In* "Protein Folding" (R. Jaenicke, ed.), pp. 53–61. Elsevier, Amsterdam.

Schiffer, M., and Edmundson, A. B. (1967). *Biophys. J.* **7**, 121–135.

Schiffer, M., Girling, R. L., Ely, K. R., and Edmundson, A. B. (1973). *Biochemistry* **12**, 4620–4631.

Schulz, G. E. (1980). *J. Mol. Biol.* **138**, 335–347.

Schulz, G. E., and Schirmer, R. H. (1974). *Nature (London)* **250**, 142–144.

Schulz, G. E., Elzinga, M., Marx, F., and Schirmer (1974a). *Nature (London)* **250**, 120–123.

Schulz, G. E., Barry, C. D., Friedman, J., Chou, P. Y., Fasman, G. D., Finkelstein, A. V., Lim, V. I., Ptitsyn, O. B., Kabat, E. A., Wu, T. T., Levitt, M., Robson, B., and Nagano, K. (1974b). *Nature (London)* **250**, 140–142.

Schulz, G. E., Schirmer, R. H., Sachsenheimer, W., and Pai, E. F. (1978). *Nature (London)* **273**, 120–124.

Shaw, P. J., and Muirhead, H. (1977). *J. Mol. Biol.* **109**, 475–485.

Sheridan, R. P., Lee, R. H., Peters, N., and Allen, L. C. (1979). *Biopolymers* **18**, 2451–2458.

Silverton, E. W., Navia, M. A., and Davies, D. R. (1977). *Proc. Natl. Acad. Sci. U.S.A.* **74**, 5140–5144.

Spiro, T. G., and Gaber, B. P. (1977). *Annu. Rev. Biochem.* **46**, 553–572.

Sprang, S., and Fletterick, R. J. (1979). *J. Mol. Biol.* **131**, 523–551.

Srinivasan, R., Balasubramanian, R., and Rajan, S. S. (1976). *Science* **194**, 720–721.

Steinrauf, L. K., Peterson, J., and Jensen, L. H. (1958). *J. Am. Chem. Soc.* **80**, 3835–3838.

Steitz, T. A., Fletterick, R. J., Anderson, W. A., and Anderson, C. M. (1976). *J. Mol. Biol.* **104**, 197–222.

Stenkamp, R. E., Sieker, L. C., Jensen, L. H., and McQueen, J. E., Jr. (1978). *Biochemistry* **17**, 2499–2504.

Sternberg, M. J. E., and Thornton, J. M. (1976). *J. Mol. Biol.* **105**, 367–382.

Sternberg, M. J. E., and Thornton, J. M. (1977a). *J. Mol. Biol.* **110**, 269–283.

Sternberg, M. J. E., and Thornton, J. M. (1977b). *J. Mol. Biol.* **110**, 285–296.

Sternberg, M. J. E., and Thornton, J. M. (1977c). *J. Mol. Biol.* **115**, 1–17.

Stroud, R. M., Kay, L. M., and Dickerson, R. E. (1974). *J. Mol. Biol.* **83**, 185–208.

Stuart, D. I., Levine, M., Muirhead, H., and Stammers, D. K. (1979). *J. Mol. Biol.* **134**, 109–142.

Subramanian, E., Swan, I. D. A., Liu, M., Davies, D. R., Jenkins, J. A., Tickle, I. J., and Blundell, T. L. (1977). *Proc. Natl. Acad. Sci. U.S.A.* **74**, 556–559.

Sugeta, H., Go, A., and Miyazawa, T. (1972). *Chem. Lett.* 83–86.

Sugeta, H., Go, A., and Miyazawa, T. (1973). *Bull. Chem. Soc. Jpn.* **46**, 3407–3411.

Swanson, R., Trus, B. L., Mandel, N., Mandel, G., Kallai, O. B., and Dickerson, R. E. (1977). *J. Biol. Chem.* **252**, 759–775.

Sweet, R. M., Wright, H. T., Janin, J., Chothia, C., and Blow, D. M. (1974). *Biochemistry* **13**, 4212–4228.

Sygusch, J., Madsen, N. B., Kasvinsky, P. J., and Fletterick, R. J. (1977). *Proc. Natl. Acad. Sci. U.S.A.* **74**, 4757–4761.

Tanaka, S., and Scheraga, H. A. (1977). *Proc. Natl. Acad. Sci. U.S.A.* **72**, 3802–3806.

Taniuchi, H., and Anfinsen, C. B. (1969). *J. Biol. Chem.* **244**, 3864–3875.

Timkovich, R., and Dickerson, R. E. (1973). *J. Mol. Biol.* **79**, 39–56.

Tsernoglou, D., and Petsko, G. A. (1977). *Proc. Natl. Acad. Sci. U.S.A.* **74**, 971–974.

Tulinsky, A., Vandlen, R. L., Morimoto, C. N., Mani, N. V., and Wright, L. H. (1973). *Biochemistry* **12**, 4185–4192.

Vainshtein, B. K., Melik-Adamyan, V. R., Barynin, V. V., and Vagin, A. A. (1980). *Dokl. Akad. Nauk S.S.S.R.* **250**, 242–246.

Van Wart, H. E., and Scheraga, H. A. (1976). *J. Phys. Chem.* **80**, 23–32.

Van Wart, H. E., Lewis, A., Scheraga, H. A., and Saeva, F. D. (1973). *Proc. Natl. Acad. Sci. U.S.A.* **70**, 2619–2623.

Venkatachalam, C. M. (1968). *Biopolymers* **6**, 1425–1436.

Walkinshaw, M. D., Saenger, W., and Maelicke, A. (1980). *Proc. Natl. Acad. Sci. U.S.A.* **77**, 2400–2404.

Wang, B.-C., Yoo, C. S., and Sax, M. (1979). *J. Mol. Biol.* **129**, 657–674.

Ward, K. B., Hendrickson, W. A., and Klippenstein, G. L. (1975). *Nature (London)* **257**, 818–821.

Warme, P. K., and Morgan, R. S. (1978). *J. Mol. Biol.* **118**, 289–304.

Watenpaugh, K. D., Margulis, T. N., Sieker, L. C., and Jensen, L. H. (1978). *J. Mol. Biol.* **122**, 175–190.

Watenpaugh, K. D., Sieker, L. C., and Jensen, L. (1979). *J. Mol. Biol.* **131**, 509–522.

Watenpaugh, K. D., Sieker, L. C., and Jensen, L. H. (1980). *J. Mol. Biol.* **138**, 615–633.

Watson, H. C. (1969). *Prog. Stereochem.* **4**, 299–333.

Weatherford, D. W., and Salemme, F. R. (1979). *Proc. Natl. Acad. Sci. U.S.A.* **76**, 19–23.

Weber, I. T., Johnson, L. N., Wilson, K. S., Yeates, D. G. R., Wild, D. L., and Jenkins, J. A. (1978). *Nature (London)* **274**, 433–437.

Weber, P. C., Bartsch, R. G., Cusanovich, M. A., Hamlin, R. C., Howard, A., Jordon, S. R., Kamen, M. D., Meyer, T. E., Weatherford, D. W., Xuong, N. H., and Salemme, F. R. (1980). *Nature (London)* **286**, 302–304.

Wetlaufer, D. B. (1973). *Proc. Natl. Acad. Sci. U.S.A.* **70**, 697–701.

Wiegand, G., Kukla, D., Scholze, H., Jones, T. A., and Huber, R. (1979). *Eur. J. Biochem.* **93**, 41–50.

Wierenga, R. K., de Jong, R. J., Kalk, K. H., Hol, W. G. J., and Drenth, J. (1979). *J. Mol. Biol.* **131**, 55–73.

Wilson, I. A., Skehel, J. J., and Wiley, D. C. (1981). *Nature (London)* **289**, 366–373.

Wodak, S. J., and Janin, J. (1980). *Proc. Natl. Acad. Sci. U.S.A.* **77**, 1736–1740.

Wolfenden, R. (1978). *Biochemistry* **17**, 201–204.

Wright, C. S. (1977). *J. Mol. Biol.* **111**, 439–457.

Wright, C. S., Alden, R. A., and Kraut, J. (1969). *Nature (London)* **221**, 235–242.

Wright, H. T. (1973). *J. Mol. Biol.* **79**, 1–23.

Wrinch, D. M. (1937). *Proc. R. Soc. London Ser. A* **161**, 505–524.

Wu, T. T., Szu, S. C., Jernigan, R. L., Bilofsky, H., and Kabat, E. A. (1978). *Biopolymers* **17**, 555–572.

Wüthrich, K., and Wagner, G. (1978). *Nature (London)* **275**, 247–248.

Wyckoff, H. W., Tsernoglou, D., Hanson, A. W., Knox, J. R., Lee, B., and Richards, F. M. (1970). *J. Biol. Chem.* **245**, 305–328.

Yakel, H. L., Jr., and Hughes, E. W. (1954). *Acta Crystallogr.* **7**, 291–297.

Yaron, A., Katchalski, E., Berger, A., Fasman, G. D., and Sober, H. A. (1971). *Biopolymers* **10**, 1107–1120.

Yu, N.-T., Lin, T.-S., and Tu, A. T. (1975). *J. Biol. Chem.* **250**, 1782–1785.

Zimmerman, S. S., and Scheraga, H. A. (1977a). *Biopolymers* **16**, 811–843.

Zimmerman, S. S., and Scheraga, H. A. (1977b). *Proc. Natl. Acad. Sci. U.S.A.* **74**, 4126–4129.

AUTHOR INDEX

Numbers in italics refer to the pages on which the complete references are listed.

SUBJECT INDEX

A

Acetylcholinesterase, activation volume of, 151
Acid proteases, 278
 domain structures of, 258, 297, 314
 structure–function correlation in, 319
Actin, subunit dissociation volumes of, 156
Actinidin, *see* Papain
Adenylate kinase
 structure of, 248, 251
 in domains, 257, 266, 278, 294
 function and, 319
Agglutinin(s)
 structure of
 in domains, 259, 276, 278, 307, 310, 314
 function and, 319
Albumin, subunit denaturation volume of, 156
Alcohol dehydrogenase
 activation volume of, 151
 structure of, 243, 251
 in domains, 257, 265, 272, 278, 294, 304, 314
 function and, 319
Aldolase, domain structures of, 257, 264, 278
Alkaline phosphatase, activation volume of, 151
Amino acids, protein backbone configuration and, 170–178
Antiparallel α domains, in protein taxonomy, 256–257, 283–287
Arabinose-binding protein
 structure of, 250
 in domains, 257, 266, 278, 294, 314
Argininosuccinate lyase, activation volume of, 151
Asparagine, role in protein structure, 233–234
Aspartase, activation volume of, 151
Aspartate aminotransferase, domain structure of, 257, 278

Aspartate transcarbamylase
 structure of, 241
 in domains, 257, 258, 265, 273, 278, 294, 296, 314
ATP-phosphoribosyltransferase, activation volume of, 151
ATPases, activation volumes of, 151
Azurin, domain structure of, 258, 278, 325

B

β annulus, as protein domain structure, 305–306
β barrels
 domain structures with, 258
 schematic drawings, 264, 266–267
 in protein structure, 200–203, 257
β sheets
 in protein structures, 190–203, 257
 schematic drawings, 265
Bacteriochlorophyll protein
 structure of, 244, 248
 in domains, 258, 274, 278, 305
Bence-Jones REI protein, 304 (*see also* Immunoglobulins)
 structure of, 205, 208, 212, 221, 228, 243
Buffer, ionization of, at high pressure, 143–144
Bulges, in protein structures, 216–223, 328

C

Calcium-binding protein
 structure of, 182, 234, 236
 in domains, 256, 263, 278, 288, 317
Carbonic anhydrase
 structure of, 186, 194, 220, 240, 248
 in domains, 258, 268, 278, 294
Carboxypeptidase
 structure of, 194, 198–199, 207, 248
 in domains, 257, 268, 278, 294, 295, 311, 312
 function and, 319

356

CONTENTS OF PREVIOUS VOLUMES